CORPORATE HEALTH MANAGEMENT

JONATHAN E. FIELDING, M.D.

▲

ADDISON-WESLEY PUBLISHING COMPANY, INC.
Reading, Massachusetts • Menlo Park, California • Don Mills, Ontario • Wokingham, England
Amsterdam • Sydney • Singapore • Tokyo • Mexico City • Bogotá • Santiago • San Juan

To those who have made the nonbook-writing time so meaningful:

Karin
Gertrude
Robert
Brian
Herbert
and
Phoebe Foster

Cover design by Mike Fender
Set in 10 point Caledonia by Techna Type, Inc., York, PA

Library of Congress Cataloging in Publication Data

Fielding, Jonathan E.
    Corporate health management.

    Bibliography: p.
    1. Industrial hygiene—United States.   2. Insurance,
Health—United States.   I. Title.
HD7654.F54   1984   658.3′82   84-2809
ISBN 0-201-13366-0

ABCDEFGHIJ-DO-8654
First printing, October 1984

CONTENTS

FOREWORD BY ROBERT N. BECK   *ix*

PREFACE   *xi*

---

1 THE CORPORATE COST OF ILL HEALTH   1

A BRIEF HISTORY OF CORPORATE HEALTH
MANAGEMENT   *1*
THE NEWER DEFINITION OF HEALTH   *5*
EMPLOYER RESPONSIBILITY FOR HEALTH   *11*
NOTES   *17*

---

2 WHAT'S WRONG WITH THE HEALTH CARE SYSTEM?   19

MAJOR PROBLEMS WITH THE HEALTH CARE
SYSTEM   *19*
NOTES   *29*

3 OVERCOMING INERTIA: A CHALLENGE TO EMPLOYERS   30

    KEY STEPS IN OBTAINING COMMITMENT   *36*
    NOTES   *38*

4 HEALTH INSURANCE CARRIERS AND ADMINISTRATORS: AN OVERVIEW   39

    THE "BLUES"   *40*
    COMMERCIAL CARRIERS   *41*
    THIRD-PARTY ADMINISTRATORS   *42*
    THE FUTURE   *43*
    NOTES   *44*

5 MANAGING MEDICAL BENEFITS   46

    EXISTING BENEFIT FEATURES   *46*
    WHERE THE MEDICAL HEALTH INSURANCE DOLLAR GOES   *49*
    BENEFIT REDESIGN: WHAT WORKS   *50*
    NOTES   *65*

6 MANAGING DENTAL BENEFITS   67

    GROWTH IN DENTAL BENEFITS   *67*
    DENTAL BENEFITS AND COVERAGE   *68*
    DENTAL COSTS   *71*
    DENTAL PRACTICE CHARACTERISTICS   *72*
    CONTROL OF COSTS AND QUALITY   *75*
    NOTES   *80*

7 MANAGING MENTAL HEALTH
   BENEFITS   82

   HISTORICAL PERSPECTIVE ON MENTAL
   HEALTH BENEFITS   82
   PROVIDERS OF MENTAL HEALTH SERVICES   84
   UTILIZATION, COSTS AND EFFECTS OF
   MENTAL HEALTH SERVICES   88
   A RECOMMENDED APPROACH   93
   NOTES   94

8 MANAGING OTHER IMPORTANT
   HEALTH CARE BENEFITS   97

   ALCOHOLISM AND OTHER DRUG ABUSE
   BENEFITS   97
   PRESCRIPTION DRUG BENEFITS   101
   VISION CARE BENEFITS   104
   HOSPICE CARE   105
   HOME HEALTH CARE   106
   NOTES   108

9 FROM DATA TO USEFUL
   MANAGEMENT INFORMATION: HOW
   TO ANALYZE HEALTH INSURANCE
   CLAIMS   110

   REASONS TO EXAMINE HEALTH CARE CLAIMS   110
   ANALYSIS OF HEALTH INSURANCE CLAIMS
   FROM DATA ANALYSIS TO COST-CONTROL   114
   STRATEGIES   150
   NOTES   154

10 SELF-INSURANCE 155

A BRIEF HISTORY 156
EFFECTS OF SELF-INSURING 156
SOME MAJOR SELF-INSURANCE ISSUES 159
THE FUTURE 164
NOTES 164

11 SECOND OPINION SURGERY PROGRAMS 166

THE IMPACT OF UNNECESSARY SURGERY 166
A BRIEF HISTORY 170
GROWTH OF PROGRAMS 174
DESIGNING A WORKABLE PROGRAM 177
THE FUTURE 179
NOTES 179

12 UTILIZATION REVIEW 181

A BRIEF HISTORY 182
TYPES OF REVIEW 183
THE FUTURE 191
NOTES 191

13 HEALTH MAINTENANCE ORGANIZATIONS (HMOs) 193

DEFINING THE HMO 193
ATTITUDES ABOUT HMOs 197
FINANCIAL VIABILITY 201
HMO EVALUATION 202
THE FUTURE 206
NOTES 207

14 PREFERRED ARRANGEMENTS WITH PROVIDERS   209

DEFINING PREFERRED ARRANGEMENTS   209
REASONS FOR PARTICIPATION IN PREFERRED
ARRANGEMENTS   211
TYPES OF PREFERRED ARRANGEMENTS   214
EMPLOYER OPPORTUNITIES   216
RECOMMENDATIONS   221
NOTES   222

15 EMPLOYER COALITIONS: COMBINING KNOWLEDGE AND CLOUT   224

THE MEMBERSHIP DILEMMA   225
COMMON COALITION ACTIVITIES   227
NATIONAL AND REGIONAL COALITIONS   232
THE FUTURE   233
NOTES   235

16 HEALTH CARE REGULATION: A ROLE FOR EMPLOYERS   236

EMPLOYERS AND GOVERNMENT REGULATION   236
LESSON FOR EMPLOYERS   244
POSITIVE EFFECTS OF REGULATION   248
NOTES   250

17 INCREASING COMPETITION IN THE HEALTH CARE SYSTEM   253

COMPETITIVE APPROACHES   253
COMBINING COMPETITIVE AND
REGULATORY APPROACHES   261
NOTES   262

✓ 18 DISEASE PREVENTION AND
HEALTH PROMOTION 263

PREVENTABLE HEALTH PROBLEMS 264
BRIEF REVIEW OF MAJOR HEALTH RISKS 277
GROWTH OF HEALTH PROMOTION/DISEASE
PREVENTION PROGRAMS 287
HEALTH PROMOTION PROGRAMS: KEYS TO
SUCCESS 295
EVALUATION 297
THE FUTURE 304
NOTES 304

✓ 19 MENTAL HEALTH AND
PRODUCTIVITY AT WORK 309

STRESSORS, JOB PERFORMANCE AND HEALTH 309
MINIMIZING STRESS-RELATED PROBLEMS 317
NOTES 330

20 EMPLOYEE ASSISTANCE
PROGRAMS 332

EAP RESULTS 334
EAP COMPONENTS 335
EAP MANAGEMENT 336
FEATURES OF SUCCESSFUL PROGRAMS 338
PROGRAM EVALUATION 341
NOTES 343

21 MANAGING DISABILITY
COMPENSATION 345

DISABILITY: A GROWING PROBLEM 345
CONTROLLING THE COST OF DISABILITY
COMPENSATION PROGRAMS 351
OPTIONS AND RECOMMENDATIONS 360
NOTES 366

22 WORKERS' COMPENSATION 369

OVERVIEW 369
PROGRAM FEATURES 371
ESTABLISHING WORKERS' COMPENSATION
RATES 372
MAJOR TRENDS 373
OPPORTUNITIES FOR COST CONTROL 377
PREVENTING AND REDUCING WORK-RELATED
DISABILITY 385
NOTES 390

23 WHAT SMALLER EMPLOYERS CAN DO 393

SPECIAL PROBLEMS OF SMALLER EMPLOYERS 394
STRATEGIES FOR SMALLER BUSINESSES 395
NOTES 400

✓ 24 WHERE TO START: DEVELOPING A HEALTHFUL CORPORATE ENVIRONMENT 401

REDUCING THE FREQUENCY OF SERIOUS
HEALTH PROBLEMS 401
REDUCING THE USE OF THE HEALTH
CARE SYSTEM 404
DECREASING THE PERIOD OF DISABILITY 406
KEEPING TRACK OF HEALTH 406

INDEX 409

FOREWORD

Helping employees and their families maintain or improve their physical and mental health is an important goal for enlightened employers. The need to be more competitive in the international marketplace has caused employers to examine all their costs and seek productivity improvements.

Health care costs have been consistently spiraling upward at a pace faster than most other consumer items. As a result, employers' health care costs have risen to a point that for many, they are the single largest cost of services purchased. Unnecessary health care expenses erode the bottom line and reduce the funds available to meet other employee needs. Additionally, a highly productive workforce that produces high-quality goods and services is essential. More and more evidence indicates a close relationship between employee health and productivity. Other things being equal, a healthier workforce is a more productive workforce.

Jonathan Fielding, who embodies the unusual and welcome blend of considerable corporate experience with academic credibility, has provided a comprehensive employer guide to controlling employee health-related costs. He gives practical yet often overlooked approaches to restructuring benefits to the advantage of both employers and employees. Dr. Fielding lucidly presents the different options for cost controls and explains how to develop cost control programs that don't sacrifice quality of care but, in fact, improve it.

Most important, this volume makes sense of many employee health areas about which there is much confusion. How good is the evidence that employee assistance programs can save money and what kind of program is likely to be the most cost-effective? What can be done to reduce the frequency of workers' compensation claims? What are the critical factors that spell success in a worksite health promotion program? How can an employer minimize job burnout among key employees? These are among the many questions that this book answers with clarity, authority and sensitivity.

There is another good reason why employers should explore new ideas about health care. An interest in wellness seems to be more than a passing fad or fancy by employees and their families. They are asking for more support to learn about health and wellness as well as assistance in taking actions to improve their health. Numerous employers are now responding to this employee need. They are finding a positive affect on employees' and their families' attitudes about the company for relatively little increase in total expenses.

I believe that health and productivity are inextricably linked. Yet, from my experience, there can be considerable slippage in developing effective strategies which spring from this simple truth. *Corporate Health Management* is likely to become a classic in how to simultaneously reduce health care cost inflation and improve employee health. For any corporate executive, human resources manager, personnel director, benefits planner or corporate medical director who wants to get a good start on maximizing employee productivity and slowing the increase in costs in both the short and long term this book is an indispensable first step.

ROBERT N. BECK
Executive Vice President
Bank of America

PREFACE

Rene Dubos, the famous scientist and environmentalist, defined health as "a measure of each person's ability to do what he wants to do and become what he wants to become." Health is thus an enabling function. With it we can be satisfied both in our work and in our leisure activities. But there is another essential component of good health: the ability to adapt well to a wide variety of external and internal challenges. These challenges are as diverse as a promotion, a divorce, bad smog, a virus, a child on drugs, a financial setback, a failed exam, or a heart attack. Adapting to them well means we are healthy, and that, even if we are unable to carry on in a physical sense, we can continue to think positively and to be productive. Being productive is thus part of the definition of personal health.

The same is true for a business organization. Meeting its objectives for productivity cannot be separated from the health of its workforce. A healthy organization can adapt well to changing business environments. Its employees can live with ambiguity and paradox better than employees of other companies. This adaptability is one of the eight characteristics shared by America's best-run companies, according to Thomas J. Peters and Robert H. Waterman, Jr. in *In Search of Excellence* (New York: Harper and Row, 1982). An organization's adaptability can be translated into higher returns on investment and equity, increasing profitability and greater consistency in reaching its planning objectives. A healthy corporation recognizes another of the eight characteristics of excellence: a key to productivity is to treat each employee as an individual who is valuable to the organization. There is no better way to evidence concern for those who make the company run than to do whatever is possible to help them maintain and improve their health.

Employers are often unaware of how many ways they can positively affect the health of their employees and the larger family of dependents and retirees. Naturally, some of the influences on health are unalterable, such as an individual's genetic constitution. However, most can be affected in a positive direction.

Providing adequate protection for workers against workplace injuries is a well-established employer responsibility that has led to steadily declining work-related death and accident rates. More recently, concern for the effects of chemical exposures on human health has led to the development of sophisticated techniques for limiting exposures and for epidemiologic monitoring systems to identify associations between exposures and disease at an early stage.

To an increasing degree we are finding that to minimize the toll of work-related health problems requires attention to the interactions between the health and risk characteristics of an individual and the characteristics of the job. The best way for an asbestos worker to minimize his risk of getting lung cancer is not to smoke. Being in good physical condition and not being overweight are important contributors to a reduced incidence of low-back problems in a workforce which lifts or moves heavy loads.

Accidents and lapses in quality control are not chance events. They are often the result of alcoholism, other drug use or preoccupation with personal problems. Supervisor training can help identify workers with emotional/behavioral problems that affect performance and refer them to counselors for confidential discussion and referral to appropriate expert resources.

Work can be a major source of stress, which is in turn associated with high rates of absenteeism, turnover and overall productivity while on the job. Corporate culture and management style have much to do with the degree to which work is a major source of stress for employees. Conscientious efforts to reduce the ambient stress level can directly benefit the health of employees and their productivity and creativity.

On-site or contract health programs can have equally positive health effects. For example, a voluntary hypertension screening program with referral to usual sources of care and periodic follow-up to assure that treatment has been continued can reduce one of the most common risks for ill health. The availability of an empathetic and confidential health professional can help employees get early assistance in identifying and sorting out both physical and psychological problems.

One of the greatest contributions to health that an organization can make is helping employees adopt and maintain personal habits that contribute to health. Employer-sponsored health programs that encourage nonsmoking, adequate exercise, good nutrition, better stress management, self-examination for early cancers (i.e., breast, skin and testicular cancers) can both teach employees the benefits of healthful personal habits and the skills to practice them. Employers can reduce the adverse health impact of motor vehicle accidents by requiring the use of safety belts in company vehicles and by providing monetary or other incentives for their use off the job. Incentives can also be used to reward employees for other healthful changes, such as weight loss, smoking cessation or improved fitness.

A complement to organized health improvement programs should be an environment that fosters health: a cafeteria that offers healthful dietary choices, opportunities to exercise during the lunch period or before and after work, a quiet place to sit for a few minutes and practice stress management. Ergonomic design of office furniture and production facilities can also improve health.

To achieve the maximum benefit for the organization, efforts to improve health must target not only employees but their dependents. It must also include retirees. It is in an employer's self-interest to maintain a healthy dependent and retiree population for whom it pays the majority of the health care bills. In addition, dependent health affects employee health. An employee's husband with an untreated alcohol problem is affecting her productivity on the job.

A combination of these efforts can reduce the burden of illness and improve productivity. Nonetheless, no matter how intense these efforts, some employees, dependents and retirees will get sick. Opportunities of equal scope exist for employers to minimize the personal and dollar costs for treating these health problems. Health care costs are not fixed, and they are, in fact, more manageable than many other costs. Changes in benefit packages, review of health care services to assure appropriateness before they are provided or before they are paid, second-opinion programs, preferred arrangements which offer discounts to employer and insuree, improved claims payment systems, self-insurance and other financing arrangements all offer opportunities to reduce the health care cost burden. Of equal or even greater importance, better management of the health benefit also can translate into improvements in the quality of care.

Opportunities for better management also extend to the problem of disability. The frequency of disability claims and the size of claims, both under workers' compensation and employer-sponsored disability plans, can be diminished through attention to the structure of benefits, timing and intensity of efforts toward rehabilitation, and effort made to reintegrate the employee into the workforce.

Health management as a corporate strategy also extends to employer involvement in the development of health policy. Employers are the largest payers for health care services and the source of a large portion of the taxes that go to public financing of medical care for older Americans and the economically disadvantaged. Employers have an undeniably large stake in local, state and federal legislation that affects health benefits and health care financing in both the public and private sectors. For example, employers can help encourage policies that make health care more efficient, that reduce the incentives to provide more services than may be necessary and that encourage the development of cost-effective services that will be made available to their insurees.

Participation in employer health coalitions can be an effective mechanism to better understand health care options in the area. Through this vehicle employers can compare their health problems with those of other area companies, exchange ideas for better health management, use their collective clout to improve the data available from carriers and jointly define initiatives that facilitate better corporate health management.

The following chapters more fully define the major employer opportunities to improve the health of current and past workers and their families. My hope is that it will help managers not only to better appreciate these opportunities, but to refine and add to their action plans. Many specific recommendations are offered, all with the knowledge that they will not fit the needs of every reader, and that the wisdom of some will be increasingly questioned as time makes us smarter. However, the desire to provide practical suggestions overrode the safer tendency to pose only questions and options.

Traditional occupational health and employee health service activities are intentionally not treated in this book. The reasons are simple. They are very broad subjects on which excellent works have already appeared, and they could not be adequately covered within the space constraints. Instead, emphasis has been placed on health problems and costs of ill health to employers, on which less has been written and questions have been more plentiful than answers.

ACKNOWLEDGMENTS

I am greatly in the debt of a number of expert reviewers who provided different perspectives as well as valid criticisms of my treatment of a variety of subjects. In addition, they directed me to a number of resource materials of which I was unaware. Without their comments I would have felt very insecure trying to present material on so many different aspects of a very broad field which is changing more rapidly than ever. These reviewers include Robert Beck, Executive Vice-President, Bank of America; Dr. Duane Block, Medical Director, Ford Motor Company; Willis Goldbeck, President, Washington Business Group on Health; Gary Bahr, Director of Compensation and Benefits, National Semiconductor Corporation; Gail Schwartz, Manager of Research, Washington Business Group on Health; and Dr. Leon Warshaw, Executive Director, Washington Business Group on Health.

I would also like to acknowledge the important contribution of my wife, Karin, an experienced editor who gently but firmly pointed out when I had not covered what I had promised, used jargon instead of English or tried to shoehorn four separate ideas into one incomprehensible sentence. Her careful editing greatly improved the readability of the manuscript.

Finally, and most importantly, I would like to acknowledge the extremely valuable contribution of Leslie Alexandre, M.S.P.H., who faithfully worked with me from the inception of the project to its completion. In addition to providing very helpful suggestions on both content and form, her valuable contributions ranged widely: she foraged in many UCLA libraries for hard-to-find statistics and important data sources, reviewed each chapter and improved it through constructive criticism and kept the project well organized, focused and reasonably on time, despite continuing efforts in the opposite direction by the author.

THE CORPORATE *1*
COST OF ILL
HEALTH

*I*n early 1984 the management of a large midwestern manufacturer
decided to figure out how much the company was paying for the ill health
of its employees. They added together all their payments for the direct costs
of health care services, including health insurance benefits to dependents
and retirees, and the indirect costs of illness of their workers. They expected
a big number, but nothing like $6,482 per active employee in that year.
This figure dwarfed expenditures for all other benefits, even pension costs,
and prompted two questions: "To what degree are the costs of ill health
manageable?" and "How can employers be most effective in bringing management techniques to the problem?"

A BRIEF HISTORY OF CORPORATE HEALTH MANAGEMENT

Until the last hundred years, we were quite pessimistic about man's ability
to intervene in disease processes, and with good reason. We didn't know
the origins of diseases, and our erroneous speculations led to treatments that
were occasionally helpful, often useless and more frequently harmful. The
likelihood that a patient would get better as a result of treatment was probably
no greater than the chances he would be worse off from the ministrations.

1

Hospitals were where the poor went to die when they couldn't afford the better treatment at home. Several advances changed the odds:

1. *Anesthesia.* Operations could be performed without pain. Surgery was greatly improved and could lead to better health through repair, drainage, and/or removal of nonvital tissues.
2. *Discovery that germs caused diseases.* This opened the door to diagnosis of what germs are responsible for specific diseases, and led to the expansion of laboratories to make these diagnoses. Acceptance of the germ theory ushered in new ways to prevent diseases, especially through the use of vaccines—for such diseases as typhus, diphtheria, whooping cough, and in more recent memory, polio, measles, German measles and mumps.
3. *Better diagnostic tools.* Laboratory identification of bacteria taken from patients permitted better therapeutic approaches. With x-rays, many problems could be diagnosed without surgery. When surgery was indicated, it could be more exact. Physicians could figure out if bones were broken and whether they were healing. And they could see whether some medicines were working; for example, if prescribing digitalis for congestive heart failure made the heart smaller and more efficient.
4. *Specific therapies.* Medicines were developed that could cure some diseases, especially infectious diseases, and palliate other problems, improving the ability of patients to function even in the absence of cure. Sulfonamide, streptomycin, penicillin and tetracycline were among the first of these "wonder drugs."

These and related advances contributed to lowering the toll of ill health. But surprisingly, the precipitous decline in mortality over the past century came less as a result of better medical care than from improvements in the general standard of living and specific public health measures. Improved nutritional status enhanced the ability to withstand many infections, and the percentage of Americans with these problems who succumbed to them decreased. Of the total decline in death rates from tuberculosis over the period 1900–1973, 92 percent occurred prior to 1950 when streptomycin, a specific therapy, became available (Table 1.1). Scarlet fever, influenza, whooping cough, typhus, typhoid and other infectious diseases showed significant declines long before specific immunizing agents or chemotherapy were available. Those factors which had the greatest impact on the frequency and severity of these diseases, in addition to improved nutrition, were water and food purification, refrigeration, improved waste disposal systems, and better housing conditions.[1]

Due to all these efforts, both public health and medical, death rates, the standard but crude indication of the health of our population, have

TABLE 1.1
THE CONTRIBUTION OF MEDICAL MEASURES (BOTH CHEMOTHERAPEUTIC AND PROPHYLACTIC) TO THE FALL IN THE AGE AND SEX-ADJUSTED DEATH RATES (S.D.R.) OF TEN COMMON INFECTIOUS DISEASES, AND TO THE OVERALL DECLINE IN THE S.D.R., FOR THE UNITED STATES, 1900–1973

Disease	Medical intervention and year became available	Fall in S.D.R. per 1,000 population, 1900–1973	Fall in S.D.R. after intervention as % total fall for the disease
Tuberculosis	Izoniazid/ Streptomycin, 1950	2.00	8.36
Scarlet fever	Penicillin, 1946	0.10	1.75
Influenza	Vaccine, 1943	0.22	25.33
Pneumonia	Sulphonamide, 1935	1.42	17.19
Diphtheria	Toxoid, 1930	0.43	13.49
Whooping cough	Vaccine, 1930	0.12	51.00
Measles	Vaccine, 1963	0.12	1.38
Smallpox	Vaccine, 1800	0.02	100.00
Typhoid	Chloramphenicol, 1948	0.36	0.29
Poliomyelitis	Vaccine, Salk/ Sabin, 1955	0.03	25.87

Source: Adapted from McKinlay, J.B. and McKinlay, S.M. "The Questionable Contribution of Medical Measures to the Decline of Mortality in the United States in the Twentieth Century," *Milbank Memorial Fund Quarterly/Health and Society,* Summer 1977, Vol. 55, No. 3, p. 418.

declined as longevity has increased. The annual death rate since 1900 has dropped from 17 per 1,000 persons to less than 9 per 1,000.[2] Perhaps even more striking is the tremendous gain in life expectancy over the past 60 years, from 54.1 years for babies born in 1920 to 73.8 years for those born in 1979, an increase of more than 35 percent.[3] All this helps to make the case that, at least to some degree, health can be managed. When appropriate resources are introduced, the burden of ill health declines.

Cardiovascular Disease: A Modern Success Story

A more recent and striking example of the ability to manage our health is the precipitous drop in heart disease death rates from the mid- to late-1960s through the present. The risk of a middle-aged man dying of a heart attack during a subsequent ten-year period declined by over 25 percent during this period. It's not magic; the majority of the decline is due to concurrent changes in those factors known to increase risk for acquiring heart disease: high blood pressure, cigarette smoking, high blood total cholesterol, and

TABLE 1.2
ESTIMATED PERCENTAGE CONTRIBUTION OF THE FOUR
ELEMENTS OF THE HEALTH FIELD TO PREMATURE
MORTALITY

Cause of death	Health system	Life-style	Environment	Human biology
Heart disease	12	54	9	28
Cancer	10	37	24	29
Cerebrovascular disease	7	50	22	21
All other accidents	14	51	31	4
Influenza and pneumonia	18	23	20	39
Motor vehicle accidents	12	69	18	0.6
Diabetes	6	26	0	68
Cirrhosis of the liver	3	70	9	18
Arteriosclerosis	18	49	8	26
Suicide	3	60	35	2
Average for ten causes	10.8	48.5	15.8	26.3

Source: Center for Disease Control, Public Health Service, 1979.

lack of sustained vigorous physical activity. During this period there was a 14 percent decline in adult male smokers and a 5 percent decline in adult female smokers,[4] a 5 percent reduction in blood cholesterol due to declines in the consumption of saturated fats and cholesterol,[5] and a substantial increase in the percentage of hypertensives whose blood pressure was adequately controlled.[6] Coincidentally, many millions of Americans began to undertake exercise programs, especially running. There may also have been a minor contribution from higher quality emergency services and cardiopulmonary resuscitation (CPR) training programs. But the main point is that health can be managed.

As illustrated by the heart disease example, health management today is not simply a question of improved public health measures or better therapy, although both are important. Most important is the role individuals play in determining their own health. It is estimated that of excess deaths in the United States today, almost 50 percent of the cause is to be found in individual health habits, about 15 percent in the environment, 25 percent based on biological factors, and only 10 percent due to lack of dissemination or use of current medical knowledge.[7] (Table 1.2)

Medicine's Contribution to Ill Health

Obviously, in some instances the work environment is one of the most powerful influences on health. Worksite injuries remain common. Many occupationally related diseases have been uncovered and the synergy of health habits and occupational exposures (such as asbestos and smoking) clarified in a few instances. But all in all, the toll of medicine is probably much higher. Many hospital admissions are due to an undesirable secondary effect of medical care. For example, a 1979 study of 815 new admissions to a medical service of a university hospital found that 290 patients (36 percent) had one or more iatrogenic (medicine-caused) illnesses, with 76 (9 percent of all admissions) suffering from major complications.[8] A significant portion of iatrogenic illness is due to drug reactions, but complications of medical and surgical procedures also make important contributions to this largely avoidable problem. Minimizing these effects by helping individuals understand that unnecessary medical care is not without dangers is a part of managing health.

All this is not to discourage or disparage appropriate health care. We have found that some forms of cancer, heart disease and many other common problems, including alcoholism, can be effectively treated. Surgery can sometimes make the difference between life and death or between being able to function in a job and being on permanent disability. Improved drugs can help to bring people out of depressions, reduce pain, forestall the otherwise debilitating effects of some diseases and substitute for vital chemicals which the body usually manufactures. The difficulty is that current financing approaches not only cover appropriate care but often provide incentives for inappropriate care, thus creating another opportunity for health management. Redirecting incentives away from conducting too many tests and spending more time than necessary in the hospital or nursing home are important components of health maintenance.

Finally, health is in part determined by the environment in which we live. Air pollution can adversely affect health, especially in persons with cardiac and respiratory problems. If cars don't have seat belts which automatically lock into place, it is unlikely that drivers and passengers will use them to reduce their risk of serious trauma. Where handguns are plentiful, deaths and injuries from them are higher than where they are scarce. If nuclear proliferation continues unabated, the likelihood of a worldwide nuclear catastrophy rises.

THE NEWER DEFINITION OF HEALTH

"Health" is of course more than the absence of disease or environmental calamity. A physician can examine a patient carefully, perform a full battery of expensive and sometimes uncomfortable diagnostic tests and conclude that he or she is healthy. The same patient can protest that he or she is very depressed, hates to get up in the morning, has been chronically unemployed, has poor relationships with family, has few friends and constantly feels tired. Surely the patient is not healthy. Without subscribing to a definition of

TABLE 1.3
LEADING CAUSES OF DEATH—UNITED STATES

Rank	1981[1] Cause of death (rate)*	1900[2] Cause of death (rate)*
1	Heart disease (336.0)	Pneumonia and influenza (202.2)
2	Cancer (183.9)	Tuberculosis (194.4)
3	Stroke (75.1)	Diarrhea, enteritis and ulceration of the intestines (142.7)
4	Accidents (46.7)	Diseases of the heart (137.4)
5	Chronic obstructive pulmonary diseases and allied conditions (24.7)	Intracranial lesions of vascular origin (106.9)
6	Pneumonia and influenza (24.1)	Nephritis (88.6)
7	Diabetes mellitus (15.4)	Accidents (72.3)
8	Chronic liver diseases and cirrhosis (13.5)	Cancer and other malignant tumors (64.0)
9	Arteriosclerosis (13.0)	Senility (50.2)
10	Suicide (11.9)	Diphtheria (40.3)

*Crude (non-age-adjusted) rate per 100,000 population.
[1]Source: Based on data from the National Center for Health Statistics, Division of Vital Statistics, 1981 (Provisional) Mortality Rates.
[2]Source: National Heart and Lung Institute Fact Book, 1975. U.S. Department of Health, Education and Welfare, Public Health Service, National Institutes of Health, DHEW Publication No. (NIH) 76–980, November 1975, p. 24.

health that is so broad as to be unworkable, agreement can be reached that health includes positive attributes, including a good self-image, self-confidence, close and rewarding personal relationships and some feeling of fulfillment. While not everyone will have all of these desirable characteristics, in many cases specific activities can increase the number and intensity of almost any of them. Positive health is thus to some degree amenable to health management.

Changing Patterns of Disease

One way to begin looking at the opportunities to reduce the burden of disease and to improve health is to consider the major causes of death in the United States. Table 1.3 compares the ten major killers for 1981, which together accounted for a full 85 percent of the deaths,[9] with a similar list from 1900. What are the major differences? First, the overall death rate is much lower, consistent with greater longevity (70 years for men; 78 years for women). In addition, there has been a drastic change in what kills us. Life-threatening infectious diseases which figured so prominently in 1900 are now rare, or at least less common, and considered controllable. Replacing them have been degenerative diseases which usually develop over a significant period

TABLE 1.4
CAUSES OF DEATH, AGES 15–44, 1980

Rank	Cause of death, age 15–24 (rate)*	Cause of death, age 25–44 (rate)*
1	Accidents (61.7)	Accidents (42.6)
2	Homicide (15.6)	Cancer (28.0)
3	Suicide (12.3)	Heart disease (23.1)
4	Cancer (6.3)	Homicide (17.8)
5	Heart disease (2.9)	Suicide (15.7)
6	Congenital anomalies (1.4)	Chronic liver disease and cirrhosis (7.6)
7	Stroke (1.0)	Stroke (5.0)
8	Influenza and pneumonia (0.8)	Diabetes mellitus (2.3)
9	Chronic obstructive pulmonary diseases and allied conditions (0.3)	Influenza and pneumonia (2.3)
10	Anemia (0.3)	Congenital anomalies (1.3)

*Rate per 100,000 population in specified group.
Source: Based on data from the National Center for Health Statistics, Division of Vital Statistics.

of time, frequently in silence, and then, at a fairly advanced stage, suddenly manifest themselves. Perhaps the most important common denominator for these slowly developing health problems is that they are preventable or their onset and complications can be forestalled.

The Burden of Ill Health to Employers

From an employer's point of view, primary focus is on the major problems in age groups represented in the workforce. Accidents, particularly motor vehicle accidents, are the single greatest cause of death and prolonged disability through about age 40 (Table 1.4). After that age, the chronic diseases, particularly diseases of the heart and blood vessels, and cancer, become the largest killers and continue through age 65 and beyond (Table 1.5). Of course, much of the burden of disease comes from nonfatal problems, such as mental health disorders, back problems, gynecological maladies, respiratory ailments, and alcoholism and other drug abuse.

Employees account for a significant portion of employer direct costs and almost all the indirect costs (absenteeism, reduced productivity, etc.) of ill health. There are also important direct costs from the ill health of dependents, including spouses, and children, and of retirees. While very serious disease is fortunately rare in childhood, it is common in newborns and infants during the first year of life. The rate of infant death is over thirty times the rate of death in young children and early teenagers, primarily because of prematurity and hazards associated with birth (Table 1.6).

Among the major health care expenses in infancy are the costs of preserving the life and health of neonates (infants under 28 days). The advent

TABLE 1.5
LEADING CAUSES OF DEATH, AGES 45—, 1980

Rank	Cause of death, age 45–64 (rate)*	Cause of death, age 65+ (rate)*
1	Heart disease (333.3)	Heart disease (2330.4)
2	Cancer (304.9)	Cancer (1011.3)
3	Stroke (44.7)	Stroke (573.1)
4	Accidents (40.8)	Influenza and pneumonia (178.1)
5	Chronic liver disease and cirrhosis (36.2)	Chronic obstructive pulmonary diseases and allied conditions (170.6)
6	Chronic obstructive pulmonary diseases and allied conditions (25.9)	Arteriosclerosis (109.9)
7	Diabetes mellitus (17.9)	Diabetes mellitus (98.7)
8	Suicide (15.9)	Accidents (97.2)
9	Pneumonia and influenza (13.0)	Nephritis, nephrotic syndrome and nephrosis (50.8)
10	Homicide (9.1)	Chronic liver disease and cirrhosis (37.3)

*Rate per 100,000 population in specified group.
Source: Based on date from the National Center for Health Statistics, Division of Vital Statistics.

of neonatal intensive care units (NICUs) has greatly improved our ability to save infants of very low birth weight, as well as newborns suffering from a variety of other conditions that were not too long ago nearly always fatal. However, while NICUs are believed to have been a major contributor to recent declines in neonatal mortality, good evidence is lacking to substantiate claims that NICUs reduce long-term health problems associated with prematurity. Neonatal intensive care may actually increase the frequency of certain disabilities, such as blindness due to the administration of high concentrations of oxygen to premature infants.[10] Regardless of outcome, the costs associated with neonatal intensive care are enormous. One analysis over five years ago placed the average cost for in-hospital intensive care for infants weighing 1,000 grams or less at birth at $14,236 per nonsurvivor, $40,287 per survivor, (with or without handicap) and $88,058 per normal survivor.[11] Overall, however, the per capita cost of insurance for health care to children is much lower than for adults, since after the newborn period serious illnesses are rare.

Death during childhood is also rare. Therefore, the major causes of hospitalization among children (Table 1.7) are a more accurate reflection of the leading causes of ill health during this period. Pneumonia, fractures and congenital anomalies are the top three causes of hospitalization among youths age 1 to 15, and together account for nearly 20 percent of total hospital days for this group. Bronchitis, emphysema and asthma are also high on the list for both males and females.

Retirees represent a different picture. Although retirement age is

TABLE 1.6
LEADING CAUSES OF DEATH FOR INFANTS, 1980

Rank	Cause of death	Rate*
1	Congenital anomalies	255.2
2	Sudden infant death syndrome	152.5
3	Respiratory distress syndrome	138.1
4	Disorders relating to short gestation and unspecified low birth weight	101.0
5	Newborn affected by maternal complications of pregnancy	43.5
6	Interuterine hypoxia and birth asphyxia	41.4
7	Accidents and adverse affects	32.3
8	Birth trauma	29.3
9	Pneumonia and influenza	28.0
10	Infections specific to perinatal period	26.9

*Rate per 100,000 live births.
Source: National Center for Health Statistics, Division of Vital Statistics.

TABLE 1.7
LEADING CAUSES OF HOSPITALIZATION AMONG CHILDREN UNDER AGE 15

Diagnosis	Days of care per 1,000 population
Male	
Pneumonia	30.6
Fracture	22.0
Congenital anomalies	22.0
Bronchitis, emphysema, asthma	14.9
Inguinal hernia	6.0
Intercranial injury	5.8
Female	
Pneumonia	22.0
Congenital anomalies	13.1
Fracture	11.2
Bronchitis, emphysema, asthma	8.4
Eye diseases and conditions	3.4

Source: Health, United States, 1981.

TABLE 1.8
DISTRIBUTION OF POPULATION AND OF PERSONAL
HEALTH CARE SPENDING BY AGE GROUP, 1978

Age	Health care spending (billions)	Population (millions)	Per capita spending	Percentage distribution Health care spending	Population
All ages	$167.9	223.0	$ 753	100.0%	100.0%
Under 19	19.9	69.5	286	11.9	31.2
19–64	98.7	129.2	764	58.8	57.9
65+	49.4	24.3	2,026	29.4	10.9

Source: Fisher, C.R. "Differences by Age Groups in Health Care Spending," *Health Care Financing Review*, Vol. 1, No. 4, Spring 1980, pp. 65–90.

changing, looking collectively at the 65-and-over age group provides a reasonable picture of the major health problems for the retiree. In addition to very high death rates from many chronic diseases (Table 1.5), this age group has extremely high illness and disability rates for all types of illness, but especially arthritis, cataracts, prostate problems, hip fractures, hearing difficulties and cardiovascular disease. And of course with advancing age senility becomes an increasingly frequent problem. According to the Health Care Financing Administration, the average annual per capita medical bill for persons over 65 is about two and one-half times that of persons 19 to 64 and nearly seven times the per capita expenditures for persons under 19[12] (Table 1.8). As the cost of care for this group becomes a greater and greater financial burden under the Medicare system, employers are increasingly being required to take up the slack. The Tax Equity and Fiscal Responsibility Act (TEFRA) of 1982 requires that company-sponsored insurance plans become the first payer for covered retirees age 65 to 69. Increasing deductibles and coinsurance under Medicare are also placing greater cost burdens on many companies, who pay the premium for "Medi-gap" policies which cover these cost sharing requirements.

Disabling Conditions

While mortality statistics are the most reliable and most readily available, a better picture of the burden of ill health appears by looking at major chronic health problems. In 1980 an estimated 31.4 million individuals, 14.4 percent of the U.S. civilian noninstitutionalized population, were restricted in activity because of one or more chronic conditions. Arthritis and rheumatism, heart conditions and hypertension are among the leading causes of self-reported long-term disability. Also high on the list are visual and hearing impairments, diabetes, mental and nervous conditions and asthma.[13] In addition, some of the commonly reported causes of absenteeism from work due to long-term disability include back problems, accidents and pregnancy.

Permanent disability is on the rise in our population, and its impact is felt directly by employers in many ways. Between 1966 and 1976 there was an 83 percent increase in persons claiming to be unable to do their main activity as a result of a chronic health problem.[14] Data from the Social Security Administration indicate that 21 million American adults in 1978 were limited in their ability to work due to a chronic condition, a 15 percent increase from 1972.[15,16] Other national data show that between 1969 and 1978 there was a 40 percent increase in self-reported disability among men age 45 to 64.[17]

An increasingly disabled workforce not only results in greater productivity losses but also raises the employer's costs of disability compensation and health insurance premiums. Chapter 21 amplifies the discussion of the causes and effects of disability in the workforce and ways in which employers can manage their disability compensation programs. Chapter 18 examines worksite health promotion programs as a possible means for employers to prevent, or at least deter, chronic health problems among the workforce.

EMPLOYER RESPONSIBILITY FOR HEALTH

Given the many causes of ill health and its costs, what can be the role of the employer in health management? Can there really be a payoff on the investment required to become deeply involved? Irrespective of employer desire for involvement, employers' responsibility for the health of employees has increased markedly and relentlessly. Not so long ago, industry responsibility was limited to providing care for occupational injuries and payments for resulting disability. Today employers are held responsible for the rehabilitation, retraining, adjustment of job duties to accommodate workers whose occupational injuries have left partial handicaps, and they are required to hire the handicapped. Employer responsibility extends beyond physical injuries at the worksite not only to occupational diseases, but to diseases whose time of appearance can be accelerated by the job or whose severity is worsened as a result of work.[18]

The line between occupational and other diseases has been gradually effaced with the recognition that not only work-related exposures but job content and general work ambience can affect health. First the Occupational Safety and Health Act (OSHA) and then the Toxic Substances Control Act (TOSCA) required employers to expend increasing resources to eliminate or minimize clearcut and potential risks to employee health. While risks may be more obvious in some industries (e.g., chemical, refining, asbestos manufacturing) than others, not even predominantly white-collar organizations can escape finger pointing when an employee has a mental health or physical problem that can be considered "stress-related." Since so little is known about the interaction of stress with physical and mental processes, it is difficult to exclude work stress as a contributory cause of many diseases, including cancers and cardiovascular conditions. Employees have even successfully sued their employers for failing to provide them a smoke-free environment.[19]

While many of these costs are covered under workers' compensation, many are not. In addition, workers' compensation is becoming a significant expense to many employers, and in many jurisdictions is experience-rated (See chapter 22). Companies are expending ever-increasing sums to provide periodic examinations and to maintain epidemiological data bases to study the effects of worksite exposures to chemicals, heavy metals, products of combustion, heat, noise and other work factors. The requirements to comply with the Coal Mine Safety Act (1969), OSHA (1970), the Rehabilitation Act (1973) and TOSCA (1976) have moved the employer toward a position of guardian, analyst, and monitor of employee health in large jumps. TOSCA alone requires some form of annual screening for an estimated 20 million workers.[20] So, like it or not, employers are very responsible for the health of their employees and their responsibility does not cease at the end of the employment relationship.

Direct Costs of Ill Health

Of all employer responsibilities in the health area, none has gotten more attention than health insurance. At a recent meeting of benefits managers from different companies, typical war stories were exchanged about the staggering year-to-year increases in health insurance costs. One benefits manager lamented having to tell the chief financial officer that health benefit costs would increase 24 percent next year, whereupon the person he was addressing responded, "You think that's bad? My carrier proposed a forty-three percent increase, and I'm afraid if I bring that to my boss, after the twenty-five percent increase last year, I may not be around to tell anyone about next year's increase."

Health insurance costs have grown at a much faster rate than almost any other cost of doing business; in 1980, they rivaled pension costs as the single largest component of the benefit dollar (Table 1.9). Much of this increase is pure inflation, frequently growing at more than 150 percent of the rate of increase in the consumer price index (CPI). A smaller yet significant part of the increase in health insurance costs is due to a growth in health care benefits. Almost all large employers and many smaller employers pay all or the lion's share of the premium costs of health insurance for their employees. Benefits have been expanded: major medical, prescription drugs, dental care, care for drug abuse and alcoholism, vision care, podiatric services, home health services, and more recently, hospice care and use of

TABLE 1.9
EMPLOYEE BENEFITS AS PERCENT OF PAYROLL, 1981

Type of Benefit	*Percent of payroll*
Total employee benefits	37.3
1. Legally required payment (employer's share only)	9.0
a. Old-age, survivors, disability and health insurance (FICA taxes)	6.3
b. Unemployment compensation	1.2
c. Worker's compensation (including estimated cost of self-insured)	1.4
d. Railroad retirement tax, railroad unemployment and cash sickness insurance, state sickness benefits insurance, etc.	0.1
2. Pension, insurance and other agreed-upon payments (employer's share only)	12.7
a. Pension plan premiums and pension payments not covered by insurance-type plan (net)	5.2
b. Life insurance premiums, death benefits; hospital, surgical, medical and major medical insurance premiums, etc. (net)	6.0
c. Short-term disability	0.4
d. Salary continuation or long-term disability	0.2
e. Dental insurance premiums	0.4
f. Discounts on goods and services purchased from company by employee	0.1
g. Employee meals furnished by company	0.2
h. Miscellaneous payments	0.2
3. Paid rest periods, lunch periods, wash-up time, travel time, clothes-change time, get-ready time, etc.	3.4
4. Payments for time not worked	10.0
a. Paid vacations and payments in lieu of vacation	5.0
b. Payments for holidays not worked	3.4
c. Paid sick leave	1.3
d. Miscellaneous payments	0.3
5. Other items	2.2

Source: Reprinted with the permission of the Chamber of Commerce of the United States of America from *Employee Benefits 1981.*

birthing centers have been added to what was twenty years ago a much leaner offering.

Collective bargaining has led to very rich health care benefit packages. For example, one U.S. automobile manufacturer estimated that its 1982 medical benefits costs were more than $2,500 per participant.[21] In many cases new benefits have been negotiated without careful analyses of the cost

TABLE 1.10
HEALTH-RELATED EMPLOYEE BENEFITS, EXPENSE PER WEEK, 1981

Type of benefit	Employer's expense per 40-hour work week ($)
Old-age, survivors, disability and health insurance (FICA taxes)	21.76
Insurance premium (life, hospital, surgical, medical, dental, etc.)	20.80
Worker's compensation	4.96
Paid sick leave	4.64
Short-term disability	1.24
Salary continuation or long-term disability	.80
Total health-related benefits	54.20
Total employee benefits	127.44
Health-related benefits as a percentage of total benefits	42.5%
Average weekly salary	341.67
Health-related benefits as a percentage of average weekly salary	15.9%

Source: This table was developed based on several pieces of data contained in *Employee Benefits 1981* (U.S. Chamber of Commerce).

effects over a several-year period and without provisions that would tend to make employees and their dependents discerning purchasers of health care services. For example, early institution of unlimited outpatient psychiatric benefits led to companies spending tens of thousands of dollars for an employee to undergo psychoanalysis. Unlimited podiatric care benefits have led to a remarkable growth in total cost of this benefit in states where podiatrists are numerous.

Health care benefits aren't the only direct health-related costs. Disability payments are high, both under worker's compensation, now twice as large as a percentage of payroll as fifteen or twenty years ago, and under private disability insurance programs, both short-term and long-term. The frequency of these claims has increased markedly and there is a growing pattern of claims that are much more difficult to adjudicate—the disability based on excess stress from the job, the lower back problem which is hard to diagnose but claimed to prevent an employee from returning to work. And employers also pay a tax which goes towards Social Security Disability Insurance.

Health-related benefit expenses, which together constituted 15.5 percent of the average weekly salary in 1981, are summarized in Table 1.10. However, these identifiable direct costs do not constitute the majority of total employer costs of employee ill health.

Indirect Costs of Ill Health

It is estimated that the indirect costs of illness are frequently on the order of two-thirds of the total illness costs.[22] A major indirect cost is illness-related absenteeism. Turnover, which is expensive, is another consequence of ill health. Costs of recruiting, hiring and training a new person to assume the responsibilities of someone who left due to ill health is expensive both in dollars and in managers' time. Depending on the nature of the job and the labor market for those skills, the cost can sometimes equal six months to two years of pay. One recent investigation of the cost of turnover to employers found the approximate replacement cost for an experienced systems programmer in a data processing department to be $67,000. By way of comparison, another study found the same cost for industrial engineers in a government setting to be $20,000. The cost for replacing bank tellers, a position requiring significantly less training than the other two positions, has been estimated at $1,000.

Reliable estimation of the costs of turnover is difficult because these costs vary widely among employers, even those with similar characteristics. More difficult to estimate, however, is the cost of lower productivity from employees in less than good health but who remain on the job. A manager whose lunch is largely alcohol is operating at greatly reduced capacity, and a line worker with a similar problem is at very high risk of having a serious accident on the job. In either case production can be adversely affected. More insidious yet is the cost of ill health in terms of reduced service or quality of product. The quality control manager worried about his marital problems is more likely to let a defective product slip by. The retail clerk whose obesity and lack of exercise have contributed to chronic lower back pain will be less agreeable toward shoppers and is more likely to have put the wrong merchandise in the box, much to the surprise of the shopper upon arriving home. Even the loss of state and federal personal income taxes due to illness-related unemployment leads to higher tax burdens on companies.

An important part of the employer cost burden comes from dependents who make up the majority of individuals covered by company health care policies. Although the company may not shoulder the full burden for the premium costs for dependents, it is usually picking up the majority. The 1983 Hay-Huggins survey of noncash compensation practices in industry found that of 854 participating employers, 40 percent paid fully the cost of dependent medical coverage and another 30 percent contributed at least 75 percent of the cost.[23] In many companies, one-half to two-thirds of claims costs is for dependents. This argues for the importance of employers involving dependents in programs that can lead to improved health and more efficient and effective use of health care services.

Many employers pick up the cost of health insurance policies for retirees to supplement Medicare coverage. The cost of this coverage is growing rapidly as Medicare benefits are eroded while costs of health care escalate. Rapid growth of the pool of retirees is illustrated by the dramatic experience of a major automobile manufacturer. In 1978, each active employee at the company was supporting benefits for 0.5 inactive persons (i.e., retirees and surviving spouses). By 1982, the ratio was 1.0 inactive person for each active employee due to a swelling of the number of retirees during a period of high structural unemployment. The result of these shifts is higher health-related benefits costs regardless of the inflation rate.

Aging of our population is reflected not just in a growing pool of retirees, but also by an increase in the average age of employees. One manufacturer saw the average age increase from 38.5 to 43.5 in only five years. Even a few years' difference in average age of a workforce can lead to a very large increase in the frequency of many diseases. For example, the risk of a heart attack for a man with the same risk characteristics can increase 63 percent during the five-year period from 40 to 45. This translates into much higher utilization of health care services, higher disability costs, and a corresponding increase in indirect costs. Thus, there is a large price for doing nothing.

On the optimistic side of the aging problem, there is some evidence that survival curves are assuming a more rectangular shape, in which more and more of the nontraumatic, premature deaths have been eliminated. The major chronic health problems of later life which remain can be postponed by changes in lifestyle. This postponement appears to be longer than the period that the average life span can be extended. Employers thus have the opportunity, through lifestyle-change programs, to extend the period of high quality life and raise the average age at which a serious infirmity develops. In turn, the time between infirmity and death will be reduced. If this occurs, employers, employees and retirees are all winners. Quality of life will be improved while health care utilization and attendant costs will be decreased.[24]

Employers also have opportunities to change the health care system in ways that improve health and reduce costs. Until recently, providers of health care largely have gotten their way in terms of growing technology, increases in prices and expansion of services. Some definite improvements have resulted, but some of what has occurred may not be necessary or desirable. Involved private payers can help the health care system evolve in ways that maximize efficiency and focus clearly on the outcome of care rather than the science and the process. Corporate health management in this sphere involves becoming active and even aggressive in arranging for health-effective care for employees, either alone, through business coalitions, through insurance carriers, or health care claims administrators.

The chapters which follow take fuller account of these opportunities and report and analyze employer health management efforts to date. Suggestions are offered for how to make health management an even more effective set of activities and how to decide on the effects and value of these efforts. In addition, these chapters treat what is likely to happen to employee health-related benefits and costs in the future.

NOTES

1. McKinlay, J.B. and McKinlay, S.M. "The Questionable Contribution of Medical Measures to the Decline of Mortality in the United States in the Twentieth Century," *Milbank Memorial Fund Quarterly/Health and Society*, Vol. 55, No. 3, 1977, pp. 405–428.
2. U.S. Department of Health and Human Services. Public Health Service. *Health United States 1981*. DHHS Pub. No. (PHS) 82–1232. Hyattsville, Maryland: U.S.G.P.O., 1981.
3. U.S. Department of Commerce. Bureau of the Census. *Statistical Abstract of the United States: 1981*. Washington, D.C.: U.S.G.P.O., 1981.
4. Op. cit. (*Health United States 1981*).
5. U.S. Department of Health, Education, and Welfare. Public Health Service. National Institutes of Health. *Proceedings of the Conference on the Decline in Coronary Heart Disease Mortality*. NIH Pub. No. 79–1610. Washington, D.C.: U.S.G.P.O., 1979, pp. xxiii–xvii.
6. Borhani, N.O. "Mortality Trend in Hypertension, United States, 1950–1976," in *Proceedings of the Conference on the Decline in Coronary Heart Disease Mortality*. Ibid., pp. 218–235.
7. Center for Disease Control, Public Health Service. *Ten Leading Causes of Death in the United States, 1975*. Atlanta: Center for Disease Control, 1979. As cited in Hettler, B. "Wellness Promotion and Risk Reduction on a University Campus," in *Promoting Health Through Risk Reduction*. Faber, M.M. and Reinhardt, A.M. (editors). New York: Macmillan, 1982, pp. 207–238.
8. Steel, K., Gertman, P., Crescenzi, C. and Anderson, J. "Iatrogenic Illness on a General Medical Service at a University Hospital," *New England Journal of Medicine*, Vol. 304, No. 11, 1981, pp. 638–642.
9. U.S. Department of Health and Human Services. National Center for Health Statistics. *1979 Revised Mortality Rates* (Unpublished).
10. Sinclair, J.C., et al. "Evaluation of Neonatal Intensive Care Programs," *New England Journal of Medicine*. Vol. 305, No. 9, 1981, pp. 489–494.
11. Pomerance, J.J. "Cost of Living for Infants Weighing 1,000 Grams or Less at Birth," *Pediatrics*. Vol. 61, No. 6, 1978, pp. 908–910.
12. Fisher, C.R. "Differences by Age Groups in Health Care Spending," *Health Care Financing Review*, Vol. 1, No. 4, 1980, pp. 65–90.
13. Health Insurance Association of America. *Source Book of Health Insurance Data, 1981–1982*. Washington, D.C., 1982.
14. Colvez, A. and Blanchet, M. "Disability Trends in the United States Population 1966–1976: Analysis of Reported Causes," *American Journal of Public Health*, Vol. 71, No. 5, 1981, pp. 464–471.
15. U.S. Department of Health and Human Services, Social Security Administration. *Work Disability in the United States: A Chartbook*. Washington, D.C.: U.S.G.P.O., 1980.
16. U.S. Department of Health and Human Services, Social Security Administration. *Disability Survey 72—Disabled and Nondisabled Adults*, Research Report No. 56, Washington, D.C.: U.S.G.P.O., 1981.

17. Sunshine, J. "Disability Payments Stabilize After Era of Accelerating Growth," *Monthly Labor Review*, Vol. 104, No. 5, May, 1981, pp. 17–22.
18. Collings, G.H. "Health—A Corporate Dilemma, Health Care Management—A Corporate Solution," in *Background Papers on Industry's Changing Role in Health Care Delivery*, edited by R.H. Egdahl, New York: Springer-Verlag, 1977, pp. 16–17.
19. *Employee Health and Fitness.* Vol. 2, No. 4, 1980, p. 43.
20. Goldbeck, W.B. *A Business Perspective on Industry and Health Care.* New York: Springer-Verlag, 1978, p. 37.
21. Unpublished company data.
22. Hartunian, N.S., Smart, C.N. and Thompson, M.S. "The Incidence and Economic Costs of Cancer, Motor Vehicle Injuries, Coronary Heart Disease, and Stroke: A Comparative Analysis," *American Journal of Public Health*, Vol. 70, No. 12, 1980, pp. 1249–1260.
23. Hay Associates. *1983 Noncash Compensation Comparison.*
24. Fries, J.F. "Aging, Natural Death and the Compression of Morbidity," *New England Journal of Medicine.* Vol. 303, No. 3, 1980, pp. 130–135.

WHAT'S WRONG 2
WITH THE HEALTH
CARE SYSTEM?

At a lunch of three friends there was a discussion over who would pay the check. Steve, a self-employed businessman, offered to take it, knowing he could write it off as a tax-deductible business expense. Mary, a lawyer, graciously offered to pay it with her firm's credit card. Her firm could consider the expense as part of its overhead. It would eventually be part of the expenses considered in upwardly adjusting her hourly billing rates. But Marvin brushed away these offers and insisted on paying the check. As a hospital administrator, he could put it on the hospital's credit card. The cost would go into the base rate and a share would be paid by the various insurers who pay hospital bills. More advantageously, the amount would become part of the hospital cost base upon which year-to-year allowable cost increases are predicated. As a result, the hospital would continue to be paid for that lunch, again and again, in greater amounts in the years to come.

MAJOR PROBLEMS WITH THE HEALTH CARE SYSTEM

This tale hits close to the truth of how hospitals have been reimbursed for services rendered. More than $150 billion goes to the United States hospitals every year. Yet, with few exceptions, they do not provide those who are about to use their services a binding or even nonbinding estimate of what their services will cost. Rather, they bill and get paid based on what they feel is appropriate. In some cases they are required by regulatory authority or payers to justify the bill based on their costs; in many other cases they are not. An analogous situation in business would be suppliers who did not quote prices but simply got the ordering company to agree to pay whatever the supplier billed them. If the way hospitals have been paid were normal business procedure, a company needing raw materials or services would call up a supplier at random, and without getting competitive bids or even asking the price the supplier would charge, would place an order. While reimbursement approaches are changing, due in large measure to new payment principles for Medicare recipients who use over one-half of hospital services, private insurance still pays primarily based on charges or costs.

Physicians and other health workers are professionals who charge fees based on what they feel their services are worth, based on an expectation of a certain annual income and what those who pay for their services are willing and able to pay. Few consumers of these services either ask about prices in advance or shop around to make sure that the price is in line with others providing similar services.

Based on these approaches to setting prices for most health care institutions and most health professionals, it should not be surprising that the charge for similar services in the same community can vary considerably. Sometimes the most expensive providers charge more than double that of the lower cost providers. How can it cost twice as much to have a gallbladder out or knee cartilage removed in one hospital as in another? The first place to look for an answer is payment mechanism.

Provider Reimbursement

Over 90 percent of hospital bills are paid by insurers, companies and public sources, not by individuals. Individuals pay only about one-third the cost of physician services. Public sources pay for the vast majority of nursing home, kidney dialysis and other long-term care services.

Under private insurance, hospitals are reimbursed primarily based on their costs. Whatever a hospital says the care costs, within limits which are not always well defined, they receive in reimbursement. Some hospitals rebuild or add to their physical plant more often and at greater expense per square foot than others. Some are fully occupied while others are only half-full; the full hospital has twice as many patient days over which to spread its fixed costs as the other.

The reimbursement approaches have been based on costs or charges, not on whether or not the hospital is cost competitive, whether it has an appropriate rate of utilization or whether it has good internal management practices. This principle of retrospective reimbursement based on costs is built into the original legislation that established Medicare, the largest single payer of hospital services.

Blue Cross/Blue Shield generally pay the lesser of costs or charges, and commercial insurers pay based upon charges, which are almost always above costs as defined by Medicare or Blue Cross/Blue Shield. Commercial insurers, until some recent changes, have been precluded by law from negotiating contracts with hospitals to obtain a lower charge. Most new hospital services do not require special review or authorization by insurers to have these services paid as part of the hospital bill. Medicaid, the state-federal program which pays for the health care of many who can't otherwise afford it, uses a number of mechanisms to reduce how much it pays. These efforts, however, generally lead to the costs they are excluding being transferred to the other payers through higher prices. Lack of a unified set of principles of hospital reimbursement has frequently led to costs not being allowed for payment by one payer being transferred to the others. When attempts have been made to reduce public spending for health services, the costs to private payers have risen even faster due to this cost shifting.

Traditional principles of hospital reimbursement are inflationary, since the more services used, the more increased costs can be justified. With hospitals trying to obtain maximum reimbursement, neither physicians nor hospital administrators have had incentives to attempt to minimize cost increases by concentrating on efficiency in the delivery of services. Hospital expansion has been aided by a predictable revenue stream from third-party payers. Guaranteed revenues have led not-for-profit hospitals to borrow money for expansion rather than to continue the traditional approach of waiting for money in hand before building. Reflecting the unique way health care is viewed by legislators, the interest paid by nonprofit hospitals to their bond holders has been exempted from both federal and state taxes, allowing these hospitals to obtain very favorable interest rates. For-profit hospitals have been able to secure funding both from debt markets, although not usually at tax-exempt rates, and equity markets, due to very rapid growth and high return on equity.

More than half of all physician reimbursement is based on "usual and customary" fees, which are set by each physician based upon what he or she believes to be appropriate, modified by the profile of fees for his or her area and specialty. This system allows two physicians with similar training, skills and energy to practice in the same community and set very different rates for their services. Under the "usual and customary" approach, how much insurers will pay *next* year is based upon what was charged by a physician *this* year. It is, therefore, in a physician's long-term financial interest to charge more, even if insurance won't fully reimburse it this year. The more physicians increase their fees in a given year, the more the area profile for physician fees will increase in the future.

Physicians As Consumers

If a consumer is defined as the person who makes purchasing decisions, physicians—not their patients—are the primary consumers of health care.

If you go to a doctor for relief from your stomach pain or because you've discovered a lump, you are not likely to decide what services you will receive. You could wind up having a number of blood tests, be sent for an x-ray or even be told you need to be in the hospital without being actively involved in the decision. It is the physician who usually has the greatest role in deciding what services you receive. Of course, patients are becoming more active and insisting on a role in making those decisions after understanding their health care options. Nonetheless, even informed consumers will properly rely on a trusted physician's judgment about what their condition warrants.

Most physicians are paid on a fee-for-service basis. Reimbursement is based on the number of services provided. This reimbursement system provides few incentives to minimize the services provided. There are no monetary rewards for efficiency. There are also no good reasons for physicians to strongly resist patient requests for services which the physician believes are not indicated. Insistent patients will simply go elsewhere to get these services. For example, a person with classical stress-related headaches may insist on a CAT (computerized axial tomography) scan to relieve his anxiety that he has a brain tumor. Relief of anxiety is important, but should the health care system be burdened with several hundreds of dollars of expense reimbursed through insurance for this purpose alone?

Demand for *caring* services as opposed to *curative* services is very price-elastic. When insurance reduces the direct cost of a physician's visit to the patient by several dollars, demand for this care will always increase, irrespective of whether cure is possible. Need, in most situations, is impossible to define, so that few objective criteria exist to place enforceable limits on demand.

Reimbursement principles under most policies strongly favor technology over personal services, such as education and counseling. Any primary care physician can generate a much greater return per hour from routinely ordering a full panel of blood tests, a stress EKG, a resting EKG, pulmonary function studies and additional tests conducted by technicians or nurses than by spending time listening, responding to questions and providing advice. Reimbursement is in general more liberal for surgical than medical procedures, and medical specialists are paid much more than primary care physicians for the same services, regardless of their respective experience.

Misconceptions About Health Care

Public misconceptions about the relationship of health care to health have also contributed to what's wrong with the United States health care system. Some of the common but largely incorrect perceptions include:

Health care is the major determinant of health.   Regular checkups and access to the best doctors and hospitals are the most important ways to safeguard one's health.

Correction: The major determinants of health are health habits, environment and biology (genetics). Annual physical exams are no longer recommended for the "well" (i.e. asymptomatic) population. Routine chest x-rays and electrocardiograms appear to have little if any value in the "well" population. However, periodic check-ups, based on age, sex and medical condition, are important.

More health care is better health care.   The best medical care requires full use of the ever-expanding array of diagnostic equipment. Up-to-date doctors are those who are comfortable using CAT (computerized axial tomography) or PET (positron emission tomography) scans, ultrasound and radioactive isotopic techniques routinely to assess body functions. A physical exam without graded exercise electrocardiography or without a battery of lab tests is not really thorough. Not receiving a prescription from a physician one has seen may mean that the physician is not doing everything he or she should be.

Correction: More health care is not always better. In many cases it is worse, leading to undesirable side effects. Many experts believe that the marginal benefit of more health care for the majority of the population, given the amount currently provided, is either very small or negative. The Rand Health Insurance Experiment, a major study of the effects of cost-sharing on consumers' use of medical care services, found that adults whose care was free used many more services but did not generally demonstrate a better health status than adults required to share in the costs of their care.[1] Overtreatment is associated with as many hazards as undertreatment. Many diagnostic procedures and treatments have the potential for significant adverse side effects.

The newest equipment is necessary for high-quality care.   Care cannot possibly be provided as well in a thirty-year old building as in a brand-new facility. An older x-ray machine can't take pictures as well as a new one. Newer equipment leads to better diagnoses and therapies which translate into improved results.

Correction: While good equipment is important, many of the new machines and procedures bring insufficient improvements to justify their additional expense, which is usually substantial. There has been no convincing demonstration that providing services in an older building leads to a poorer result than in a more modern one.

What is needed to provide quality health care is best assessed by health professionals. If a health professional says that a certain piece of equipment is required to maintain high-quality care, it should be purchased. Health care professionals agree on what constitutes high-quality care and this care gives the best outcome.

Correction: For any consumer, the quality of health care is both difficult to define and to assess. While there are some bad results which can be traced to patently substandard care, in most cases it is difficult to relate different diagnostic and treatment processes to differences in health outcomes.

Health professionals are in the best position to decide what medical care is needed for each patient. While patients' wishes should be respected, in general their role in deciding what diagnostic or therapeutic procedures are appropriate will naturally be limited due to lack of expertise about medicine.

Correction: Certainly the patient is in the best position to assess if the human aspects of the process of medical care are of acceptable quality. And patients have the right to make decisions about what medical care they receive, after understanding their options and the risks and benefits associated with each option.

One commonly held perception is correct: the quality of health care in the United States is good. Health professional training programs are generally very good, if not excellent, and applicants for them are generally well prepared. With the large number of physicians trained in both medical and surgical specialties, up-to-date equipment, diagnostic tests and therapies are available in almost all parts of the country. More research on better ways to diagnose, care for and cure problems is undertaken in the United States than in any other country, and probably more than in the rest of the world combined. Hospitals and most other health care facilities are well maintained. If you have a serious illness, there is no better place to seek medical care than in the United States. But despite these excellent features, the health care system itself is still seriously ill. The illness is one of excess, with little thought to what the national health care priorities should be or what the incremental value is of each added health professional, new technique or new hospital.

Failure to Set Limits

How much medical care is too much? When kidney dialysis first came under Medicare reimbursement, those patients put on dialysis were generally younger and had primarily kidney problems. Increasingly, the indications for dialysis have broadened. People with end-stage diabetes who are blind, have had a leg amputated and have very advanced cardiovascular disease are sometimes now being put on dialysis when their kidneys fail as well. The result is that a program originally estimated to cost several hundred million dollars is expected to cost the federal government more than two-and-a-half billion dollars in 1985.[2] We are spending more and more for health care during the last year of life. For example, the Medicare program spent $3,351 per enrollee who died in 1976 during their last year of life, more than six and a half times the $509 spent per surviving enrollee.[3] Often painful and unwanted medical treatment is being substituted for support that makes dying easier and more dignified for both the patient and family.

As a nation we have been understandably reluctant to set limits on what is appropriate care. There is always a chance that any additional treatment might help in the treatment of a particular condition, that a new piece of equipment or a very expensive drug could help some group of patients have a better outcome. An implicit assumption in the decisions that are usually made about health care services is that human life has an infinite value. We have decided to ignore the fact that we have to make choices. While we spend over $1,000 per day to keep a 93-year-old person with very little chance of recovery on life support in an intensive care unit, many children remain inadequately immunized. While the dispute rages whether liver transplants at up to $300,000 each should be covered under insurance, inexpensive preventive or therapeutic services of proven effectiveness, such as well-child care and hearing aids, are frequently not covered by insurance. What is and is not covered by insurance is not dictated by what works. Rather, insurance protects against expensive medical care becoming a financial burden for patients or a barrier to receiving care, by covering best the most expensive types of services.

The public, government and health care payers are recognizing that there are limits. It is untenable and unaffordable to continue saying that whatever medical service could conceivably improve health care should be bought, regardless of price or impact. Other industrialized nations have addressed the problem squarely and set up criteria for when particular forms of care will be provided and when they will not. Kidney transplants, for example, are limited in Great Britain to those under a certain age. In Sweden, heart transplants have not been made available unless the municipality (county) makes an explicit decision to provide funds for the individual to receive a heart transplant.

Other Problems with the Market for Health Care

In general, patients do not make the decision of which hospital to enter based on price. Instead, they choose the hospital where their doctor has admitting privileges. Even if they tried to make a decision based upon price, it would be difficult. Hospitals have different approaches to pricing. An item which is part of the basic room and board charge in one hospital may be a separate charge in another. To compare bills for the same services in several hospitals, it would be necessary to ask the price for each of tens or hundreds of items. And few hospitals are willing to agree in advance to a negotiated price for a hospital stay.

Similar issues cloud the determination of the costs for physician office visits and related services. While one physician may charge somewhat less for a basic office visit, he or she may tend to order more tests, which would lead to a higher total cost. Few people even attempt to price-shop for physician services. If they have a good relationship with a physician, they are unlikely to change for a lower price, especially when insurance pays the lion's share, and the same percentage despite considerable disparity in charges.

Fragmentation

Another problem with our health care system is that it encourages fragmentation. If a person is not a member of a health maintenance organization, he or she is buying health care à la carte. Specialist attention is sought for problems that could just as readily be handled by family physicians or general internists. Expensive emergency units and newly available urgent care centers appear to be replacing the family doctor for many Americans, especially during evenings and weekends. Continuity of care frequently suffers. Physicians, always sensitive to the possibility of malpractice suits, are even more cautious when patients are seeing them only once; they tend to order more tests for patients they don't know.

Employer-Sponsored Health Insurance Benefits

Health insurance, developed originally to prevent financial disaster due to hospital bills, has achieved its purpose but has also created some new problems, including benefit packages that employers can no longer afford.

Health care benefits have been shaped primarily by union negotiations, employer desire to maintain competitive benefit packages, very optimistic estimates about future costs of new or expanded benefits and the general rate of increase in health care costs. They have also been strongly influenced by the ability of employers to deduct health insurance premiums as a business expense, thus reducing their tax burden. In 1982 the health care tax expenditure from employer contributions to health insurance premiums amounted to $23 billion, $16.5 billion from income tax deductions and $6.5 billion from payroll tax deductions.[4]

Until relatively recently it was assumed that any health care provided under insurance coverage was necessary and appropriate. Broadening health benefits on a regular basis became one measure of an enlightened company.

Workers and their dependents were successively given greater access to the medical, dental and mental health care of their choice. Giving in on a union health-care coverage demand was preferred over spending the same money by accepting a new work rule or by giving more paid holidays.

Several important ramifications tended to be overlooked or their effects underestimated as insurance coverage became increasingly liberal:

- Increasing insurance might lead to too much care.
- Broadening third-party coverage would spur health costs to increase at a faster rate than almost all other goods and services.
- Almost all the benefits were to provide care after a problem had appeared rather than to prevent or postpone it.

When initial improvements in health insurance coverage and employer contributions were negotiated or given by management, health care insurance costs to employers were on the order of 1 to 2 percent of payroll.[5] The full effect of broadening third party coverage in health care costs wasn't seen until the latter half of the 1960s, spurred on by the Medicare principles of reimbursement. Currently health insurance vies with pension plans in accounting for the largest single portion of the benefit dollar, frequently averaging in the range of 4 to 6 percent of payroll.[6]

Too Many Providers

In many parts of the health care system, oversupply is a serious and expensive problem. Due to federal incentives provided through increased capitation grants, medical schools greatly increased their enrollment in the late 1960s and 1970s. As a result, the number of doctors per 100,000 rose from 142 in 1960 to 202 in 1980.[7] And the numbers are expected to continue to increase rapidly. Between 1981 and 1990 it is estimated that the number of active physicians will grow 27 percent, or three times as fast as the United States population. In 1990 there will be an estimated 591,000 active physicians.[8] Already there is virtual consensus that a surplus exists. Physicians in most specialties have great difficulty establishing independent practices in most large urban communities. Getting a job with an existing group is equally difficult. Newly trained physicians are frequently accepting salaries lower than those offered two to three years ago. Dentists in many parts of the country are having great difficulty attracting sufficient patients to maintain the income they received in prior years. Despite this, the number of active dentists is projected to increase from 129,000 in 1981 to 155,000 in 1990, a 20 percent increase.[9] Psychologists also frequently experience problems similar to dentists in attracting enough patients.

Oversupply is also present in hospital facilities. Many areas have average hospital occupancy rates of 60 to 70 percent, below an efficient level.

Overall, the average national hospital bed utilization rate is only 74 percent.[10] Reimbursement based on what a hospital costs to run, even if it has high fixed costs and low occupancy, is a major contributor to this problem. And if every hospital had the age-adjusted low number of bed days per 1,000 population that HMOs experience, an additional 30 to 40 percent of hospital beds might not be utilized.

The Future

In the short term, a growing surplus of health professionals, each of whom orders health care services for his or her patients, coupled with liberal reimbursement through employer-sponsored health insurance, should contribute to a continued increase in per capita health care expenditures for employers. In the short term it is very likely this will continue to happen at a faster rate than for most other products and services.

However, the winds of change are already being felt, and it is likely that their intensity will increase. The last half of the 1980s will probably be remembered as one of the periods of greatest change in the delivery of health care. Unless the attitude of government changes drastically over the short term, it is likely that competition will become increasingly fierce. Reimbursement will increasingly be based upon a health problem rather than directly related to the specific services provided to an individual patient. Insurers will contract with providers for a package rate, and employers will provide strong incentives for enrollees to use providers with whom there are contracts. Government subsidies for health professional education will either diminish or be contingent upon significant reductions in number of students, especially in medicine and dentistry.

An increasing number of health professionals will work for a number of large health care organizations that pay them a salary, with incentives for efficient use of services for their patients. Health care will come more to be regarded in much the same way as a number of other services and will lose some of its special, almost mystical veneer. More health professionals will become unionized and physicians may strike with increasing frequency as the large differential between their salaries and those of others with professional education is slowly eroded.

Prepayment for health care services of many types will increase. Fewer and fewer physicians will be able to rely on fee-for-service patients. Payment for services will become more uniform between providers, and advertising will increase so that consumers are more aware of the prices they are confronting in the market. More health professionals will have to accommodate to patients' leisure hours, providing services on evenings and weekends. Hospitals, physicians and other health care service providers will come together more and more to constitute systems appropriate for prepaid care.

Employers are likely to band together to increase their purchasing power. The largest such aggregations will get the best prices.

Public debate over what health care we can afford as a nation is likely to intensify. Guidelines for withholding care should be recognized increasingly as rational and humane. Whether poorer Americans will receive care from the same systems that provide care to their more affluent compatriots is an open question, determined in part by the degree of health care cost inflation, the perspective of the political party in power both in Washington and in the states, and industry's perspective. Quality of care, always a concern, will be increasingly monitored by sophisticated data systems which look at both process and outcome, backed up by respected professionals. As incentives grow for efficiency and economy in the use of services, concerns about the provision of insufficient care may gain in legitimacy.

If most of this vision is correct, discussions of what is wrong with the health care system a decade from now will be much different than the summary in this chapter.

NOTES

1. Brook, R.H., Ware, J.E., et al. "Does Free Care Improve Adults' Health?" *New England Journal of Medicine*, Vol. 309, No. 23, pp. 1426–1434, 1983.
2. U.S. Department of Health and Human Services. *End-Stage Renal Disease Second Annual Report to Congress*. Washington, D.C.: U.S.G.P.O., 1980.
3. Lubitz, J., Prihoda, R. *Use and Costs of Medicare Services in the Last Years of Life*. Office of Research and Office of Statistics and Data Management. Health Care Financing Administration. June 29, 1982.
4. Congressional Budget Office. *Containing Medical Care Costs Through Market Forces*. Washington, D.C.: U.S.G.P.O., 1982.
5. U.S. Chamber of Commerce. *Employee Benefits Historical Data 1951–1979*. Washington, D.C., 1981.
6. U.S. Chamber of Commerce. *Employee Benefits 1981*. Washington, D.C., 1982.
7. U.S. Department of Health and Human Services. Public Health Service. *Health United States 1981*. DHHS Pub. No. (PHS) 82–1232. Washington, D.C.: U.S.G.P.O., 1981.
8. U.S. Department of Health and Human Services. Health Resources Administration. *Third Report to the President and Congress on the Status of Health Professions Personnel in the United States*. Washington, D.C.: U.S.G.P.O., 1982.
9. Ibid.
10. Op. cit. (*Health United States 1981*).

3 OVERCOMING INERTIA: A CHALLENGE TO EMPLOYERS

"Health care benefit costs are escalating at about twenty percent per year, but luckily labor is only twenty percent of my total costs, and my total benefits are only thirty percent of payroll." So spoke the president of a 2,000-employee manufacturing firm who wasn't sure that health management should be a top priority. The following conversation ensued between him and a consultant.

CONSULTANT(CT): What is your after-tax net on sales?

PRESIDENT(PT): It varies between 2 percent and 4 percent, with 3 percent a reasonable average figure.

CT: How much of your benefit costs are going toward your health care premiums?

PT: About one-quarter, say 7 percent of payroll.

CT: If you could reduce your rate of inflation in these costs from 20 percent to 10 percent for each of three years, your compounded increase would be 33 percent instead of 73 percent. Assuming you started with health benefit costs at 7 percent of payroll, you would be saving 40 percent (73 to 33 percent) of 1.4 percent (20 percent × 7 percent) or .6 percent of sales. This would increase your post-tax net by 30 percent (2.6 percent ÷ 2 percent) in the bad years and by 15 percent (4.6 percent ÷ 4 percent) in the good years.

PT: Actually, my financial vice-president is more worried about worker's comp. He projects it will rise at about 35 percent per year based on what the insurance company is saying about our claims pattern. Worker's comp is now about 2 percent of payroll.

CT: Once again there are some things that can be done to help reduce the rate of rise. Nothing is likely to keep it where it is, but you may be able to keep those increases closer to the general inflation rate. What about disability? How are your costs there?

PT: Until recently there wasn't a problem, but now with some of our older guys getting up there in years we're seeing more disability retirements: heart attacks, nervous diseases, and a lot of our younger workers are having horrendous accidents. Some will never get back to work.

CT: Unless some effort is made to structure the plan to provide incentives for returning to work and to provide more rehabilitation options, it may be very difficult not to see those costs rise faster than any other. But even more important, how much do you think ill health, whether physical or mental, costs you in terms of absenteeism and reduced productivity on the job?

PT: I don't know, but a lot—maybe another 10 percent of payroll. And some of that is due to the alcohol and drugs. You should see the parking lot, littered with beer bottles.

As this only slightly fictionalized account points out, health-care benefit costs are merely the tip of the iceberg. The cost of ill health is much greater, but too often the various cost components are not combined and considered as a total problem.

Management of health-related benefits and services is frequently also fragmented. Health insurance may be managed by the benefits department, worker's compensation by the insurance department, health promotion by the medical department, the employee assistance program by the training department and safety by its own separate department. The line manager is left alone to wrestle with the problems of absenteeism and many other elements of productivity. In larger companies, the point of convergence of all these responsibilities is usually the CEO or a senior executive.

It is rare that companies make an effort to bring together all the components of the cost of ill health and examine the full dimensions of the problem. Yet all the pieces are inextricably interrelated. A heart attack and associated medical expenses might have been avoided by attention to a worker's high blood pressure. After the heart attack occurred, more concerted efforts at rehabilitation might have prevented a disability retirement that required the manager to start looking for a replacement and to take time to train a new person. The twenty-three year old with a drinking problem had high absenteeism, but also wasn't pulling his weight when he *was* at work. Predictably, he got a finger caught in a machine, leading to a large worker's compensation award. Two months after returning to the job he was injured in a car accident, probably associated with excess alcohol intake, and at twenty-four he is a paraplegic, on disability, with little likelihood of ever returning to work. He now represents an over-$1 million liability to the company. For every employer the health problems are there, the costs are there, but the recognition of both how the problems interrelate and how much can be done to reduce their frequency and magnitude has often lagged.

As recognition of the dimensions of the health cost problem has grown, many corporations have actively sought better methods to manage their health care costs. Still it has often been hard to sell top management the idea that the problem deserves attention and resources in proportion to its size. One reason for a cautious attitude is that health care management requires expertise which most corporations haven't had. To get directly involved in the delivery aspects of health care or attempt to have a major impact on the frequency of non work-related employee health problems are major commitments outside traditional employer areas of interest. In many ways it is like starting an entire new type of business, a business very foreign to most corporations and large nonprofit employers. Like other possible business ventures, it requires a significant investment both in time and in manpower. But in addition, it has a significant potential for strong adverse effects on employee relations. Protection from the soaring costs of health care are among the most prized of employee benefits.

Many strikes in recent years in auto, coal, oil and other industrial sectors have been over whether management or labor would pay the increase in the health insurance premiums or over the proposed changes in these benefits. What benefits director wants to have irate employees protesting that they should not have to get a second opinion before their elective surgery can be covered? What personnel director wishes to confront incensed employees arguing for payment of hospital bills for the two days that utilization review decided were unnecessary? It is difficult to refuse to pay for a major health problem which is marginally covered if the company wishes to be sensitive to its employees' problems. In one company the head of human resources wanted to reduce outpatient mental health benefits, which were virtually unlimited and very costly, but found that the largest users were the vice-presidents and their families. He demurred.

Good health care coverage is also a potent recruiting tool. When engineers are needed by a defense contracting firm, recruiting them away from other potential competitors is likely to appear much more important than shaving a few percentage points off the costs of health insurance. Economic conditions also influence the willingness to attack the costs of ill health. When times are good, the potential savings seem small compared to the opportunities for enhanced profits from investments in improving existing company products or services. When times are bad and layoffs are contemplated or occurring, alternate investment options are less attractive and there is a willingness to consider any opportunity for belt-tightening that can reduce losses and the need for terminating career employees.

Another reason for a passive attitude from employers has been the tax deductibility of health benefits. In general, all health-related premiums are tax deductible. It's like spending fifty-cent dollars. And there has been no cap on how much can be spent and still be deductible. On the other hand, most of what an employee spends for health care for the family is in after-tax dollars. The employee therefore pays up to double the amount for the same service. Only those expenditures over 5 percent of gross adjusted income are deductible from federal income tax. Reducing health benefits in essence transfers a greater cost to the employee than what the company saved. Given these circumstances, some employer reluctance to make benefit cutbacks is understandable, as is the strong reaction of employees to any such proposals.

No one questions whether more employers are active today in seeking solutions to the high costs of ill health. More cost-sharing by employers has been a clear trend, and second opinion programs, utilization review, encouragement to join HMOs and employee expense accounts are all being adopted to a greater degree than in the past. Yet there is still legitimate question about the priority accorded the health cost problem by top line managers. An effort to assess corporate attitudes toward health care costs was undertaken by a group of health policy researchers through interviews with about 250 executives (almost always including the CEO or other board-level officer) of major U.S. companies that were: a) randomly selected from various *Fortune* lists; b) selected either from firms in cities that have a reputation for business involvement in health affairs (e.g., Minneapolis, Rochester) or from industries in which there was widely reported special interest in health care costs (automotive and steel); or c) selected from a sample of small firms in the Boston area. The interviewed group also included senior representatives of major carriers selling group health insurance, insurance brokers, health benefit consultants, knowledgable and relevant state and local officials, representatives of health provider associations concerned about the topics being explored, labor union officials and the federal officials responsible for the design of federal employee benefit programs.[1] Among the relevant findings of the survey, which was published in 1981, were:

1. Firms appear willing to provide superior benefits to retain the managerial freedom of a nonunionized work force.

2. The usual locale of benefit design and management under the vice-president for personnel or human resources reinforces the tendency to be generous with benefits, because the typically overriding concern of these groups is recruiting employees and maintaining the workforce morale.

3. Many firms develop benefits to meet their broader compensation goals, and frequently these goals are derived from surveys of compensation practices of "peer firms." Therefore, "major benefit improvements

implemented by the nation's richest or most unionized firms diffuse throughout the economy by means of a chain of interfirm comparisons."[2]

4. Most firms are very sensitive to complaints about poor benefits yields or reports that the employer offers fewer benefits than similar firms. Benefit improvements are generally viewed in a favorable light.
5. Top management rarely expressed a deep interest in health care costs, and were generally more concerned about whether their benefit package was up to the industry average.
6. For many companies, concerns about disability costs or complaints from retirees about inadequate pension benefits loomed as larger problems than health insurance costs.
7. Most firms knew little about the details of their health care claims experience.

The main conclusion of the authors was that "most of the firms we visited felt little impetus to seek changes in health benefits."[3] In other words, top managers were reluctant to take aggressive action to attempt health care cost containment. Strong disagreement with some of these findings has been voiced by business coalitions and others working with employers on cost containment activities. While there is little question that there is currently much greater willingness to cut back and restructure benefits and to offer employees trade-offs within benefits than in 1980 when this survey was conducted, it is probably still true that health care cost containment is only a priority for some top executives. Those companies with the slimmest profit margins or very high personnel costs are the most likely to place strong pressure on the human resources units to do something about health care costs.

It is still a rarity for large companies to put all the pieces of the health problem together and map their interrelationships. Yet in recent years we are seeing the appearance of health specialists within benefits or personnel departments. These individuals may have their primary training in benefits or management but more and more have some knowledge of health or health systems. Health planners and public health analysts are increasingly being recruited into important corporate positions for their special expertise. Effective programs to manage health and minimize the costs of ill health will require not only this expertise in the management of the health benefit but an understanding of how to improve health and prevent avoidable illness. In the future it is likely that there will be a team of health care and benefit professionals in larger corporations who together develop a coordinated health strategy. The corporate medical director, health and safety manager, occupational health nurse, a consulting or full-time mental health professional and those with special expertise in prevention and in rehabilitation would be logical team members to add to benefits and insurance staff.

Line managers hold an important key to minimizing the costs of ill health. As they realize the tremendous dollar cost of ill health to their bottom line and the abundant opportunities to reduce these costs, they will become more active. They have always been concerned about productivity, absenteeism and turnover. But now they are starting to realize that these problems can be responsive to efforts to minimize ill health, physical and mental. Line managers are also starting to find themselves being charged by their corporate office for health insurance, disability insurance and even worker's compensation, based upon the experience of their workforce and dependents. This gives them a much stronger incentive to take active steps to control these costs. Often they need training in how to evaluate their experience and options.

As the problems of ill health are recognized by management as controllable, employers are establishing better systems to determine the costs and are holding a specific senior manager responsible for meeting predetermined health and cost containment objectives. Even more basic is developing a set of general objectives which help guide all corporate health management efforts. The following general objectives were formulated by a concerned company with over 75,000 domestic employees:

1. Develop, promote and maintain a comprehensive, integrated strategy to manage health care expenditures in the most efficient manner on an ongoing basis.
2. Offer incentives to management, providers, employers and carriers to improve the efficiency of medical services while maintaining the quality of care.
3. Optimize the equity and balance of coverage among various groups of employees, including:
   union/nonunion
   salaried/hourly
   field offices
   multidivision offices
   geographic areas
   labor markets
4. Delineate corporate and business responsibilities in order to optimize centralized and decentralized strengths.
5. Develop a corporate environment and set of programs to improve and maintain the health of employees, dependents, and retirees.

As measurable objectives are established, corporate health managers will increasingly be judged by whether they attain the objectives, and bonuses and promotions will depend on this performance. Examples of measurable objectives that might be established for evaluating the performance of a corporate health manager include:

1. Reduce the number of sick leave days used by _____ percent per year for the next three years.
2. Reduce the rate of rise in direct medical care costs to no more than the rate of increase in the medical care component of the Consumer Price Index.
3. Reduce the rate of rise in dental care costs to no more than the rate of increase in the dental care component of the Consumer Price Index.
4. Increase employee awareness and comprehension of cost containment features in the health insurance plan so that by _(date)_ at least _____ percent of employees respond correctly to at least 8 out of 10 questions on a survey regarding features of their health insurance plan.
5. Increase participation in the employee health promotion program so that during _(year)_ , _____ percent of the workforce will participate in at least one program activity and _____ percent will participate in two or more activities.
6. Reduce the frequency of on-site injuries by _____ percent by _(date)_ .

Line managers, also a beneficiary of these changes since most improvements will accrue to their cost center profits, will be more anxious to cooperate in effecting strategies to reduce the employer costs of ill health.

Unions will naturally be opposed to any reductions in benefits or cost containment efforts that limit the choices of health care services or providers by their members. However, most unions recognize that health care services are frequently overutilized, reducing dollars available for other benefits. In addition, cost containment strategies can save money for employees and their families as well as for the employer. Programs designed to reduce preventable diseases and secure early confidential help for personal problems are also in the employee's interest as well as the employer's. It is worthwhile, therefore, for unionized companies to devote substantial effort toward building a shared feeling for the value of health improvement and cost containment programs, and to structure programs so that their value to employees is clear and significant.

KEY STEPS IN OBTAINING COMMITMENT

Whether an employer decides to adopt as a priority the management of the costs of ill health depends on many factors, among them:

- overall profitability
- the general economic picture in the industry

- the percentage that labor represents of the total costs
- the employer's own experience with these costs over the past several years
- the degree of concern about the employee relations aspects of possible action

However, a commitment is more likely to be made and sustained if the following steps are taken:

1. Bring together direct and indirect costs of ill health to the employer, explain them, display them in graphic form and compare them to other types of company costs and rate of cost increases.
2. Make a presentation to top management, including the chief executive officer, that includes:
   a. costs of ill health over a several-year period
   b. projected costs if nothing is done over the next several years
   c. concrete examples of how other companies have been able to reduce some of these costs and the techniques they have used
   d. estimates of cost savings and effects on bottom line in short- and long-term
   e. estimates of the resources required to improve health care management
   f. discussion of the strong interrelationship between health and productivity
   g. discussion of how improved cost-effectiveness can also improve the quality of health care services to insurees
   h. suggested objectives and time frames for their achievement
3. Reach agreement among senior management of long-term goals and short-term objectives after discussion of alternatives. To the degree possible, objectives should be quantifiable, both in dollars and in other relevant terms (e.g., improvements in job satisfaction based on employee surveys).
4. Give a senior person within the organization clear responsibility for health care management. If that person does not have all the pieces of that function reporting to him or her, make explicit coordination mechanisms and a mechanism for resolving disagreements. Over time, improved management usually requires increasing centralization of overall responsibility for health care management within the company.
5. Make available staff with knowledge of the health care system, the company health-related problems, data systems, and suppliers (e.g., insurers or plan administrators). In larger companies it is probably worth the investment to have at least one full-time person assigned to this task.
6. Give line managers of larger units financial responsibility for the health-

related costs of their employees and their dependents. They also should be provided education and training in the health-related cost problems, what they can do as line managers to reduce the magnitude of these problems, and how the corporate staff will help them in these efforts.

If a commitment is obtained from top management, there are a number of ways to increase the likelihood that it will be sustained. Two crucial ways are:

a)   Incorporate health care management as one of the factors upon which a line manager's performance will be assessed.

b)   Invest in collecting and carefully analyzing accurate data on health status, health risks, health care costs and utilization, merge the different data systems whenever possible and have a system, either internal or external, that can provide timely and accurate information to identify problems and help assess the success of health care management activities.

NOTES

1.   Sapolsky, H.M., Altman, D., Greene, R. and Moore, J.D. "Corporate Attitudes Toward Health Care Costs", *Milbank Memorial Fund Quarterly/Health and Society*, Vol. 59, No. 4, 1981, pp. 561–585.

2.   Ibid.

3.   Ibid.

HEALTH INSURANCE *4*
CARRIERS AND
ADMINISTRATORS:
AN OVERVIEW

*T*he largest measurable health-related employer cost is paying for the health care of employees and their dependents. There are four major decisions which employers have to make about providing this benefit:

1. Who will take the financial risk?
2. Who will provide the administrative services and make the payments to providers?
3. What services will be covered and to what extent?
4. What additional approaches, if any, will be used to help control costs paid under the benefit?

This chapter provides an overview of questions 1 and 2. Self-insurance is covered more extensively in Chapter 10. Health care benefit design is covered in Chapters 5 through 8, and other cost containment mechanisms are discussed in Chapters 11 through 14.

As company size increases, so does the likelihood that the company carries all or the majority of the financial risks for the health care services covered under its health insurance plan. In these instances, insurers, who once bore all or the majority of the risk that the claims paid and related administrative costs would exceed the premiums, function as administrative agents of the employer. In these cases they estimate and periodically receive funds from the employer to pay the claims. In many cases they simply write checks against an account maintained by the employer, who puts in additional funds as necessary during the year.

Many companies, however, still prefer to have another organization underwrite all or much of the cost of paying for health insurance claims. Although in general this approach is more costly than self-insuring, the creative mechanisms devised by insurers to reduce the net cost to employers have helped to maintain third-party insurance as a viable option. An additional benefit to letting an insurance company take the risk is that an employer can budget for what his or her costs will be, and not face a large increase in the middle of a fiscal year which requires an explanation to top management about why costs are above projections.

Employers utilize two types of organizations to provide insurance, Blue Cross/Blue Shield and commercial insurers. A more recently formed third type of organization, third-party administrators, join the first two in providing administrative services under self-insurance. While these organizations provide similar services, the approach that each takes to performance of its responsibility differs. Understanding the history that conditions their respective approaches helps to explain the current advantages and disadvantages of each type of organization.

THE "BLUES"

Blue Cross plans were founded under the auspices of the American Hospital Association in the 1930s to help make health care services in hospitals affordable and to provide a steady source of revenue for hospitals.[1] Blue Cross plans were established as a special type of nonprofit institution exempt from usual insurance laws and from payment of federal income tax in exchange for the promise of open enrollment and community rating. Under community rating risks are shared equally among members, with all paying the same amount, regardless of medical condition or prior use of health care services.

Blue Cross has always been closely aligned with hospitals, and the American Hospital Association owned the Blue Cross name and logo until 1972.[2] Traditionally, Blue Cross negotiated discounts with hospitals in return for prompt payment and first dollar coverage, which assured that the entire bill would be paid. Large market share ensured that the discount remained.

Blue Shield plans were initiated in the mid-1940s under the auspices of the American Medical Association primarily for the same basic reasons as the Blue Cross plans. However, ties between physicians and Blue Shield plans have never been as tight as those between the hospitals and Blue Cross. In 1960 the American Medical Association formally severed its connection with the Blue Shield plans, which currently number about sixty-eight. In at least twenty-eight states Blue Cross and Blue Shield plans are merged, as is their national association.[3]

The "Blues" are the most common carriers for plans which provide first dollar coverage (i.e., no deductible), a concept which many plans strongly pushed until quite recently. In fact, many plans did not offer other coverage for many years. To meet the challenge from the commercial carriers, in recent years the Blues have retreated from community rating and have based premiums on the company's experience. They have also offered indemnity plans, not requiring first dollar coverage. While they still offer enrollment to individuals in the community, the prices are extremely high, much above those available to group subscribers.

The Blues are the favored carriers for medium and small businesses, many of which shoulder only a portion of the insurance risk. Overall, the Blues are the single largest carrier in the market. They are the first-ranked insurer for companies with 100 to 499, 500 to 999, and 1,000 to 4,999 employees, according to an analysis of over 20,000 businesses with over 100 employees that must report to the Department of Labor under ERISA.[4] They are also the second-ranked contractor for businesses with 5,000 to 9,999 workers and with over 10,000 workers. Overall the Blues had 37 percent of the contracts reported to the Department of Labor and these contracts accounted for 42 percent of the total dollars.* The major advantage of Blue Cross hospital coverage has traditionally been the discounts that Blue Cross receives from hospitals based upon a negotiated contract. In some states this can be as great as 10 to 15 percent. They have thus enjoyed a considerable cost edge over commercial carriers who have been legally precluded from contracting with providers and forced to pay based on what the hospital charged. Another advantage of Blue Cross and Blue Shield coverage is that physicians and hospitals are more likely to be willing to bill these carriers rather than give the initial payment responsibility to the patient and require the patient to be reimbursed by the insurer. Finally, the Blues have been exempt from paying the taxes on premiums which commercial carriers have had to pay.

COMMERCIAL CARRIERS

Prior to World War II Blue Cross dominated what was a very small market for health insurance. World War II spurred rapid growth of private health insurance due to the exemption from wage and price controls allowed for nonmonetary compensation (i.e., fringe benefits). Commercial insurance carriers began to enter the market in a big way during the war and postwar period. Commercial carriers sold insurance directly to employers and unions, generally comprised of healthier than average individuals. Unlike the Blues,

---

*Information furnished to the Department of Labor applies to either calendar year 1980 or fiscal year 1981.

they were under no obligation to offer community rating or open enrollment. In addition, the commercial carriers generally offered indemnity plans (paying a fixed dollar amount or percentage for each service), not service plans as did the Blues, who were thus obligated to provide first dollar coverage. As a result the commercials undercut the Blues and gained a substantial market share, about 41 percent by the late 1970s, roughly equalling the Blues.[5]

Commercial carriers have favored the use of cost sharing to reduce overall utilization and the rate of health care cost increases. Their plans have traditionally included co-insurance and deductibles. They have been strong supporters of rate setting in a number of states and have fought hard to be allowed to contract with providers. While together they probably account for more total business than Blue Cross/Blue Shield, the business is spread over a large number of carriers.

Together, the commercial carriers are the preferred insurers of larger corporations. Based on the ERISA 1981 survey, the carriers with the greatest premium volume were: Prudential, $1.55 billion; Aetna, $1.36 billion; Travelers, $1.29 billion; CIGNA (formed by merger of Connecticut General and INA), $1.17 billion; Equitable, $1.12 billion; and Metropolitan Life, $1 billion. No other carrier had more than $500 million.[6]

THIRD-PARTY ADMINISTRATORS

Self-insurance has grown quickly and is the most common financing mechanism for companies with over 5,000 employees, the second most common form for those with 1,000 to 4,999 workers and 100 to 499, and is in third place for employers with 500 to 999 workers.[7] Both the Blues and commercial carriers have tried hard to get a large share of the growing business of providing administrative services for self-funded plans. The need for efficient claims processing has spawned the establishment of a new group of companies, called third-party administrators (TPAs), who believe that they can do this more efficiently and effectively than the traditional insurers.

Due to the enormous financial investment of establishing a data processing system able to process thousands of claims daily, many insurance companies are tied to existing computer systems which are far from state of the art. Third-party administrators, much smaller than most insurance companies and without large investments in older, more limited data systems, have often been quicker to take advantage of newer systems which have more flexibility and the ability to supply increasing amounts of data which are demanded by companies. They were among the first companies to go to completely on-line claims processing. They do not need a large underwriting staff and don't have to conform to many of the fairly onerous financial and reporting requirements promulgated by state insurance commissioners.

Their generally small size also gives them an advantage in being able to change rapidly in response to changing markets for their services.

Some TPAs, such as U.S. Administrators, (U.S.A.) have pioneered in performing much tougher prepayment claims review than is usually practiced. Each medical claim processed by U.S.A. is subject to a comprehensive set of screens designed to assure that the services provided to the patient based on the stated diagnosis were medically necessary and appropriate and that charges are reasonable. Medical appropriateness is determined by U.S.A.'s Model Treatment Program which has incorporated a model treatment screen for each ICD-coded diagnosis. These screens consist of parameters for treatment including such services as physician visits, laboratory and x-ray. Reasonableness of charges for physician services is determined based on U.S.A.'s internally developed relative value scale. Physician claims found to be medically appropriate with reasonable fees are paid within 48 hours of receipt; undisputed hospital claims are paid within 24 hours. Procedures and services considered to be medically unnecessary or inappropriate are not reimbursed, and excessive fees are reduced. Providers are notified in writing by U.S.A. the reasons for reduced or denied fees and are given the opportunity to furnish U.S.A. with more information that might justify additional reimbursement. To avoid employee relations problems caused by a physician requesting additional payment directly from the patient, U.S.A. always holds the patient harmless in a disputed case.

These features have made TPAs an attractive alternative during periods of hyperinflation of health care costs. Union-administered plans, which generally have seen a fixed-employer contribution buy less and less medical care, were among the first to be attracted to TPAs. However, TPAs are now effective competitors for employer contracts for administrative services. Some insurance brokerage firms have formed their own TPAs, going into competition with the insurers whose services they sell.

THE FUTURE

Reaction to a loss of business by both the Blues and the commercial carriers has been to become more competitive and to develop the internal capacity or external arrangements to help employers with cost containment. Two of the largest commercial carriers bought out fledgling companies that provide employers, based on analysis of hospital records and/or claims tapes, with greatly improved data reports over what the insurers had been providing. Other large insurers are contracting with firms who analyze health insurance claims. All of the major insurers and the Blues are involved in the formation of preferred provider organizations (see Chapter 14). Some have developed agreements with professional review organizations to cooperate in the review of claims or have developed their own utilization review programs. Some of the largest insurers have developed multidisciplinary teams to deal with patients who have high claims and could benefit from a case management approach.

What is clear is the increasing sophistication of employers as purchasers of administrative services. They are becoming as concerned with the process of claims review and the sophistication of the data systems as with the efficiency of claims processing, the turnaround time for claims payments, and employee satisfaction with the bill-paying process. Increasingly, detailed requests for proposals to administer the claims payment system are being developed by employers and sent to a large number of potential bidders.

Winning a health insurance contract used to mean being willing to shave off a few tenths of a percentage point from the usual percentage fee for processing claims and having a good minimum premium plan that maximized employer cash flow and avoided the payment of insurance taxes. Today and tomorrow, winning a contract should be based on:

- helping subscribers refine their benefit package to facilitate cost containment;
- possessing a demonstrated capacity to perform very careful claims review that can exclude payment for inappropriate or overpriced services, especially in the hospital;
- providing accurate coding of medical diagnoses so that the nature of insuree health problems can be analyzed;
- furnishing management reports which clearly identify problems in the costs and utilization of health care services under the benefit plan;
- having a flexible data-based system that permits analysis of claims information both routinely by the carrier and in response to ad hoc requests by the subscriber;
- performing other cost control activities, such as preadmission and concurrent hospital review and case management of patients with serious problems requiring extensive use of services under the benefit plan.

In the future it is likely that many more carriers will either tie in with firms that can do extensive analysis of their data for them or that will make major investments in developing the capability internally. Larger employers will be given computer access to the claims tape file, with the possible exception of names of individual claimants, and will manipulate the data-base to generate a variety of reports on costs, utilization, and the nature of health problems.

NOTES

1. Law, S.A. *Blue Cross: What Went Wrong?* New Haven, Connecticut: Yale University Press, 1976.

2. Ibid.
3. Health Insurance Association of America. *Source Book of Health Insurance Data 1981–1982*. Washington, D.C., 1981.
4. Drury, S.J. "Blues Dominate Group Benefit Market: Study." *Business Insurance*. July 18, 1983, pp. 1, 43.
5. Carroll, M.S. and Arnett, R.H. "Private Health Insurance Plans in 1978 and 1979: A Review of Coverage, Enrollment and Financial Experience." *Health Care Financing Review*. September, 1981, Vol. 3, No. 1, pp. 55–86.
6. Op. cit. (Drury).
7. Op. cit. (Drury).

5 MANAGING MEDICAL BENEFITS

EXISTING BENEFIT FEATURES

Today, almost any combination of medical coverages is available from insurance carriers or from benefits administration service companies. Nonetheless, fairly standardized arrangements have been developed over time.

Basic, Major Medical and Comprehensive Coverage

Virtually all employees have a health insurance plan that covers inpatient hospitalization. Traditionally, this benefit was provided under a "basic" plan which also usually covered some physician services while in the hospital. "Major medical" policies cover primarily ambulatory services although they may also cover expenses which exceed "basic" plan benefit limitations. Major medical insurance often requires co-insurance and deductibles, especially if underwritten by a commercial carrier. "Comprehensive" plans combine the coverage provisions of basic and major medical plans with usually a single deductible and a uniform co-insurance percentage.

Between 1970 and 1980 the number of persons protected by major medical coverage increased by approximately 50 percent to a total of 154 million.[1] By 1980, 90 percent of all employees and virtually all employees of larger companies had some form of major medical benefit.[2,3] The 1983 Hay-Huggins survey found that two-thirds of 854 participating companies provide major medical benefits which supplement basic benefits limitations or which cover categories of services not covered under the basic benefit. About one-third of companies surveyed have comprehensive coverage.[4] The trend to combine basic and major medical into comprehensive coverage is continuing to gain momentum with more than half of the large employers having adopted this approach by 1983.[5] It is likely that in the near future most employees and their dependents will be covered under comprehensive plans. What is covered and not covered is clearer in a comprehensive plan. Additionally, moving to comprehensive plans is usually accompanied by an increase in employee cost sharing, which is a primary aspect of many company cost containment programs.

Deductibles

Deductibles are usually a fixed dollar amount per insuree, often with a family maximum at two to three times the individual deductible. A handful of companies have set deductibles as a percentage of earnings (ranging from 1 to 3 percent). Xerox Corporation, for example, at the start of 1984 changed their deductible from $100 per individual and $200 per family to 1 percent of each employee's annual earnings.[6] During 1984 Olin Corporation implemented a plan whereby employee deductibles increase with higher salary and larger family. Single employees pay 0.5 percent of their annual salary; employees with one dependent pay 1.0 percent of their salary; and employees with two or more dependents have a deductible equal to 1.25 percent of pay.[7] More companies can be expected to move to this type of arrangement in the future.

In general, the level of deductibles has not been increased regularly to reflect even inflation in the overall economy, let alone the hyperinflation of medical care costs. In the 1983 Hay–Huggins survey, 94 percent of the deductibles for supplemental major medical plans were $100 or less per individual.[8] A similar conclusion was reached by a 1982 Hewitt survey which found only 4 percent of 659 major employers required an annual deductible of $150 or more per individual for their group health insurance plan.[9] Many deductibles have not changed in twenty years. To have the same impact as a $100 deductible twenty years ago, the level today would have to be set at $400 to $500.

Co-insurance

According to the 1980 Employee Benefits Survey, nine in ten participants in major medical plans pay 20 percent of their medical bills through a co-insurance provision, one in twenty pays 15 percent, and the rest have other co-insurance arrangements.[10] As the chances of an individual or family with a co-insurance provision being financially devastated by a serious injury or illness have significantly increased due to higher medical costs, a growing

TABLE 5.1
HEALTH INSURANCE: PERCENT OF FULL-TIME PARTIC-IPANTS IN PLANS WITH MAJOR MEDICAL COVERAGE BY EMPLOYER PAYMENT, PRIVATE INDUSTRY, 1980

Final employer paid	Total	Initial employer paid portion		
		80%	85%	Other
Total	100	89	5	5
Final employer paid portion changes to 100% (stop-loss)	55	48	4	2
When covered expenses within a year reach:				
$0–2,000	7	7	—	(*)
$2,001–4,000	15	14	(*)	(*)
$4,001–6,000	20	18	1	1
$6,001–8,000	4	1	2	(*)
$8,001–10,000	6	5	1	(*)
More than $10,000	3	3	(*)	(*)
Final employer paid portion changes to other than 100%	1	1	—	(*)
Employer paid portion unchanged	44	40	1	3

*Less than 0.5 percent.
Note: Because of rounding, sums of individual items may not equal totals. Dash indicates no employees in this category.
Source: Employee Benefits in Industry, 1980 (modified). U.S. Department of Labor, Bureau of Labor Statistics, September 1981, p. 19.

number of major medical plans have adopted a "stop-loss" feature. Under stop loss, which is also called catastrophic coverage, insurance pays 100 percent of all covered expenses after a dollar limit is reached during a 12-month period. In 1983 about 70 percent of supplemental and 90 percent of comprehensive major medical plans sponsored by larger employers included such arrangements.[11] The levels of stop loss, based on a 1980 survey, are summarized in Table 5.1.

Total Coverage Limits

In addition to deductibles and co-insurance provisions, most major medical plans set a maximum limit on total coverage over a lifetime. Eighty-five percent of companies with supplemental major medical plans and 84 percent with comprehensive plans indicated to Hay–Huggins that their plan establishes a maximum coverage.[12] The most common figure in 1983 was $250,000 for supplemental major medical and $1,000,000 for comprehensive plans. Although a significant minority of the limits are in the $100,000 to $250,000 range, the general trend is to raise limits, often to $500,000 or $1,000,000. In most cases the plans are reinstatable.

As expensive organ transplants continue to be taken off insurers' lists of experimental procedures, many employers who do not specifically exclude these transplants from their plans will find the need to increase their lifetime coverage maximums. A recent survey by the Health Insurance Association of America found that 82 percent of the nation's largest insurance companies are willing to pay for liver transplants and 85 percent are willing to pay for heart transplants. The vast majority are also reimbursing for kidney and bone marrow transplants.[13] Liver transplants can require up to six weeks of hospitalization, costing $70,000 to $100,000. Heart transplants have an average cost of $70,000 and a combination heart and lung transplant brings the total to $100,000. The cost of a transplant can run as high as $300,000.[14] Transplant patients will usually continue to have high bills for health care services for the rest of their lives.

WHERE THE MEDICAL HEALTH INSURANCE DOLLAR GOES

The largest portion of the health insurance dollar goes to hospitals, which now account for about one-half of all expenditures under employer-sponsored health insurance plans. Hospital expenses are virtually uniformly covered as a basic benefit, regardless of length of stay or level of expenses, and, according to recent national employee benefit surveys, between 90 and 95 percent of employees are covered for room-and-board expenses up to the semiprivate room rate.

Under basic and comprehensive benefits, most plans limit total exposure, either by restricting the total number of days per confinement (usually between 120 and 365) or the total dollar amount per confinement, with major medical benefits usually picking up coverage for the excess. The 1983 Hay–Huggins survey found that coverage for hospitalization under basic medical benefits is limited to a maximum number of days in 82 percent of plans (24 percent at 120 days; 51 percent at 365 days), by aggregate dollar expenditure in 3 percent, and no maximum is specified in 13 percent of plans. Eighty-four percent of basic medical plans set no maximum for miscellaneous hospital charges, while most of the others set an overall maximum dollar figure or a maximum percentage of total charges.[15]

Surgical benefits are usually provided under the basic benefit provision, although they may be reimbursed under major medical. About two-thirds of employees are covered by plans that pay "usual and customary" charges for the procedure performed, with the rest covered by a schedule of maximum fees for each procedure.[16]

Inpatient visits and outpatient x-rays and laboratory tests also tend to be covered in most instances by the basic medical benefits. According to the Hay–Huggins survey, in 55 percent of plans, in-hospital doctor visits are covered by the basic benefit and in 4 percent by major medical coverage. These visits are reimbursed on the basis of reasonable and customary fees in 70 percent of plans and on a dollar limit per visit (or per day) basis by the remainder. All companies provide coverage for outpatient laboratory and x-ray services, with the majority (84 percent) including this coverage as part of the basic medical benefit. The vast majority (95 percent) of plans reimburse outpatient x-ray and laboratory services on a reasonable and customary basis (full or partial), with the remainder applying a fee schedule. About half of the plans have a maximum coverage for outpatient x-ray and laboratory tests, most frequently a yearly maximum in the range of $100 to $200. [17]

Unlike surgical fees and inpatient physician visits, physician office visits are primarily covered under major medical benefits (84 percent of plans). [18] Over seventy percent of plans reimburse a percentage of reasonable and customary charges (usually at 100 percent, less frequently at 80 or 90 percent), while the others pay a flat dollar amount per visit.

BENEFIT REDESIGN: WHAT WORKS

For many years most companies looked at their medical insurance as a keystone of their entire benefits package. A "progressive" employer would increase the degree of protection employees and their dependents had against large hospital bills. Blue Cross prided itself on offering service benefits rather than indemnity coverage which only paid a portion of the hospital or physician bill. First-dollar coverage became virtually uniform for basic benefits, and was frequently extended to major medical. When the total health insurance benefit was one or two percent of payroll, the price of buying very broad coverage did not seem too high to secure employee satisfaction with what many considered to be the most important benefit.

Providing such broad medical coverage violated the fundamental principle of insurance, which is to spread the risk of very infrequent events over a large number of individuals at risk for the events. Medical coverage has been designed to include reimbursement for relatively small, routine bills. Few health benefits plans cover only large expenses that could cause serious financial hardship. The combination of extremely broad coverage and willingness to pay whatever providers charge has made insurance a major contributor to the meteoric rise of health care costs. All incentives have been to utilize services with little concern about their costs. As a result, many employers are revamping their medical benefits.

The major focus of most benefit plan redesigns is reducing the rate of rise of employer health care costs. For example, one large midwest-based employer sought to achieve three goals: 1) prevent any rise in health insurance benefit costs per employee in year one; 2) have per employee increases no greater than the CPI in year two; and 3) reduce the increases to less than the CPI starting in year three. But the constraints faced by this company (like most others attempting to meet similar objectives) are considerable. Among them are:

- need to maintain competitive recruitment position
- desire to maintain good employee relations
- union contracts
- desire to retain image as progressive employer
- concern about possible union organizing efforts
- concern about cost shifting from *employer* pre-tax dollars to *employee* after-tax dollars

Altering Incentives

A general principle to consider in plan redesign is the need to provide incentives for the insured to use the health care system in a cost-effective manner. Education about how to use health care services more judiciously is very important but insufficient to effect major changes in the absence of altering the incentives. There are at least five approaches to altering incentives:

1. Increase cost sharing by the employee.
2. Provide money to those who are low dollar users of services and/or allow employees to benefit directly from improved cost experience of the entire group.
3. Require increased or even full employee shouldering of the costs for services not judged to be appropriate or necessary.
4. Provide opportunities for employees to opt for different coverage levels with corresponding variation in employee contribution to costs of plans.
5. Establish opportunities for employees to pay for a variety of health benefits and/or services with pre-tax dollars and to share in cost savings.
6. Encourage entry into systems which have inherent incentives to minimize costs and utilization, i.e., prepaid plans.

Approach 1: Increase cost sharing by employees The fastest way to reduce employer health care expenditures is to increase the portion of costs borne by their employees. The arguments in favor of this approach are:

1. Employee contributions have tended to decrease during the time of greatest health care cost inflation in our nation's history. There is a need to reverse the disincentives which have accompanied this trend.
2. Increased cost sharing is one of the few options available for which clear savings can be obtained for the employer very quickly; the savings start almost immediately.

TABLE 5.2
ANNUAL PROBABILITY OF ONE OR MORE VISITS TO PHYSICIANS OR HOSPITAL ADMISSIONS IN NINE SITE-YEARS*

Plan	Visits to Physicians	Hospital Admissions
Free care	0.84 ± 0.02†	0.102 ± 0.013
25% co-insurance	0.78 ± 0.03†	0.081 ± 0.014‡
50% co-insurance	0.75 ± 0.05†	0.072 ± 0.021†
95% co-insurance	0.69 ± 0.04†	0.076 ± 0.014†
Individual deductible, 95% co-insurance§	0.73 ± 0.04†	0.090 ± 0.016

*95 percent confidence intervals are presented for all plans. Standard errors are corrected for intrafamily and intertemporal correlations.
†$P < 0.01$ as compared with the free plan by the one-tailed t-test.
‡$P < 0.05$.
§This plan has zero coinsurance (free care) for inpatient services.
Source: Newhouse, J.P. et al., "Some Interim Results from a Controlled Trial of Cost Sharing in Health Insurance." Reprinted, by permission of *The New England Journal of Medicine*, Vol. 305, No. 25, 1981, pp. 1501–1507.

3. Greater financial participation will make employees and their dependents more conscious of how to buy cost-effective health care services; it will increase their willingness to ask questions in advance about the need for services and the charges they will be facing.

4. There is evidence that cost sharing, especially increases in co-payments, reduces utilization of some services.

Economists have long been at work to quantify the effects of increased cost sharing on utilization. While results are far from uniform, one review of major studies in this field concluded that a fully insured patient would utilize approximately twice as many physician services as a consumer without insurance.[19] Preliminary results from the Rand Health Insurance Experiment, where families were randomly assigned to different insurance coverage, which varied all the way from entirely free care to 95 percent co-insurance up to a certain percentage of a family's yearly income, indicated that persons completely insured for medical services spend about 50 percent more than those who only have catastrophic coverage.[20] More co-insurance was also associated with a reduced hospital admission rate (Table 5.2), but no reduction in costs per hospitalization. Increased co-payments for ambulatory medical care did not appear to increase hospitalization rates, as some experts had predicted.

Most of the experiments in this area have dealt with major changes in the proportion of costs borne by individuals and families, changes much greater than those commonly contemplated by employers. Therefore, the degree of change in utilization is difficult to predict, particularly with respect to deductibles, where few careful experiments have been reported.

Although increased cost sharing can provide a quick fix for employers wishing to control health benefit costs, there are problems associated with choosing this approach. Among them are:

1. An increase in deductibles, or in payroll deductions, can make employees feel that since they have paid a considerable amount for the benefit, they are going to make sure they get their money's worth by using it to the greatest extent possible. Both deductibles and stop-loss coverage leave employees with little or no incentive for cost consciousness after these limits have been reached.

2. The employees will be spending more of their after-tax income on health care which may lead to demands for enough additional pre-tax salary to pay for the increases. While in recessionary periods employers may not have to pay great attention to this demand, it is difficult to ignore when the job market is more competitive.

3. Employees frequently don't take account of why they are not receiving reimbursement on their submitted claims. They may not realize that they are required to pay a larger part but instead assume that plan inadequacies in terms of payment rates to providers is the reason they are paying more. Thus, employees may not realize that they have the ability to reduce their out-of-pocket expenditures by making more careful use of services and asking questions about charges in advance.

4. The largest single expense under health insurance plans is for hospital-related services. It is almost always the physician who makes the decision about hospitalization. Particularly in urgent situations, the employee or his/her family is unlikely to compare charges of the hospital recommended by the doctor to the charges of other nearby hospitals.

Cost sharing arrangements (how much and what type) will be different for each employer. However, a few general recommendations are warranted:

1. Employee contributions through payroll deductions are unlikely to have a moderating effect on utilization or claims costs. Therefore, effects of this approach should only be considered the pure shifting of some existing employer costs to the employee. Making employees pay a reasonable share of the premium for dependent coverage is a good mechanism to minimize double coverage and to reduce overall costs. It is more effective to have the employee pay a separate charge per month for each dependent covered than to pay a single charge regardless of covered dependents.

2. Fixed deductibles, while easy to communicate, are less desirable from a cost-containment standpoint than deductibles which increase based on changes in the CPI or index of medical cost inflation. With fixed deductibles, each change is remembered as a takeaway. With floating deductibles, employees can be educated that they are not being asked to increase the percentage they contribute to health care benefits but merely to keep pace with inflation just as the employer is required to do.

3. A good scheme for deductibles is to have them vary with salary levels. This approach helps to insulate the low income employee from paying a large percentage of disposable income for health care while asking the higher income employees to absorb greater but readily affordable costs before the company plan begins to pay. Several insurance carriers have this type of plan in place for their own employees.

4. Deductibles and co-insurance should be waived for preventive services that are known to improve health and prevent more serious future health problems, such as immunizations, well-baby care, periodic (but not annual) health examinations, some cancer prevention techniques (e.g., Pap smears and mammography), and prenatal care (including amniocentesis).

5. Co-payments should be used to provide appropriate incentives for cost-effective use of services. For example, co-insurance can be waived for some surgical procedures when performed on an outpatient basis. The rate of co-payment can be increased over the usual percentage if these procedures are performed on an inpatient basis in the absence of circumstances requiring that setting. Likewise, an insured person who does not get a second opinion on the need for elective surgery can be required to pay a much higher percentage of the charges than if a second opinion had been obtained (see Chapter 11).

Approach 2: Provide money to low dollar users of covered health services Increasingly, employers are experimenting with plans that return money to employees whose total reimbursement under the health insurance plan is below a certain amount in a given year. The rationale for this strategy is that employees are able to see tangible dollar rewards for being efficient in their use of health care services. In general, this approach is combined with an increase in the deductible so that employees with higher utilization are paying a larger share of the costs for health care.

The original plan based on this model was the Mendocino County Stay Well Program, developed by Blue Shield of California and the local Teachers Association. In brief, the first $500 of expenses are self-insured by the employer, and all costs in excess of this amount are covered by Blue Shield. If the family unit spends less than $500 a year, the unused portion is credited to an account and is redeemable when the employee terminates employment or retires.[21] Bank of America offered a group of employees an experimental plan whereby the bank would pay the employees half of the health insurance premium (at that time $35.14 per month per family, $14.48 per month for an individual) for one year if the Bank of America program did not have to pay a single claim for that employee and dependents during the pilot year.[22]

To date, the degree to which this approach is effective in reducing utilization is unclear.

The Bank of America program's preliminary findings suggested that the pilot test group experienced fewer injury claims and doctor visits at a lower per capita cost than the control group.[23] However, a number of problems are associated with this type of plan. It rewards low utilizers, but those rewarded were probably always the low utilizers. Employees who are low utilizers tend to be young, healthy and without children. By contrast, once they have reached the cut-off point for refunds, the high utilizers, who comprise a small percentage of the total insured population, have no incentive to be more cost conscious under this approach. Any hospitalization and almost any significant outpatient diagnostic work-up will exceed the cut-off point. As a net result, the employer may find itself giving more money back to the majority of employees who tend to be low utilizers than they save by the higher deductibles for the high utilizers. Another potentially deleterious outcome can be that employees will postpone necessary preventive services in order to get their refund, unless these services are exempted from the costs on which refunds are based.

A related approach is to share cost savings with all employees if aggregate costs and utilization fall below the projected level for the year. It is felt that this provides an incentive for all employees and their dependents to be cost conscious in using health care services. It has the advantage, especially in dealing with unions, of assuring employees and their representatives that the company is not trying to reduce health care expenditures "on the backs of employees" and is willing to share savings with employees. It could be particularly useful, for example, in situations where a number of significant changes were to be made in the benefit plan and the net result was impossible to predict, but the company was committed to sharing the savings with employees. A company could tell employees, "In the absence of these changes we would expect our per employee costs to increase fifteen percent next year, and any reductions we will split fifty/fifty with employees. In subsequent years, to the extent that increases in total expenditures per employee are less than the percentage increase in the CPI, these savings will also be shared." Union acceptance of any incentive approach may require the employer to share savings with the employees.

An important negative feature of this type of collective approach is that it provides very weak incentives for reducing utilization or being much more cost conscious in purchasing services. When the refund is not to be based

FIGURE 5.1
TWENTY COMMON PROCEDURES WHICH CAN GENERALLY BE PERFORMED ON AN OUTPATIENT BASIS

1. Application of leg or body casts
2. Arthroscopy
3. Aspiration of breast cyst
4. Biopsy (mouth, breast, skin, prostate, cervix, uterus, lymph node)
5. Bronchoscopy
6. Colonoscopy
7. Diagnostic Laparoscopy
8. Dilatation and currettage (D&C)
9. Esophagogastroduodenoscopy
10. Excision of ingrown toenails
11. Heart catheterizations
12. Inguinal herniorraphy
13. Laryngoscopy
14. Myringotomy (tympanotomy)
15. Nasal polyp(s) removal
16. Podiatric surgery
17. Sigmoidoscopy/colonoscopy
18. Simple hemorroidectomy
19. Tenotomy
20. Vasectomy and tubal ligation by laparoscopy

on an individual's behavior but that of hundreds or thousands of employees, the predictable employee reaction is "The results are out of my control, and what I do will make very little difference."

Approach 3: Increase employee costs for unnecessary/inappropriate services or settings All plans limit their coverage to services that are medically necessary and appropriate. However, trying to determine from a claim form what services were either unnecessary or inappropriate is at least very difficult and frequently impossible. A better approach, where possible, is to specify those procedures or circumstances that are excluded from payment under the plan. For example, a number of procedures where effectiveness has not been demonstrated should be listed as exclusions so that the employees know in advance they will have to bear the full costs. A second approach is to specify procedures which generally do not require hospitalization and to reimburse for them only at the outpatient rate unless there is prior approval or retrospective justification of special circumstances necessitating hospitalization. A reasonable list of such procedures is given in Figure 5.1. As a general rule, it is important to provide all employees and their covered families with lists of these procedures so that they know in advance, if they take the time to look, what is explicitly excluded from coverage.

Approach 4: Provide several plan options at different levels of coverage To accommodate employee preferences for different coverage levels, several levels of health insurance benefits are being offered by an increasing number of companies. While in most cases these differ primarily in deductible, co-insurance and stop-loss features, there is no reason that plans could not differ in other aspects, such as coverage for maternity, reimbursement based on a fixed dollar schedule versus reasonable costs, etc. One reason to go to a two- or three-tiered set of options is to be able to continue to offer the prior plan, even at much higher prices to employees, while also providing less expensive options.

DuPont, for example, began giving its 28,000 Texas employees a choice between their original coverage and a second option in 1982. Under the old plan, which is comprised of basic and major medical coverage with a $30 monthly employee premium for the latter, the basic plan pays 100 percent of hospitalization and 80 percent for physician services. Major medical coverage is optional for employees; it requires a $100 deductible and 80 percent co-insurance. The new option is a comprehensive plan which has no employee premiums, a $1 million yearly maximum, and a $1,000 out-of-pocket maximum. Co-insurance is 90 percent for hospitalization and 80 percent for physician services, and there is a $200 individual deductible and $400 family deductible. The new option also provides incentives for ambulatory services, including outpatient surgery.[24]

Some companies have agreed to pay a fixed percentage of each plan (often 65 to 75 percent). However, minimizing both equity problems and long-run costs could probably best be accomplished by contributing the same dollar amount towards each coverage. If employee relations concerns preclude this approach initially, it can be achieved in steps over a several-year period.

The major problem with this approach is estimating initially and maintaining the relative employee contributions in line with the cost experience of the various plans. Estimating experience under each plan requires not only taking account of potential differences in utilization due to different levels of cost sharing, but also guessing the differences in the characteristics of employees and their dependents who will opt for each coverage. Experience with the Federal Employee Health Benefit Program has shown that adverse selection can occur and over time have a significant impact on claims experiences. Healthy employees often choose the low-cost plan until they have a medical problem and then switch at the first opportunity to the broader coverage. Over time, the difference in claims experience between the broader and narrower coverage plans increases. This result can be mitigated by making sure that on a yearly basis the relative cost to the employer and to the employee is reflective of the actual costs under each plan. Changing plans can be prohibited for one to two years after the initial choice has been made to limit switching when a health problem is uncovered.

A further extension of the concept of providing more options to employees is the offering of flexible benefit packages. These allow employees to opt for more health coverage instead of extra paid vacation days, for example, or a legal services benefit. Where flexible benefits are offered, some companies are finding that employees will opt for less broad health care coverage than the company previously provided because it precludes too many other benefits that the employee wants.

Approach 5: Establish pre-tax employee expense accounts with incentives for health-related cost savings   In most cases the idea of paybacks has been merged with an employee expense account, under which each employee is given a specified amount of money in a flexible spending account, to be used to pay for deductibles and co-insurance as well as otherwise uncovered services. For example, Chemical New York Corporation, parent of Chemical Bank, deposits $300 for each of more than 20,000 employees in an Employee Spending Account and allows employees to allocate up to 50 percent of their profit-sharing revenues to the account. During the year, employees have the option of using the account to reimburse themselves for benefit expenses—medical, dental, legal, etc.—or to buy into the bank's health insurance plan toward which Chemical pays 85 percent. At year's end employees can pocket what is left (at which point it becomes taxable) or return it to their profit-sharing accounts, which are tax-free.[25]

One of the more innovative approaches which combines the employee expense-account concept with sharing savings based on the experience of the entire group is that of the Quaker Oats Company, whose program is summarized in Figure 5.2. It provides a very flexible account and shares savings below what the company estimated health costs would be with employees. During 1983, per employee health insurance costs under the Quaker Health Incentive Plan were nearly 15 percent below the figure projected under the old plan, resulting in significant rebates to employees.[26] The Berol Corporation reported that as a result of its $500 cash incentive program, they achieved a first year savings of $110,000 on 300 salaried headquarters employees.[27] In addition to limited information on results, these approaches harbor a number of potential problems, in many cases similar to those of Approach 2 (providing money to low dollar users). Low utilizers under these plans were probably perennial low utilizers. Now the company is spending more on them, while the high utilizers may be well over the limit for possible rebate and have no incentives to modify their health services-seeking behavior. In addition, efforts to avoid submitting claims may mean that employees and dependents forego important preventive services which can avoid more serious health problems later on. Finally, employees may be paying more out-of-pocket for their health insurance, not submitting claims in order to get the rebate. What the employer saves the employee pays, and more, since it is not tax deductible to the employee.

FIGURE 5.2
QUAKER HEALTH INCENTIVE PLAN

Health Expense Account:
Each eligible employee is credited with a tax-free $300 annual expense allowance
that (s)he can use to: a) cover his/her health insurance deductible; b) pay for a
wide range of other health care expenses not covered by the group policy—physicals, eyeglasses, immunizations, etc.; or c) receive as cash at the end of the year.

Comprehensive Medical Benefits:
Employees must pay a $300 deductible and 15 percent of the next $5,000 in covered expenses for hospitalization, surgery or other illness or injury-related expenses. However, the co-payment is capped at $750 for individuals and $1,500 for
families.

Annual Dividend:
If the company's overall health care costs for the year are less than the amount
budgeted, the savings is returned to employees in the form of a dividend that is
added to their health expense accounts for the next year.

Economic Adjustments:
As the health economy changes, periodic adjustments in guarantees and coverage
levels will be made.

Source: Cain, Carol. "Quaker Oats is Cooking up Health Incentives," *Business Insurance,*
November 22, 1982, p.1.

In early 1984 the IRS indicated opposition to the deductibility of monies
provided to employees under many types of flexible spending accounts,
including health care "expense accounts." Barring a final unfavorable ruling
on all of these arrangements by the IRS, it is likely that expense accounts
will grow and provide increasing flexibility, enabling employees to purchase
health insurance of various types, use it for health or personal health enhancement services (spa memberships, smoking cessation programs, etc.),
and either draw it out or transfer it to a tax-deferred profit-sharing or other
account. It is not as obvious how health expense accounts will affect overall
utilization of services. The approach is likely to have the greatest impact on
those whose total utilization consists of ambulatory services. Since hospital-
related expenses can constitute up to two-thirds of health insurance costs,
it will take a significant reduction in nonhospital expenses to make a dent
in the claims experience.

*Approach 6: Encourage enrollment into prepaid plans and use of lower
cost providers* Group model prepaid plans (i.e., HMOs) provide strong
incentives to minimize health services utilization, particularly hospital services (see Chapter 13). Groups of independent practitioners that work under

umbrella prepaid arrangements (i.e., IPAs) have more limited incentives, since they are generally paid based on their usual fees or a percentage thereof. The individual risk of these physicians is small. While plan solvency requires moderation by *all* practitioners in their ordering of services and referrals to specialists, the level of concern about the plan by an individual practitioner may be low.

From the insured's point of view, there is little or no incentive to minimize utilization in an HMO since all payment is in advance, with the possible exception of very small co-payments for some ambulatory services and medications. The organization has an incentive to be efficient; the plan members do not. In general, employees and their families can get a broader range of services while enjoying reduced out-of-pocket costs in a prepaid plan. However, as discussed in Chapter 13, the cost to the employer may be higher than that for other plans offered to employees, especially given the trend towards increased cost sharing. Therefore, the employer may not have an incentive to encourage enrollment. A clear exception is when the employer-sponsored plan is so rich in benefits that HMO premiums are significantly less than the plan cost. This is true in industries that have strong unions and high rates of pay, such as the automobile, steel and oil industries. In these industries, some employers offer cash incentives for new enrollment in HMOs, in essence sharing the company savings with the employees.

Preferred provider arrangements (see Chapter 14) offer services at lower than usual charges and therefore are inherently attractive. However, whether they can really save employers dollars in the short run, and whether they will continue to do so compared to other cost containment alternatives, has yet to be determined.

One of the problems associated with restructuring medical benefits is the difficulty of accurately forecasting what costs and utilization effects will result from a number of simultaneous changes in benefit design. While the impact of increasing the deductible or increasing co-payments can be estimated within narrow limits, no reliable estimates are available about the cost and utilization effects of several simultaneous actions, such as:

1. increasing the deductible from $100 to $200;
2. requiring preadmission review of hospital stays by a local physician review organization;
3. instituting a mandatory surgical second-opinion program;
4. requiring any insured who is hospitalized to pay the first $50 each day of hospitalization for the first five days; and
5. increasing the employee co-payment on outpatient psychiatric services from 20 to 40 percent, while raising the annual limit on payments for such services from $500 to $1,200.

Impacts of such a set of changes would vary considerably depending on a number of factors, including:

- average family income
- employee salaries
- prior rates of utilization
- previous rates of cost increases for various health care services
- general economic climate
- organization of health care services in the area
- effectiveness of physician review organizations
- age/sex distribution of the insured population

This difficulty is frustrating when a benefits manager or financial analyst is trying to determine how much to budget or what combination of changes are required to save a certain number of dollars.

Adding Lower Cost Alternate Services

Many additions to traditional benefits have been made to reduce reliance on, and incentives for, the use of high-cost services. Among newer benefits for which reimbursement is provided, usually at the same level as inpatient services are:

- outpatient testing as a substitute for inpatient testing, including pre-operative work-ups
- nursing home care as a substitute for hospital care
- home health care as a substitute for hospital or nursing home care
- hospice care as a substitute for hospital or nursing home care
- outpatient surgery as a substitute for inpatient care, etc.

The rationales for the addition of coverage for these services are clear, including potential cost control and the availability of newer services which are humane and more appropriate for the patient. However, from a cost-containment point of view, they have frequently been less effective than anticipated. For example, some hospitals have priced outpatient surgery so that it is no less expensive than having inpatient surgery performed with an overnight stay. Hospices may or may not reduce the overall cost for a protracted illness. The total number of days in the hospices are usually much greater than the number of days that would have been spent in a hospital. Even the lower per diem price may not compensate for the extra utilization in achieving economies. Preadmission testing has not been definitively shown to reduce total costs of a surgical procedure. Frequently, hospitals charge the same for outpatient and inpatient testing, much more than the cost of the same procedures performed in a free-standing laboratory of equal quality. Reduced hospital days due to shorter preoperative stays may lead to increases in hospital prices in order to cover the high fixed costs with fewer total patient days.

Therefore, in most cases, additions to the benefit package mean additional utilization. While there may be some reduction in the utilization of the particularly high cost service, some insureds who would never have used the high cost service avail themselves of the newly covered lower cost service so that overall expenses often rise. For example, free-standing ambulatory surgery centers, usually at much lower-cost than the hospital, may start to do a number of procedures which previously would have been performed in doctors' offices at a much lower charge. Changes in the benefit package in the absence of incentives rarely appear to serve the aims of cost containment. By contrast, paying women to leave the hospital sooner after childbirth does work to reduce average length of stay and total costs.

Alternate Reimbursement Methods and Shopping for Lower Price Services

As summarized in Chapter 2, most health insurance policies reimburse hospitals based on charges and physicians based on U&C (usual and customary) or U&P (usual and prevailing) fees. Hospitals thus get what they charge or a fixed percentage thereof from the insurer, assuming the services are covered and treatment is necessary. Physicians receive an amount based on a profile of charges in the area as well as their usual billed charges during the previous year(s). Usually the policy will reimburse the physician an amount no greater than the 90th percentile although some companies are paying at the 75th percentile and others pay at up to the 100th or even 125th percentile. The point is that reimbursement principles for both hospital and physician payment are inherently inflationary and are not moderated by competition from other providers.

Alternatives to usual hospital reimbursement under an employer-sponsored plan are difficult unless the employer participates in a preferred provider arrangement in which they contract directly or indirectly for special discounts or flat rates per day or per stay (see Chapter 14). Since hospitals usually price their services à la carte, it is difficult for a prospective patient to know what a particular stay will cost in advance. In general, however, employers can use either information available from the local health planning agency, state rate setting organizations, a carrier's data base or the employer's own experience with different hospitals to determine which facilities are likely to be associated with lowest overall costs per stay. The employer can then make this information available to employees who must share part of the cost of hospitalization for themselves and their dependents. Unless the insured participate in cost sharing to a meaningful extent, and do not have a low stop-loss limit, there is virtually no incentive for them to use this information or shop themselves for lower-cost hospital care.

With respect to physician services, one practical approach is to employ reimbursement schemes which are less inflationary than usual and customary. For example, a company can institute a payment schedule for particular types of services and make it available to employees. Armed with this information employees and dependents can ask around to be sure that they receive care whose cost falls within what the schedule will pay, or at least minimize the excess for which they will be responsible. Even under usual and customary payment approaches in the presence of significant co-insurance and deductibles insurees can be taught to compare prices of hospitals, physicians and other services in advance for elective problems.

An innovative approach that a growing number of employers are adopting to help their insurees make cost-effective decisions regarding their use of the health care system is that of a health information resource service. TELANSWER, a product of U.S. Corporate Health Management, is an example of such a system. Applying somewhat of a "gate-keeper" approach, TELANSWER provides information to employees and their dependents on how to select a provider, what questions to ask their physician about a recommended procedure or treatment, lower cost health care alternatives, relative charges of providers in their community, and how to take advantage of the cost-containment features in their benefit plan.

Another example of a health information service is the Blue Cross "medical service adviser" system, which was adopted by Zenith Corporation at the start of 1983. Zenith requires 2,600 of their salaried employees to contact a utilization review specialist prior to entering the hospital for non-emergency care. Using Blue Cross cost data and regional length of stay figures, this specialist informs patients what their likely out-of-pocket costs will be at different hospitals in the area. In addition, the specialist provides patients with the names of physicians who give second opinions and informs them about lower cost options for care. If a patient is reluctant to speak with his or her physician about cost saving options, the utilization review specialist will call the physician. At the end of two months of using the medical services adviser system, Zenith estimated their savings to be on the order of $4,500 per week.[28]

Employee Education Programs

An educational program can be effective in helping employees and their families understand why the company is becoming increasingly aggressive in dealing with the problems of health care costs. It can also explain the inherent economic incentives, within the health care system, the lack of direct relationship between cost and quality of health services in most cases and what questions to ask to help assure oneself that recommended procedures and services are necessary. Changes in health benefit plan design, reimbursement principles and administrative procedures also need to be communicated in straightforward fashion without an excess of bewildering detail.

A mistake frequently made is to expect that one educational exposure is sufficient for employees to understand how their health insurance package works. Even for health professionals, reading a plan summary can be confusing, appear unnecessarily complex, and be impossible to remember in details as to what is covered and how much the company plan will pay. It is therefore essential to distribute information on health benefits on a periodic basis, even though the employee probably has the basic benefit plan stuck away in some drawer at home. Face-to-face meetings of employees with company benefits experts and representatives of the insurance carrier(s) or plan administrator are also important to clarify how to effectively use the plan and should be considered a high priority activity, with a push made for high attendance.

Possible Trade-off of Medical for Other Benefits

The dollar cost of specific benefits may not bear a close relationship to their value as perceived by employees. For example, prescription drugs and vision care are generally perceived as being quite valuable, despite their low costs. Legal benefits and group insurance benefits, such as automobile or homeowners, may also be generally considered to have value out of proportion to their costs. It is therefore possible to trade off some increase in employee co-insurance and/or deductibles under a comprehensive health insurance policy for the addition of dental care, vision care, etc. The Hewitt survey of 582 major employers released in 1982 showed that only 60 percent of employers offered dental plans, 17 percent vision care, 6 percent group automobile insurance and 3 percent legal services.[29] Therefore, the opportunity exists for most employers to trade off these new services for some increased cost control measures in administration, plan design or reimbursement areas.

Union Concerns

Unions are very proud of the broad health care coverage which they have negotiated over a period of many years on behalf of their members. Unions often point to the broader benefit package and larger employer contribution in unionized companies as one of their key accomplishments and a reason for unorganized employees to establish a union (Table 5.3). A number of unions have gone on strike rather than accept any take-aways in health benefits or an agreement to share in any of the costs. Despite the general hard line, however, there is a recognition that the large increases in the cost of these plans have eroded the ability of the union to secure increases in salary and other desired benefits. Therefore, there is room for constructive proposals

TABLE 5.3
**TRADE UNIONS AND THE PROVISION OF HEALTH BEN-
EFITS: MEAN PREMIUMS AND EMPLOYER CONTRIBU-
TION PER ELIGIBLE EMPLOYEE IN FIRMS WITH HEALTH
INSURANCE BY FIRM SIZE AND UNIONIZATION, 1977**

	Number of employees in firms with plans* (000s)	Mean per eligible employee		Percent of premium paid by employer
		Total annual premium	Employer contribution	
Firm size and unionization				
More than 1,000 employees	38,795	$627	$552	84
Union	27,051	675	615	89
Nonunion	10,468	500	384	73
251–1,000 employees	12,903	587	490	79
Union	6,296	735	661	88
Nonunion	6,196	447	327	71
26–250 employees	22,954	529	416	76
Union	7,301	689	595	87
Nonunion	15,182	455	333	75
25 or less	11,090	595	517	80
Union	1,535	821	761	86
Nonunion	9,237	569	485	84
Total	85,473	591	502	80

*Totals include employees with unknown labor union status.
Source: National Center for Health Services Research, National Medical Care Expenditure Study.

from management which either trade off some reduction in medical benefits for other benefits the union wants or which promise to share the savings with the union members. For example, the plan developed by Quaker Oats summarized in Figure 5.2 has been successfully sold to at least three unions because it provides more money per employee for health services and shares savings with the employees while building in incentives for more cost-effective use of the benefit over time. A large national manufacturing company recently was able to negotiate with several unions to accept a new hospital deductible, a higher general deductible and an increase in employee contributions to medical plan premiums.

NOTES

1. Health Insurance Association of America. *Source Book of Health Insurance Data 1981–1982*. Washington, D.C., 1982.
2. U.S. Department of Labor, Bureau of Labor Statistics. *Employee Benefits in Industry, 1980*. Washington, D.C.: U.S.G.P.O., September, 1981.

3. Hay Associates. 1983. *Noncash Compensation Comparison*.

4. Ibid.

5. Health Research Institute. *Health Care Cost Containment Third Biennial Survey Participant Report*. Walnut Creek, California, Winter, 1983.

6. "Xerox Teams Up with Union to Control Health Care Costs," *Business Insurance*. October 24, 1983, p. 29.

7. Drury S.T. "Firms Basing Health Deductibles on Pay," *Business Insurance*. January 16, 1984, pp. 1, 23.

8. Op. cit. (Hay Associates).

9. "Few Workers Pay High Health Deductibles: Study," *Business Insurance*. January 24, 1983, p. 8.

10. Op. cit. (U.S. Department of Labor).

11. Op. cit. (Hay Associates).

12. Op. cit. (Hay Associates).

13. "Liver Transplants Covered by Most Insurance Companies," *Hospitals*, Vol. 57, No. 19, 1983, p. 24.

14. "Coverage of Expensive Transplants Takes Them Off 'Experimental' Lists," *Modern Healthcare*, Vol. 13, No. 9, pp. 70, 74.

15. Op. cit. (Hay Associates).

16. Op. cit. (U.S. Department of Labor).

17. Op. cit. (Hay Associates).

18. Op. cit. (Hay Associates).

19. Newhouse, J.P. "Insurance Benefits, Out-of-Pocket Payments, and the Demand for Medical Care: A Review of the Literature." Santa Monica, California: Rand Corporation, May 1978.

20. Newhouse, J.P., et. al. "Some Interim Results from a Controlled Trial of Cost Sharing in Health Insurance," *New England Journal of Medicine*, Vol. 305, No. 25, 1981, pp. 1501–1507.

21. Kiefhaber, A. "Consumer Incentives—A Cost Management Strategy?," *Washington Business Group on Health*. June 1982, p. 1.

22. Ibid.

23. Ibid.

24. "DuPont Has Good Results with Optional Health Plans," *Employee Benefit Plan Review*, Vol. 37, No. 2, 1982, p. 22.

25. "Paying Employees Not to Go to the Doctor," *Business Week*, May 21, 1983, pp. 146–148.

26. Washington Business Group on Health. March 3, 1984, Memorandum on Flexible Spending Accounts, Washington, D.C.

27. Op. cit., (*Business Week*).

28. "Zenith Plan Taps Blues' Cost Data," *American Medical News*. December 2, 1983. pp. 1, 11–12.

29. Hewitt Associates. *Salaried Employee Benefits Provided by Major U.S. Employers*. 1982.

MANAGING DENTAL *6*
BENEFITS

GROWTH IN DENTAL BENEFITS

During the 1970s, no health-related benefit grew as rapidly as dental insurance. In 1970 very few employees were covered for dental services and the Chamber of Commerce annual survey of employee benefits did not even mention dental coverage. By 1976 even the most conservative estimates indicate a 500 percent increase in dental coverage, with larger employers taking the lead.[1] And by 1983 approximately 80 to 90 percent of medium-sized and large employers sponsored dental plans.[2,3]

Employer size greatly affects likelihood of dental coverage. Metropolitan Life, for example, reported that in 1982 80 to 90 percent of the accounts with 5,000 or more employees and dependents had dental coverage, as opposed to 33 percent for those clients with 50 to 200 employees.[4] The director of marketing for group insurance operations at Connecticut General predicts that at least 100 million people will be covered by group dental plans by the mid-1980s.[5]

Dental coverage grew so rapidly for two major reasons. The first was the formation of statewide dental service plans. In 1955, in response to a request for a dental insurance plan made by officials from the International Longshoremen's and Warehousemen's Union-Pacific Maritime Association, organized dentistry in California formed the California Dental Service Corporation. Since then virtually all state dental societies have formed statewide nonprofit dental service corporations, usually known as Delta Dental Plans or Dental Service Corporations. In 1965 the American Dental Association created the Delta Dental Plans Association as a national coordinating agency for these state plans.[6]

Between 1970 and 1980 the number of enrollees in Delta plans grew about five-fold to an estimated 15 million. Commercial carriers have about a 50 to 60 percent market share followed by Delta plans (20 to 25 percent) and the Blue Cross/Blue Shield plans (10 to 15 percent). Many third party administrators have increased their market share in the past five years for some of the same reasons as their growth in administering medical benefits.[7]

A second major reason for growth of coverage was union demands. The 1973 United Auto Workers contract brought dental benefits to an estimated three million people. The following year the United Steel Workers asked for and negotiated dental benefits for its members, and in 1976 almost one million workers secured this coverage under the Bell System contract.[8]

DENTAL BENEFITS AND COVERAGE

Services commonly provided under dental plans are listed in Table 6.1. Of nearly 670 dental insurance plans surveyed by the 1983 Hay-Huggins Survey, virtually all covered preventive, basic, restorative and reconstructive services, while 70 percent also covered orthodontic services.[9] Another survey of major employers' benefits found that 78 percent of the dental insurance plans covered orthodontia in 1981.[10]

Reimbursement Principles

Participants in dental plans are usually covered in one of three ways: 1) payment of a proportion of the reasonable and customary charge for a procedure; 2) payment based on a scheduled cash allowance up to a certain amount for a given procedure; or 3) payment in full based on a reasonable and customary charge for each procedure. A variation which combines some elements of both 1 and 3 and is unique to dental insurance is the use of an incentive schedule. Under this type of arrangement the percentage of dental expenses paid by the plan increases each year if the participant is examined by a dentist or has regular preventive services performed.

Approximately four-fifths of dental participants are covered under plans which reimburse dentists at various percentages of the reasonable and customary fee. Routine exams are frequently paid at 100 percent, minor restorative work most commonly at 70 to 85 percent and orthodontia and prosthetics usually at 50 percent.[11,12] Delta dental plans are more liberal, generally paying 100 percent for primary and secondary preventive care and 75 percent for most restorative care without deductibles.[13]

TABLE 6.1
SERVICES COMMONLY PROVIDED UNDER DENTAL PLANS

Diagnostic and preventive services
  Mouth examination
  X-rays
  Cleaning
  Fluoride treatments
  Space maintainers
  Consultation
Restorative services
  Routine fillings
  Inlays
  Onlays
  Gold fillings
  Crowns
  Outpatient oral surgery, e.g., root resections, removal of impacted teeth
  Root canal therapy
  Gum treatments
Prosthodontic services
  Installation, repair and replacement of removable and permanent dentures and bridgework
Orthodontic services
  Prevention or correction of irregularities in position of teeth; related x-rays, surgery and
    braces.

Source: Bell, Donald R., "Dental and Vision Care Benefits in Health Insurance Plans," *Monthly Labor Review,* Vol. 103, No. 6, 1980.

Some of the reluctance on the part of both insurers and employers to pay higher percentages of restorative, prosthodontic and orthodontic procedures is their high unit costs. As displayed in Table 6.2, services related to crowns and bridges are used by 6.4 percent and 3.5 percent of patients during a year's period, but together account for over 30 percent of the total charges.

During the period from 1977 to 1979, two-thirds of reviewed large plans with scheduled cash allowances allowed $100 to $200 for the most expensive restorative procedure and nearly three-fourths had limits of $100 to $300 for the most expensive orthodontic procedure. In general, scheduled cash allowances were greater for orthodontic treatment. For example, Dow Chemical Company employees were allowed $50 for preliminary x-rays and diagnostic costs, $225 for the initial month of active orthodontic treatment and $30 for each succeeding month of treatment. Under this plan, the amount payable for a two-year course of treatment would be $965, compared to Dow's allowance of $205 for the most expensive prosthodontic procedure.[14]

While commercial insurers and Blue Cross/Blue Shield plans pay based on either usual, customary and reasonable charges or a table of allowances, dental service corporations usually have contracted agreements with dentists

TABLE 6.2
THE PERCENTAGE OF DENTAL PATIENTS RECEIVING DIFFERENT SERVICES AND THE PERCENTAGE OF TOTAL CHARGES FOR EACH SERVICE IN AN INSURED POPULATION

Service	Percentage of patients receiving one or more services per year	Percentage of total charges*
Examinations	77.7	4.0
Radiographs	82.0	7.3
Prophylaxes	79.7	7.1
Amalgams and composites	64.1	25.7
Extractions	11.6	2.1
Full and partial dentures	2.5	6.7
Crowns	6.4	11.6
Bridges	3.5	18.8

*Approximately 17 percent of total charges are for services not included in this table.
Source: Bailit, H.L., et al. "Controlling the Cost of Dental Care," *American Journal of Public Health*, Vol. 69, No. 7, 1979, pp. 699–703.

who agree to provide their services to covered beneficiaries at prefiled fees. In many cases the full cost of routine restorative services is covered by the dental service corporation, and the overall cost to the insured of services through these plans may be less expensive than more traditional insurance plans. However, not all dentists are plan members and dental service plans may pay less than other insurers for visits to a nonplan dentist.

Plans with Incentives for Prevention

About 5 percent of dental plans use an innovative incentive schedule for examinations, x-rays and routine restorative care, 2 percent for major restorations and none for orthodontia.[15] Typical of this group are the plan negotiated by the Machinist's Union and the one offered by Prudential Insurance Company to its insurance workers. In both, benefits for preventive and restorative services are provided at 70 percent in the first year an individual is covered, and the percentage of customary and reasonable charges reimbursed increases by 10 percent per year if the insured receives annual dental checkups, until reaching 100 percent coverage, which can be obtained in the fourth year of the plan.[16] Underlying this plan is the assumption that provision of incentives for preventive care, although increasing utilization and costs in the short run, will start paying dividends in reduced need for more extensive and costly procedures at a later time. Since prophylaxis, examinations and radiographs together account for less than 20 percent of total charges (Table 6.2), the investment has a limited downside risk.

Higher Payment for Surgery

A number of plans pay a larger proportion of reasonable and customary charges than might be expected for surgical procedures such as tooth removals, root resections and gingivectomies. The reason for this is historical. These procedures were originally covered on an inpatient basis under basic health insurance that usually paid full charges. Providing a smaller proportion of reimbursement for less expensive outpatient surgery would appear inequitable and create incentives for the surgery to be performed in the more expensive setting.[17]

Deductibles, Maximums and Employee Contributions

About one-third of all dental plans have deductibles for all services and an additional two-fifths sometimes require a deductible, which is, however, usually waived for preventive services.[18,19] Annual deductibles generally range from $25 to $50 per covered person per year, although a small number of plans provide for one lifetime deductible rather than annual deductibles.

Another method used to control overall utilization and costs, especially those associated with attempting to remedy the ravages of a lifetime of neglect of teeth and gums, is to set maximum dollar amounts that can be paid. Of plans offered by large employers, 92 percent had annual maximums in 1981, with the median amount set at $1,000. Fourteen percent of plans (some with annual maximums) had lifetime maximums, with $5,000 the median figure. Only 6 percent had no maximum.[20] Orthodontia maximums are usually separate and rarely exceed $1,000 to $1,500.[21]

In 1981, among dental plans sponsored by large employers, 31 percent required employee contributions toward their own coverage and that of their dependents, and 20 percent toward dependent coverage only; the remaining 49 percent was fully paid by the employer.[22] Although a smaller proportion of large employers contribute 100 percent of employee and dependent premiums for medical benefits (39 percent),[23] most of the difference is probably explained by the significantly lower premiums for dental insurance. On the average, in 1983 monthly dental premiums were about $7.50 to $8.50 for individual coverage and $18.50 to $25.50 for family coverage (including the employee).[24]

DENTAL COSTS

First-year claims under a new dental insurance plan can be expected to be high due to previously delayed work. Rapid increases over the first several years may occur as insurees become more familiar with the benefit. However, over time, the rate of rise has been slower than for medical claims because prices for dental services have risen at a slower rate than those for most

TABLE 6.3
NATIONAL DENTAL EXPENDITURES 1960–1980

Year	Payments (billions)	Payments per capita	Percentage paid by insurance
1980	15.9	68.42	N/A
1979	13.5	58.95	22.3
1975	8.2	37.46	13.0
1970	4.7	22.77	5.4
1965	2.8	14.19	1.5
1960	2.0	10.75	N/A

Sources: Gibson, R.M. and Waldo, D.R. "National Health Expenditures, 1980," *Health Care Financing Review*, Vol. 3, No. 1, 1981 and Carroll, M.S. and Arnett, R.H. "Private Health Insurance Plans in 1978 and 1979: Coverage, Enrollment, and Financial Experience," *Health Care Financing Review*, September, 1981.

other health care services (Table 6.3). During the period from 1967 to 1980, fees for dental services increased at an average annual rate of 10.9 percent, about the same rate of rise as the overall consumer price index (11.3 percent per year) and much more slowly than hospital costs and physician fees (Table 6.4). However, the growth of expensive procedures, especially peridontia, could lead to a much faster rate of rise, particularly as cost control measures are targetted primarily to medical rather than dental care.

Insurance coverage for dental care exhibits a strong positive influence on demand. Based on a comparison of dental premiums for two policies, one with zero co-insurance and one with a 20 percent co-insurance, working backward from a premium that was 63 percent higher for the zero co-insurance policy, Phelps and Newhouse estimated that demand was 30 percent greater without co-insurance than with 20 percent co-insurance.[25] When a free comprehensive dental benefit was added for New York Teamsters, who previously had no coverage, visits for plan enrollees rose from 2.5 in the year prior to plan initiation to 4.9 in the year after, nearly a 100 percent increase in demand. Other studies have found a similar demand responsiveness to dental insurance.[26] One impact of increasing coverage has been a lessening in the traditional fluctuation in demand for dental services due to changing national economic cycles.

DENTAL PRACTICE CHARACTERISTICS

Greater success has been obtained in controlling the costs of the dental benefit than the costs of most other health-related benefits. In part, this is due to the nature of dentistry and dental practice. Approximately 78 percent of the 130,000 active civilian dentists in 1981 were working in a solo practice

TABLE 6.4
U.S. ANNUAL AVERAGE CONSUMER PRICE INDEX FOR SELECTED HEALTH CARE ITEMS:
1967–1980 (CPI for 1967 = base 100)

Year	All consumer items	All health care items	Dental services	Physician services	Hospital semiprivate room rates
1967	100.0	100.0	100.0	100.0	100.0
1970	116.3	120.6	119.4	121.4	145.4
1973	133.1	137.7	136.4	138.2	182.1
1976	170.5	184.7	172.2	188.5	268.6
1980	247.0	267.2	242.3	274.3	416.3

Sources: U.S. Department of Health and Human Services, Public Health Service, *Dental Manpower Fact Book*, March 1979, p. 98, *Health 1980*, p. 209, and *Health 1981*, p. 198.

on a fee-for-service basis. Ninety percent of dentists are general practitioners. A typical solo dental practice sees in the range of 65 to 90 patients in an average week. Two-thirds of dentists work without a chair assistant or with only one such assistant.[27] Thus, traditional private solo practice dominates the field of dentistry. General dentistry is not a technology-intensive practice. It has undergone very little of the transformation that has affected medical care practice, especially in the hospital setting. It is also not a system that has been shaped to an appreciable degree to date by the widespread availability of third-party insurance. In 1981, of the $17.3 billion spent on dental services, 72 percent came from direct payments (i.e., directly from patients to dentists; without third-party involvement). This compares to less than 40 percent direct payments for physician services and only 11 percent for hospital care.[28]

In a 1979 survey of dental practice by the American Dental Association, 25 percent of practicing dentists stated that they were not busy and would like more patients.[29] Yet an increase of almost 20 percent in the number of dentists is expected between 1980 and 1990.[30] It seems reasonable based on the increase to anticipate greater numbers of dentists who do not perceive themselves to be sufficiently busy. One possible offshoot of this trend is a redistribution of dentists from metropolitan areas to nonmetropolitan areas. Another possible ramification of the impending dental glut, and one which is increasingly likely as the growth of third-party payments for dental care continues, is the provision of more dental services of questionable need and/or marginal value.

Alternative Settings for Dental Practices

A number of trends in the practice of routine dentistry are likely to maintain downward pressure on prices. In 1977, Sears Department Stores opened an in-house dental practice in its El Monte, California store. In the following three years, ten other retailers got into the dental business at thirty-five sites. Montgomery Ward, not to be outdone by Sears, had opened eighteen units in their stores throughout the United States by 1981 and had announced plans to more than double this number in a short space of time. Several large companies operating vision centers, most notably Searle and Sterling Optical, have either begun offering dental care or have announced plans to do so. Care is generally available with or without an appointment and the average facility is open ten to eleven hours per day, six or seven days per week. Fee schedules, which are posted, display prices that are usually 20 to 40 percent below the usual and customary rates in the area. These facilities are larger than the usual dental practice office, frequently including five to ten examination and treatment rooms.[31]

A few corporations have developed dental care practices. The first in-house dental facility opened at Stockham Valves and Fittings, Inc. in Birmingham, Alabama in 1918. Although employer provision of on-site dental services has never become a common practice, the ability to control utilization and costs and reduce time off for dental care are attractive features which may induce more companies to proceed in this direction. Another source of competition to the traditional mode of dental practice is franchised practices. In return for an initial investment by an interested dentist, the franchising company will find an appropriate location, build or rent the necessary space, furnish all the equipment and supplies, provide staff training, recordkeeping and accounting systems and let the dentist participate in a volume-purchasing agreement for supplies. These types of businesses usually market to dentists who are interested in establishing practices in shopping malls and department stores with a number of other dentists working for them.[32]

Increasingly, medical care organizations are entering the dental care field. In 1979, 27 of the 99 federally qualified health maintenance organizations included dentistry.[33] In the same year, of 98 nonfederally qualified prepaid health plans responding to a survey, 21 offered comprehensive dental services.[34] The ability to prevent the two major oral diseases, dental caries and periodontal disease, and reduce their high incidence and prevalence suggests that a prepaid approach is very appropriate and promises high future savings for present investments in preventive care.[35] Studies have not yet been reported to validate or refute this premise in the adult population, although fluoride treatments for children reduce caries by about 60 percent[36,37] and plastic sealants are also effective in preventing dental decay.

Hospitals are also increasingly offering dental services as part of ambulatory care. Whether these services will prove profitable at a time when hospitals are investigating all possible sources of new revenue is not yet known. The high overhead usually associated with hospital-run services may put them at a competitive disadvantage. However, the desire to build patient volume and to keep as many people as possible as part of the hospital system to preserve market share may spur considerable additional experimentation in this area.

CONTROL OF COSTS AND QUALITY

Concerted attempts to develop procedures for assuring that proposed dental work meets professional standards for necessity and appropriateness and that the work performed is of acceptable technical quality have been undertaken by a number of insurance carriers and dental plan administrators. Payers have adopted these activities to counteract some of the financial incentives which accompany the broader availability of reimbursement from third parties. Coverage under a dental insurance plan usually leads to a significant increase in the utilization of the more expensive elective dental services, including crowns, bridges, periodontia and orthodontia.[38]

In general, dentists who wish to be paid under a specific insurance plan are required to submit in advance a plan of treatment and x-ray for services that exceed a fixed dollar figure, usually in the range of $150 to $300. Claims that are covered by pre-authorization account for 20 percent of the number of claims filed but 80 percent of total service expenditures.[39] Occasionally models are required, and more rarely a consulting dentist examines the patient as part of an assessment of a submitted treatment plan. Dentists, either employees of or consultants to the reviewing organization, conduct the assessment and deny benefits if the rationale for treatment is not sufficiently well established or if less expensive services could adequately respond to the dental needs as outlined.

Estimates of savings from pre-authorization are three to five percent of total gross premiums from denials or acceptance of less expensive treatment plans proposed by the reviewer. However, the dampening effect of knowing that treatment plans will be reviewed is probably also responsible for significant savings. In new plans, initial rejection rates are higher than in established plans where dentists have a better idea of what is acceptable to the reviewers. Operating costs for pre-authorized programs are one to two percent of premiums.[40]

Post-treatment review can also serve to assure that the accepted plan of treatment is followed. Of greater importance, it can serve as an audit of the quality of care that has been provided. This review consists of having the dentist submit post-treatment radiographs with the bill. Usually this is required only after extensive crown and bridge treatment, and may be limited to those dentists who have a history of many denials or shoddy work.[41]

Table 6.5 provides a breakdown of the $5.5 million in claims revenue saved by Pennsylvania Blue Shield, which provides dental coverage for 1 million subscribers, as a result of professional pre- and postpayment review in 1979. Of 1,029,432 claims received in 1979, 22,085 (2.15 percent) were reviewed. Ninety-five percent of the savings was from prepayment review.[42]

While peer review can affect both quality and cost, it can cause antagonism among dentists and may result in the patient experiencing higher out-of-pocket expenses. A related but possibly more efficient approach establishes periodic profiles of the pattern of dental practices based on claims and administrative review data. Dentists who have a pattern of overproviding are identified through computer analysis, and pre-authorization and post-treatment review are concentrated on them. One estimate of possible savings from this method is 8 percent of premium dollars with administrative and computer-related costs of two or three percent. However, these estimates are based on computer simulations, not actual experience utilizing such a system.[43]

In addition to identifying dentists who appear to be overproviders,

TABLE 6.5
PENNSYLVANIA BLUE SHIELD PROFESSIONAL DENTAL REVIEW

	Number of services denied (%)		Amount denied (%)	
Prepayment				
Overtreatment	7,898	(16.4)	$ 999,801	(19.1)
Undertreatment	424	(.9)	67,193	(1.3)
Quality	28,068	(58.3)	2,434,711	(46.5)
Optional treatment	10,089	(20.9)	1,652,923	(31.6)
Misreporting	1,684	(3.5)	78,326	(1.5)
TOTAL	48,163	(100.0)	$5,232,954	(100.0)
Postpayment				
Quality	3,610	(9.0)	$ 57,653	(18.9)
Misreporting	36,572	(91.0)	247,478	(81.1)
Patient fraud	2	(.0)	79	(.0)
TOTAL	40,184	(100.0)	$ 305,210	(100.0)
			TOTAL	$5,538,164

Source: "$5.5 Million in Savings Claimed for Dental Review," *Employee Benefit Plan Review*, Vol. 35, No. 8, 1981, pp. 30, 87.

establishing computer profiles based on claims data can highlight those dentists engaged in overbilling, a practice whereby dentists tack on the patient's cost-sharing requirement to their bill and then agree to accept whatever the carrier payment is as payment in full. There are variations in overbilling schemes but the bottom line remains the same—overbilling has a negative effect on the group's experience and will result in higher future premium costs to the purchaser. Table 6.6 demonstrates the impact overbilling had on the experience of three subscriber groups covered by California Dental Service (CDS).

CDS has extensive administrative procedures and membership agreements with all of its 14,000 participating dentists which allow it to conduct dental office audits to verify that fees submitted are accurate and that patients have been billed for co-payments as appropriate. Sanctions which can be taken against dentists in violation of CDS terms include reduced reimbursement, recovery of funds, termination of membership in CDS and even legal action.[44] Many insurers do not have such extensive capabilities for monitoring claims and controlling overbilling.

Pre- and post-treatment reviews, while mitigating the natural effects of insurance coverage on utilization, do not prevent overproviding. As stated by a dental consultant with extensive experience, ". . . the indirect review

TABLE 6.6
AVERAGE CLAIMS PAYMENTS

	Non-overbilling dentists	Overbilling dentists	Average differences	Combined average*	Net effect of overbilling on claims payments
Group A	$79.14	$176.99	+123.6%	$87.52	+10.6%
Group B	74.28	137.33	+ 84.9	92.62	+24.7
Group C	55.86	242.49	+334.1	87.34	+56.4
Average	70.66	189.56	+168.3	88.31	+25.0

*Combined average was reached by using all individual claims payments of overbilling and non-overbilling dentists.
Source: "Avoidance of Co-payment Feature in Dental Plan Subverts Plan Design and Adds to Premium Costs," Employee Benefit Plan Review, Vol. 35, No. 12, 1981, pp. 14–16, 34.

utilized by pre-authorization cannot prevent the 'nickel and dime' exploitation of dental insurance by way of needless radiographs, filling of noncarious pits and fissures, replacement of functional restorations, or even the remaking of existing bridges and dentures that could be repaired or relined just as effectively. And, few dental consultants appear willing to stand up to those oral surgeons in the United States who submit inflated claims statements."[45] Among the other effects of insurance noted by this dentist are tremendous growth of full crowns as substitutes for three-quarter crowns, inlays and onlays, substitution of permanent for removable appliances, replacement of repairable appliances, routine full crowns on teeth on which root canal work has been done, more extensive and frequent periodontal treatment and routine extraction of third molars.

Although on balance the arguments for performing pre-authorization—especially that of quality control—seem to outweigh the disadvantages, not all dental insurance carriers have utilized this mechanism. A study of third-party payers revealed that 50 percent of commercial insurance companies were conducting pre-treatment reviews, as were 66.7 percent of Delta plans, 71.4 percent of Blue Cross and Blue Shield plans and 93.5 percent of Medicaid programs. Commercial carriers also had the lowest percentage of post-treatment review procedures, such as patient examinations, submissions of radiographs, checking of adherence to treatment plans and provider fee audit.[46] Commercial carriers appear to rely instead on cost-sharing provisions to reduce utilization and costs. The same survey found that the average number of claims per individual covered during a one-year period were 0.54 for commercial carriers, 0.91 for private insurance covered persons in Delta Plans and 1.02 for Blue Cross and Blue Shield plans (excluding Medicaid beneficiaries).

Cost-Effective Dental Benefits

Current dental benefits appear to have been constructed to avoid or minimize the effects of problems experienced with other health care benefits. Most laudable is the emphasis on preventive services which appear to have desirable benefit:cost ratios. Tooth cleaning and scaling are currently covered in most programs and should show benefits to the individual and help control cost increases in the dental plan over time. Dental health education aimed at improving oral hygiene practices might also be considered as a worthy benefit, since many individuals, despite coverage, do not avail themselves of preventive or other important dental services. Full coverage of routine restorative procedures with increased cost-sharing for more extensive procedures appears an appropriate balance between meeting legitimate needs and reducing overproviding.

Pre-authorization review can be an important quality assurance mechanism as well as an effective means of limiting care to the least expensive, professionally adequate treatment. Post-treatment review is also an effective means of identifying poor treatment, as well as some cases of fraud. In the past, the patient frequently paid some or all of costs that were disallowed. Employers, either directly or through their carriers, should be prepared to educate employees about their dental needs in these cases, so that they can make an informed decision about whether to proceed with treatment that will not be covered, or covered to a small degree, if less expensive treatment would be equally effective. In addition, the employer should be willing to intercede on the employee's behalf when he or she is overcharged. Finally, employers should include the application of topical fluoride and sealants as a benefit for the children of employees who are not in fluoridated communities. Such coverage will assure that the cost of the dental benefit will still be reasonable ten and twenty years from now.

NOTES

1. Praiss, I.L., Tannenbaum, K.A., Gelder-Kogan, C.A. and Hale, C.B., "Changing Patterns and Implications for Cost and Quality of Dental Care," *Inquiry*, Vol. 16, No.2, 1979, pp. 131–140.
2. Hay Associates. *1983 Noncash Compensation Comparison.*
3. Hewitt Associates. *Salaried Employee Benefits Provided by Major U.S. Employers.* 1983.
4. Sherwood, S. "Employers Hope to Contain Dental Costs Too," *Business Insurance.* December 13, 1982. pp. 10–14.
5. Norris, E. "Dental Plans Remain Popular with Employers, Employees," *Business Insurance*, December 14, 1981, pp. 34, 35.
6. Douglas, C.W. and Day, J.M. "Cost and Payment of Dental Services in the United States," *Journal of Dental Education*, Vol. 43, No. 7, 1979, pp. 330–348.
7. American Dental Association, Council on Dental Care Programs. Fact Sheet: Dental Prepayment Plans. June 1982.
8. Op. cit. (Praiss, et al.).
9. Op. cit. (Hay Associates).
10. Hewitt Associates. *Salaried Employee Benefits Provided by Major U.S. Employers.* 1982.
11. U.S. Department of Labor, Bureau of Labor Statistics. *Employee Benefits in Industry, 1980.* Washington, D.C.: U.S.G.P.O., 1981.
12. Bell, D.R. "Dental and Vision Care Benefits in Health Insurance Plans," *Monthly Labor Review*, Vol. 103, No. 6, 1980, pp. 22–26.
13. Op. cit. (Douglas).
14. Op. cit. (Bell).
15. Op. cit. (U.S. Department of Labor).
16. Op. cit. (Bell).
17. Op. cit. (Bell).
18. Op. cit. (Bell).
19. Op. cit. (Hewitt Associates, 1982).
20. Op. cit. (Hewitt Associates, 1982).
21. Op. cit. (Hay Associates).

22. Op. cit. (Hewitt Associates, 1982).
23. Op. cit. (Hewitt Associates, 1982).
24. Op. cit. (Hay Associates).
25. Phelps, C.E. and Newhouse, J.P. *Coinsurance and the Demand for Medical Services.* (R–964–1–OEO/NC) Santa Monica, California, Rand Corporation, 1974, as cited in Newhouse, J.P. *Insurance Benefits, Out-of-Pocket Payments, and the Demand for Medical Care: A Review of the Literature,* Santa Monica, California, Rand Corporation, 1978.
26. Ibid.
27. Rovin, S. and Nash, J., "Traditional and Emerging Forms of Dental Practice, Accessibility, and Quality Factors," *American Journal of Public Health,* Vol. 72, No. 7, 1982, pp. 656–664.
28. Gibson, R.M. and Waldo, D.R., "National Health Expenditures, 1981," *Health Care Financing Review,* Vol. 4, No. 1, 1982, pp. 1–35.
29. American Dental Association. *Survey of Dental Practice,* 1979.
30. U.S. Department of Health, Education, and Welfare. Public Health Service. *Dental Manpower Fact Book.* 1979. DHEW Pub. No. (HRA) 80–21.
31. Op. cit. (Rovin and Nash).
32. Op. cit. (Rovin and Nash).
33. Op. cit. (Rovin and Nash).
34. American Dental Association, Council on Dental Care Programs. "Dental Components in Prepaid Health Plans," *Journal of the American Dental Association,* Vol. 101, No. 5, 1980, pp. 817–820.
35. Schoen, M.H., "Alternative Oral Health Service Delivery Systems," *Family and Community Health,* Vol. 3, No. 3, 1980, pp. 71–80.
36. Pelton, W.J., Dunbar, J.B., McMillan, R.S., et al. *The Epidemiology of Oral Health.* Cambridge, Massachusetts: Harvard University Press, 1969.
37. Cuzacq, G. and Glass, R.L. "The Projected Financial Savings in Dental Restorative Treatment: The Results of Consuming Fluoridated Water," *Journal of Public Health Dentistry,* Vol. 32, No. 1, 1972, pp. 52–57.
38. Bailit, H.L., Raskin, M., Reisine, S. and Chiriboga, D., "Controlling the Cost of Dental Care," *American Journal of Public Health,* Vol. 69, No. 7, 1979, pp. 699–703.
39. Ibid.
40. Ibid.
41. Friedman, J.W. "The Effects of Dental Insurance on Private Dental Practice," *New Zealand Dental Journal,* Vol. 77, No. 347, 1981, pp. 14–19.
42. "$5.5 Million in Savings Claimed for Dental Review," *Employee Benefit Plan Review,* Vol. 35, No. 8, 1981, pp. 30, 87.
43. Op. cit. (Bailit).
44. "Avoidance of Co-payment Feature in Dental Plan Subverts Plan Design and Adds to Premium Costs," *Employee Benefit Plan Review,* Vol. 35, No. 12, 1981, pp. 14, 16, 34.
45. Op. cit. (Friedman).
46. Op. cit. (Praiss, et al.).

7 MANAGING MENTAL HEALTH BENEFITS

HISTORICAL PERSPECTIVE ON MENTAL HEALTH BENEFITS

The President's Commission on Mental Health estimated that as much as 15 percent of the population may need mental health care at any point in time. A careful review of epidemiological literature over a quarter-century estimated that 10 percent of the population have mental health problems at any point in time and that 15 percent experience such problems over the course of a twelve-month period.[1]

In the thirty years following World War II, as insurance for hospital expenses grew rapidly, coverage was gradually extended to mental health conditions necessitating hospitalization. Usually, however, lower coverage limits were established for treatment of mental than of physical illness. For example, in the mid-1970s a survey of 148 employee health benefit plans conducted by the Bureau of Labor Statistics revealed that 68 percent of plans offered equal levels of hospital reimbursement (although frequently with dollar or service maximums) for mental and physical conditions while the remaining 32 percent uniformly offered less hospital reimbursement for mental conditions.[2]

Users of Mental Health Benefits

More so than in other areas of health care, there is strong disagreement on who needs what type of mental health care. Over the past several decades patterns of care have shifted drastically. In 1955 more than three-quarters of all care episodes at mental health facilities occurred on an inpatient basis. In large measure due to deinstitutionalization of mental health care in state facilities, by 1975 at least three-quarters of care episodes at such facilities were to outpatients.[3] Lengths of stay in state and county mental hospitals have declined rapidly. Equally striking has been the development of inpatient psychiatric services in general hospitals, where the estimated number of available beds increased more than 20 percent, from 11.2 per 100,000 population in 1972 to 13.6 per 100,000 in 1978 and have continued to increase since.[4] There are few larger communities lacking a private hospital that offers psychiatric services, usually on a strict inpatient basis but sometimes also providing day treatment and/or outpatient services.

Criteria for need of mental health services are relatively concrete only for those one or two percent of people who are diagnosed as schizophrenic or manic depressive. In these cases, treatment by a psychiatrist, including drug manipulation, is mandatory. In some circumstances, fewer now than in the years prior to drug therapy, hospitalization is essential. In these cases, most psychiatrists believe maximum benefit can usually be derived from shorter lengths of stay than were common in the past.

Much more common than these psychoses are a category of disorders, intermittent or chronic, which can be disabling but usually do not uniformly require continuing psychiatric care, such as depression without mania, alcoholism and severe neuroses. In the view of a distinguished sociologist who has studied this problem extensively, these disorders "might best be managed through collaborative efforts involving the general medical sector but with significant assistance from psychiatric consultation and community supportive groups."[5] But while the vast majority of these problems can be handled on an outpatient basis and frequently by a psychologist or social worker, current coverage usually provides financial disincentives for pursuing outpatient treatment and may discourage care by professionals other than psychiatrists.

A brief history of how these disincentives developed illustrates the problem of whether different limits should be applied for mental health services than other services provided under employer health benefit plans. When outpatient coverage initially developed, mental health services by physicians were reimbursed on the same basis as care for somatic difficulties. Generally, after a small deductible, insurers would pay 75 to 80 percent of the charges for such treatment. However, early experience with this benefit was so adverse that insurers soon began to restrict outpatient mental health benefits.

The reason for this retreat was not based on overuse by those with alcoholism, serious depression or situational problems that temporarily disabled them but could be remedied with the help of psychiatric intervention. Rather, its cause was the high utilization by individuals who, while carrying out their usual functions in society, felt a lack of well being, experienced feelings of anxiety, malaise and sadness associated with life transitions or adverse external events. Psychotherapy for those individuals often consisted of psychoanalysis, with several visits per week to a psychiatrist over a period of several years. Reimbursement to these few individuals came to constitute a very large percentage of total outpatient mental health benefits. In a sense, those with the lowest level of need were receiving the most services while many others did not appear to be receiving needed care because of lack of awareness that it could be helpful and the stigma attached to seeking care for "mental" problems.

Despite the well-established inverse relationship between socioeconomic status and mental impairment, given comparable insurance coverage, persons with the higher use rates for mental health services are characterized by higher levels of income, formal educational attainment and sophistication. While the stigma associated with having a mental health problem is still almost universal, it is felt most strongly by those with limited educational attainment. One study found that college-educated employees with coverage for outpatient mental health services were six times as likely to seek psychiatric care and had almost tenfold the number of psychotherapy office visits as fellow employees with similar coverage but only a grade school education or less.[6] In assessing reasons for retrenchment of companies with respect to coverage for outpatient mental health services, one report concluded, "the companies became concerned over the appropriateness and equity of paying out significant portions of total benefit payments to a very few individuals who were not disabled and were continuing to work or carry on their usual functions."[7]

PROVIDERS OF MENTAL HEALTH SERVICES

Mental Health and Primary Care

Overall, only a minority of those with mental health problems receive care from a mental health professional. A task force of the President's Commission on Mental Health reported that "only about one-fourth of those suffering from a clinically significant disorder have been in treatment."[8] Where are the others? The answer appears to be that the majority of these individuals bring their mental health problems to their primary care physician. Most primary care practitioners agree that for at least a substantial minority and frequently a majority of their patients, somatic problems are not the major reason for the visit. Most cases of insomnia, unhappiness, alienation, stress-induced headaches and similar complaints are addressed by the medical portion of the health care systems as part of primary health care. One study found that 73 percent of those treated for mental disorders in 1975 received care in the general health care sector, only 19 percent were treated in the specialty sector, and 8 percent were treated in both.[9] Insurance coverage, which pays a greater portion of the bill for medical rather than mental health professionals, contributes to the medicalization of care for psychological problems.

Mental Health Professionals: Who Should be Reimbursed?

Although few employers or insurers would argue that psychiatrists are the only mental health professionals who should be eligible for reimbursement, coverage for other well-qualified professionals has been slow to evolve. The two major reasons for this have been insurers' concerns about potential overutilization and psychiatric association lobbying against the inclusion of other mental health practitioners. In the United States there are about 28,000 psychiatrists, not all of whom have an inpatient practice. There are, by contrast, at least twice as many doctors of psychology, six times as many social workers, with over 25,000 working in psychiatric settings, and many thousands of nurses with additional training in mental health care.[10] Were all of these professionals to be covered under third-party insurance, it is likely that utilization and costs of the mental health benefit would mushroom, both due to increasing availability and reduced prices. During 1977 a survey of mental health claims for federal employees under Blue Cross and Blue Shield coverage in the Washington, D.C. area found coverage costs claimed per visit of $42 for psychiatrists, $38 for clinical psychologists, $30 for psychiatric social workers, $20 for psychiatric nurses and $37 for nonpsychiatric physicians.[11]

As of late 1981, thirty-two states had passed legislation mandating that consumers should have a choice of seeing either a psychiatrist or a psychologist.[12] In these states psychologists are reimbursed on the same basis as psychiatrists. As of 1983, in ten states psychiatric social workers had won required coverage for their outpatient services under mental health insurance benefits.[13] A Washington Business Group on Health survey of its members in 1978 revealed that all sixty-eight respondents accepted psychologists as providers although less than one-half accepted social workers and psychiatric nurses.[14]

Although the impacts of these additions to the roster are difficult to isolate and likely to be disputed by one or more of the involved professional groups, some results appear probable. Nonphysician mental health therapists are more likely to utilize a situational approach to problems rather than a psychoanalytic approach. They are also more likely to place strong emphasis on the social aspects of the patient's problem. They also tend to deal with the problem in a small number of sessions, determined in part by the number of sessions provided under the benefit and the financial resources of the individual. In the study of federal employees in the Washington, D.C. area who submitted outpatient mental health services claims to Blue Cross in 1977, 60 percent of the visits provided by both psychologists and psychiatric social workers were for individuals with fifty visits or fewer as opposed to only about 40 percent of the visits provided by psychiatrists.[15]

To date there is no clear evidence that for the majority of patients seeking mental health care, short-term therapy by a psychologist or psychiatric social worker is likely to yield a poorer result than therapy by a psychiatrist. However, an important part of the training of nonphysician mental health practitioners as well as primary care physicians should be the identification of those individuals who need pharmacotherapy and/or who constitute a significant threat to themselves or others. In the latter cases, referral to a psychiatrist is mandatory. In the former, the mental health professional and primary care physician may be able to team up to provide the necessary elements of care, or a psychiatrist can be involved.

For some mental health concerns, community resourses are as important, and sometimes more important, to successful outcome as medical, psychiatric or psychological care. Alcoholics Anonymous has had remarkable success. Community groups for those with special problems, such as depression, loneliness or poor self-image may be very helpful, even in the absence of direction by a mental health professional. Educational approaches, teaching people the skills to overcome phobias, anxieties or depressive episodes have also been effective, and are low-cost. However, these types of activities are not usually covered under insurance. Summarizing the effects of insurance on care provided for mental health problems, the distinguished sociologist David Mechanic concludes, "In short, we have developed insurance programs for psychiatric benefits that . . . reinforce traditional, ineffective and inefficient patterns of mental health care, inhibit innovation and the use of less expensive mental health personnel, and reinforce a medical as compared with a social or educational approach to patients' psychological problems."[16]

Where the Mental Health Benefit Dollars Go

Accurate claims information on mental health services is difficult to obtain. Many inpatient and outpatient claims have been accepted for payment with diagnoses such as ill-defined signs and symptoms. Due to the stigma associated with mental health problems, physicians often enter somatic problems as the primary diagnoses for mental health patients. Discussions with insurers and third-party administrators suggest that, depending upon the benefit package, 10 to 20 percent of total claims dollars paid are probably for mental health services.

Overall, the rate of cost increases for mental health services has varied significantly from company to company, sometimes increasing much more rapidly than aggregate health insurance benefit costs. For example, IBM reported that between 1976 and 1980 hospital costs for their employees receiving inpatient psychiatric care grew at an annual compound rate of 16.1

percent, twice the rate of costs for patients in all other diagnostic categories. In 1980 the company spent 15.5 percent of total inpatient hospital costs for mental health patients, while mental health outpatient costs accounted for 22.2 percent of major medical expenses. Over a thirteen-year period inpatient psychiatric bills tripled as a percentage of the company's costs. Per capita costs demonstrated wide geographic variation of $27 to $231 per employee for 1980 without apparent explanation for the differences. It is worth noting that insurance for IBM employees includes full coverage of psychiatric inpatient charges and 80 percent coverage of eligible outpatient mental health charges after a $150 deductible.[17]

At least one-half of all states have statutes that mandate or regulate certain kinds of coverage for mental illness. Among the most common provisions of these statutes are:

- coverage for minimum days of inpatient psychiatric care (usually thirty days or more per year)
- coverage for minimum days of partial hospitalization treatment
- coverage for specified dollar minimum of outpatient treatment (usually $500 or more per year)
- coverage for psychologists, social workers or both in the same manner as physician coverage*
- offering to employees an option to purchase certain benefits or types of coverage
- inability to exclude coverage because service delivered in hospital without surgical facilities or state-operated[18]

Despite laws in some states mandating a specified minimum level of coverage, private insurance benefits for mental health care generally are still subject to more stringent limitations than benefits for other illnesses, as displayed in Table 7.1. As of 1980 only one-half of covered workers received hospital coverage for mental health problems to the same degree as somatic problems. The most common limitation was 30 days per year for mental health care contrasted to between 120 days and unlimited coverage for other hospitalizations. Approximately 10 percent of covered workers had outpatient mental health services covered to the same degree as other illnesses with frequent limitations in reimbursement per visit, reimbursement per year and/or higher coinsurance, with an average of 50 percent.[19] A 1981 survey by the Health Research Institute on health care cost containment, responded to by more than 600 of the 1500 largest U.S. employers, found some evidence that the percentage of employers offering coverage for mental

---

*As of 1981, thirty-two states mandated consumer choice between receiving care from a psychiatrist or a psychologist. As of 1983, ten states required reimbursement for psychiatric social workers practicing outside of an organized setting, such as a hospital.

TABLE 7.1
HEALTH INSURANCE: PERCENT OF FULL-TIME PARTIC-IPANTS BY COVERAGE FOR MENTAL HEALTH CARE, PRIVATE INDUSTRY, 1980

Coverage limitation	All participants	
	Hospital care	Outpatient care
Total	100	100
With coverage	98	93
Covered the same as other illnesses	54	10
Subject to separate limitations*	44	83
Limit on days or visits	33	20
Limit on dollars	21	58
Major medical coinsurance limited to 50 percent	3	54
Other limitations	1	4
Not covered	2	7

*The total is less than the sum of the individual breakdowns because many plans had more than one type of limitation on mental health coverage.
Source: *Employee Benefits in Industry, 1980.* U.S. Department of Labor, Bureau of Labor Statistics, September 1981, p. 22.

health-related problems to the same degree as other illness increased over a four-year period from 19.0 percent to 22.1 percent for mental/nervous conditions; from 31 percent to 46.1 percent for alcohol abuse treatment and from 32.6 percent to 40.8 percent for drug abuse treatment.[20]

UTILIZATION, COSTS AND EFFECTS OF MENTAL HEALTH SERVICES

Limited studies of cost effectiveness of different mental health treatments for particular problems have been conducted, although there is clear evidence for the effectiveness of such important therapies as major tranquilizers, antidepressants and some types of outpatient care. Of particular relevance are several studies of the effect of mental health treatment on nonpsychiatric health care usage. Many studies have shown that individuals with mental health problems use not only more mental health services but also more medical services. In one study of four settings such patients had about twice as many general health care visits as other patients.[21]

Group Health Association of Washington reported that patients treated

TABLE 7.2
MEASURES OF HOSPITAL AND AMBULATORY CARE FOR ALL CONDITIONS AND MENTAL HEALTH CONDITIONS IN RESPONDENT HMOs

Type of care	All conditions	Mental health
Hospital Care		
Days/1000 enrollees (N*)	480 (68)	65 (49)
Admissions/1000 enrollees (N*)	90 (68)	6 (49)
Length of Stay/Enrollees (N*)	5.52 (68)	10.41 (49)
Ambulatory Care		
Visits/1000 enrollees	3585 (66)	125 (51)
Patients/1000 enrollees	671 (19)	17 (20)

*N = number of HMOs providing information for all three hospital care variables for all conditions (68 out of 128 respondents) and mental health conditions (49 out of 123 respondents). *Source:* Levin, B.L. and Glasser, J.H. "A Survey of Mental Health Service Coverage Within Health Maintenance Organizations," *American Journal of Public Health*, Vol. 69, No. 11, 1979, pp. 1120–1125.

by mental health providers reduced their nonpsychiatric physician usage by 31 percent and laboratory and x-ray use by 30 percent in the year after referral. Blue Cross of Western Pennsylvania found that patients using a psychiatric outpatient benefit in community mental health centers reduced their medical/surgical utilization by $9.41 per month to $7.06, well below the average cost in a control group. Both medical/surgical inpatient days per month and outpatient visits per month were down by more than 54 percent.[22] A review of twenty studies looking at whether medical utilization is reduced as the result of alcohol, drug abuse and mental health treatment intervention suggests that this effect occurs subsequent to short-term therapy in organized settings for a wide variety of patient populations.[23]

Much of what is known about the interrelationship of factors that affect utilization and costs of mental health services derive from the experience of prepaid health plans, which have generally kept close track of such data. Based on a 1978 survey to which 123 (68 percent) of 181 HMOs responded, typical mental health inpatient benefits were 45 days per enrollee per year for hospital care and 20 outpatient visits per enrollee per year (20 visits are required for federally qualified HMOs). Only 2 percent of HMOs required co-payments for hospitalization and 32 percent for ambulatory care. Hospitalization and ambulatory care rates for these HMOs as a group are summarized in Table 7.2.[24]

A comparison of published reports on utilization and costs of mental health services provided under prepaid, cost-financed and indemnity models (Table 7.3) showed that cost-financed and indemnity plans experienced admission rates three or more times higher than prepaid plans. In general,

TABLE 7.3
ANNUAL INPATIENT UTILIZATION RATES OF INSURANCE PLANS

Type of plan	Year	Average length of stay (days)	Hospital admissions per 1,000 members
Prepaid group plans			
Arizona Health Plan	1975	5.03	0.80
HIP	1975	15.80	0.70
Health Maintenance Plan of Cincinnati	1976	10.83	4.37
GHA (indemnity)	1969	12.50	0.8
GHA (service plus indemnity)	1976	7.58	1.29
Cost-financed	1970		3.00
Indemnity Plans			
Blue Cross/Blue Shield (federal employees)	1974		
High option		17.4	5.0
Low option		13.9	2.8

Source: Craig, T.J. and Patterson, D.Y. "A Comparison of Mental Health Costs and Utilization Under Three Models," *Medical Care,* Vol. 19, No. 2, 1981, pp. 184–192.

prepaid groups had as high and usually higher rates of outpatient utilization and costs as the other plans, even when the indemnity plans have low co-insurance.[25] In addition to cost incentives for outpatient treatment in prepaid plans, it has been pointed out that when mental health services are provided in the same organization as medical services, there is opportunity for closer liaison and resultant potential increases in efficiency in dealing with psychological problems.

High utilization by federal employees, one of the groups covered in this comparative study, has caused their largest insurer, Blue Cross, to seek and receive increases in co-payment rates and restriction in number of visits under outpatient psychiatric coverage. Blue Cross noted that the 1.9 percent of subscribers using the benefits accounted for 7.7 percent of total mental health benefits payments. At congressional hearings on proposed cutbacks, the director of the Office of Personnel Management of the Federal Employees Health Benefit Program testified that "mental health conditions do not lend themselves definitive diagnoses on causes of treatments, making it virtually impossible to control utilization.[26]

Psychiatrists and psychologists are increasingly concerned about shrinking mental health insurance benefits, arguing that limitations on number of visits (fifty or fewer is the norm), fees ($30–$40 per visit is the norm) and lifetime expenditures for outpatient benefits ($20,000 or less, usually) prevent needed care which would not only help more workers on the job but also lead to reduced utilization for medical services.

Although the pendulum appears to have swung away from providing 75 or 80 percent reimbursement and unlimited visits for outpatient mental health benefits, it may have swung too far in the other direction. Some of the more stringent reimbursement policies probably adversely impact appropriate utilization. Low-income employees, who have the highest frequency of serious mental health problems within the workforce, are likely to be most affected. At least one study strongly suggested that the poorest patients had their utilization of mental health services most strongly influenced by insurance.[27] Early detection and prevention are likely to be the casualties of a payment structure that requires a 50 percent co-payment for even the first several visits to assess the nature and severity of a mental health condition.

Increasing Cost Effectiveness

Experience of HMOs and two prepaid outpatient mental health plans suggests that a cost-effective and perhaps health-effective policy should minimize barriers to mental treatment but either provide economic disincentives or peer review barriers to prolonged psychotherapy.

A prototype of coverage that emphasizes short-term treatment and removes barriers to initial assistance is that of the United Auto Workers. As described in 1977, it provided full payment for the first five visits, a 15 percent co-payment was required for sessions six through ten and co-payment increased to 45 percent on additional visits, subject to a $800 annual limit on individual payments.[28]

Effects of specific changes in benefits are not always easy to predict. Nonetheless, several researchers have attempted to calculate a price elasticity of demand for outpatient mental health services. Using data on two populations that have 50 and 20 percent co-payment features for this benefit, McGuire calculated a price elasticity of demand of $-.434$. He further estimated that a national financing plan that provided 80 percent payment with unlimited visits would raise utilization by 60 percent. Finally, he estimated that if a twenty-visit limit were imposed on the federal employees with a Blue Cross high-option plan, who have no limit other than a $250,000 maximum, demand would fall 60.6 percent.[29]

The most careful study to date on the effect of cost sharing on the demand for ambulatory mental health services has been done as part of the Rand Corporation Health Insurance Experiment. Major findings from the study, in which families were randomly assigned alternative health insurance coverages were:

- Ambulatory mental health expenditures per enrollee rise by three-quarters on the same plan when co-insurance falls from 95 to 0 percent.

- Small deductibles ($150 per person per year for ambulatory care, followed by free care) have a very small effect on overall expenditures.
- Even when all services were free, the average annual per capita cost for ambulatory mental health services was only $24, 5 percent of total health expenditures (excluding dental).
- Those with different incomes and mental health statuses responded similarly to changes in insurance coverage.[30]

Several approaches in addition to benefit limitations have been attempted to improve the cost effectiveness of the mental health benefit. Such efforts are hampered by lack of agreement among experts on need and definitions of mental health problems, absence of standard treatment methods and durations, and difficulty in evaluating therapeutic outcomes. Nonetheless, several benefit programs and insurers have enlisted the American Psychiatric Association and the American Psychological Association to conduct peer review. In the CHAMPUS program, which covers civilians related to memebers of the uniformed services, review considers the intensity of therapy as related to the diagnosis and attempts to identify forms of doubtful practice such as daily visits for a depressed patient or shock treatment to an adolescent. Therapy is automatically reviewed after eight, twenty-four, forty, and sixty sessions. Although no figures have been published, CHAMPUS considers the program successful.

Aetna was the first private insurer to contract with these professional organizations, with the medical director forwarding the most suspicious cases for review. While the American Psychological Association ran into trouble with incensed members who were denied reimbursement based on peer review, the American Psychiatric Association quickly expanded its programs and by November 1981 was working with four other commercial carriers and Blue Cross's Florida federal employees program.[31] One of the problems that make review difficult is the questionable accuracy of diagnostic information submitted to insurance companies for review. Utilization review for psychiatric hospitalization, carefully selected through local professional review organizations, holds promise of considerably reducing total days of stay per 1,000 enrollees. Concurrent review with a requirement that the hospital notify the reviewing organization within twenty-four hours of admission if they want to be sure of payment can initiate the process. In many cases, lengths of stay can be drastically reduced. In others, such as admissions for some eating disorders, admission may be found to be inappropriate and payment denied or covered only long enough for the patient to be discharged.

One newer phenomenon is an increase in the number of admissions for adolescents with behavioral problems. While there is no doubt that some teenagers are sufficiently disturbed to need hospitalization, in many cases hospitalization is being used inappropriately to deal with behavioral problems which can better be dealt with through increased parental attention, understanding and perseverance. Utilization review can also help to identify these situations and suggest more appropriate alternatives.

A totally different approach to providing cost-effective mental health care is a prepaid mental health plan. The California Wellness Plan, for example, provides outpatient mental health services in the private offices of 450 California-licensed psychiatrists, psychologists and social workers. No co-payment or deductible is required for the first five visits and additional visits have co-payments gradually increasing to 50 percent. To be eligible for reimbursement for continuing treatment, the provider must submit a diagnosis, prognosis and review plan to one of the thirteen statewide professional peer review committees for evaluation. As of mid-1984 participating health professionals, regardless of degree, agreed to accept $60 per hour as a fee per session. (The usual monthly fee is $6 to $8 per employee or family.) A number of insurance companies have endorsed this approach. The originator of this concept, John Armer, claims wide satisfaction by subscriber groups and stabilization of outpatient mental health payouts. Careful studies are necessary to assess the impact of this innovative approach on costs, utilization of mental health and other health care services and mental health of the participants.[32] However, the California Wellness Plan approach, also used by the older California Psychological Health Plan, permits employers to offer a much broader benefit with greater confidence that it will not be abused. In addition it can be expected that the presence of these plans will reduce medical ambulatory visits and related costs.

The California Wellness Plan also provides a companion inpatient mental health and chemical dependency benefit, utilizing a group of well-respected psychiatric hospitals that agree to require preadmission and continued-stay review by the Plan. As employers obtain better data and realize the large percentage of total claims dollars going to mental health services, these types of arrangements should become increasingly attractive.

A RECOMMENDED APPROACH

The following is a list of recommended mental health benefit plan features to achieve high cost effectiveness:

1. Cover at least 30 to 60 days per year of hospitalization for serious mental health disorders under the same cost-sharing provisions and reimbursements principles as other hospitalization.
2. Institute pre-admission and concurrent review for mental health admissions with review organizations that have proven track records.
3. Require that all hospital admissions provide an accurate diagnosis as a condition of payment.

4. Institute prepayment review of all mental health claims to identify inappropriate and unnecessary services for which payment should be withheld.

5. Intervene on behalf of the insurees if they are billed for services for which payment has been denied.

6. Reimburse for partial hospitalization and day treatment to at least the same degree as full hospitalization, subject to concurrent review.

7. Pay for outpatient mental health visits on a sliding scale of the type utilized by the California Wellness Plan.

8. Consider contracting through a mental health plan with mental health professionals who charge reasonable rates and have stringent peer review.

9. Limit outpatient mental health service payments, except when they can be shown to be a direct substitute for hospitalization or are subject to strict peer review, to $1,000–$1,500 per year per enrollee.

10. Put a ceiling on the per visit charge at the average rate for a psychiatrist in your area, subject to usual, reasonable and customary limits, as they may be below the ceiling.

11. Allow payment to psychologists and psychiatric social workers and other health professionals licensed by their state to supply mental health services.

NOTES

1. Reiger, D.A., Goldberg, I.D. and Taube, C.A. "The Defacto U.S. Mental Health Services System: A Public Health Perspective," *Archives of General Psychiatry,* Vol. 35, No. 6, 1978, pp. 685–693.

2. Reed, L.S. *Coverage and Utilization of Care for Mental Health Conditions Under Health Insurance—Various Studies, 1973–1974.* Washington, D.C.: American Psychiatric Association, 1975.

3. McGuire, T.G. "Financing and Demand For Mental Health Services," *Journal of Human Services,* Vol. 16, No. 4, 1981, pp. 501–522.

4. Witkin, M.J. Mental Health Statistical Note No. 155. *State and Regional Distribution of Psychiatric Beds in 1978.* U.S Department of Health and Human Services. Public Health Service, Alcohol, Drug Abuse and Mental Health Administration and National Institute of Mental Health. January 1981.

5. Mechanic, D. "Considerations in the Design of Mental Health Benefits Under National Health Insurance," *American Journal of Public Health,* Vol. 68, No. 5, 1978, pp. 482–488.

6. Ibid.

7. Reed, L.S., Myers, E.S. and Scheidemandel, P.L. *Health Insurance and Psychiatric Care: Utilization and Cost.* Washington, D.C.: American Psychiatric Association, 1972.

8. President's Commission on Mental Health. Report to the President. Vol. 1, Washington, D.C.: U.S.G.P.O., 1978.

9. A Report on Mental Illness and Third-Party Reimbursement, prepared by a committee of the American Psychiatric Association, 1981.

10. Brown, B. "The Federal Government and Psychiatric Education: Progress, Problems and Prospects," *New Dimensions in Mental Health*, U.S. Department of Health, Education, and Welfare, Alcohol, Drug Abuse and Mental Health Administration. Washington, D.C.: U.S.G.P.O., 1977.

11. Towery, O.B., Sharfstein, S.S. and Goldberg, I.D., "The Mental and Nervous Disorder Utilization and Cost Survey: An Analysis of Insurance for Mental Disorders," *American Journal of Psychiatry*, Vol. 137, No. 9, 1980, pp. 1065–1070.

12. Washington Report on Medicine and Health Perspectives. "Mental Health Benefits: Shrinking Reimbursement." Washington, D.C.: McGraw-Hill, November 9, 1981.

13. Personal Communication with Tim Carr, Reimbursement Specialist. National Institute of Mental Health, 1983.

14. Webber, A. "Trends in Corporate Mental Health Insurance," in *Mental Wellness Programs for Employees*, edited by R.H. Egdahl, D.C. Walsh and W.B. Goldbeck. New York: Springer-Verlag, 1980, pp. 158–178.

15. Op. cit. (Towery).

16. Op. cit. (Mechanic).

17. "For IBM Employees, Hospital Costs for Psychiatric Care Up Sharply," *American Medical News*, June 4, 1982, p. 17.

18. Goldberg, F.D. "State Laws Mandating Mental Health Insurance Coverage," *Hospital and Community Psychiatry*, Vol. 28, No. 10, 1977, pp. 759–763.

19. U.S. Department of Labor, Bureau of Labor Statistics. *Employee Benefits in Industry, 1980*. Washington, D.C.: U.S.G.P.O., 1981.

20. Health Research Institute. *Health Care Cost Containment Third Biennial Survey Participant Report*. Walnut Creek, California, 1983.

21. Hoeper, E.W., Nyoz, G.R., Reiger, D.A., Goldberg, I.D., Jacobson, A. and Hankin, J. "Diagnosis of Mental Disorder in Adults and Increased Use of Health Services in Four Outpatient Settings," *American Journal of Psychiatry*, Vol. 137, No. 2, 1980, pp. 207–210.

22. Carr, T.J. and Sharfstein, S.S. "Insurance and Insurability for Mental Health Services," in *Mental Wellness Programs for Employees*, (op. cit., pp. 151–157).

23. Jones, K.R. and Vischi, T.R. "Impact of Alcohol, Drug Abuse and Mental Health Treatment on Medical Care Utilization, A Review of the Literature," *Medical Care*, Supplement to Vol. 17, No. 12, 1979, pp. 1–82.

24. Levin, B.L. and Glasser, J.H. "A Survey of Mental Health Service Coverage Within Health Maintenance Organizations," *American Journal of Public Health*, Vol. 69, No. 11, 1979, pp. 1120–1125.

25. Craig, T.J. and Patterson, D.Y. "A Comparison of Mental Health Costs and Utilization Under Three Insurance Models," *Medical Care*, Vol. 19, No. 2, 1981, pp. 184–192.

26. Op. cit. (Washington Report on Medicine and Health).

27. Op. cit. (McGuire).

28. Glasser, M.A., Duggan, T.J. and Hoffman, W.S. *Obstacles in the Pathways to Prepaid Mental Health Care*. National Institute of Mental Health, Department of Health, Education, and Welfare. Washington, D.C.: U.S.G.P.O., 1977.

29. McGuire, T.J. "The Cost of Private-Practice Psychiatry Under National Health Insurance," in *Mental Wellness Programs for Employees* (Op. cit., pp. 197–212).

30. Wells, K.B., Manning, W.G., Duan, N., Ware, J.E. and Newhouse, J.P. *Cost Sharing and the Demand for Ambulatory Mental Health Services.* Santa Monica, California: Rand Corporation, September 1982, p. vi.
31. Op. cit. (Washington Report on Medicine and Health).
32. Armer, J.E. "Mental Wellness and the Cost of Health Care," in *Mental Wellness Programs for Employees* (Op. cit., pp. 213–220).

MANAGING OTHER *8*
IMPORTANT HEALTH
CARE BENEFITS

W hile medical, dental and mental health are the largest benefits in terms of paid claims, how some other health benefits are structured is of increasing importance in terms of both dollars and health impact. Among these additional benefits are alcohol treatment services, prescription drugs, vision care and hospice and home health services.

ALCOHOLISM AND OTHER DRUG ABUSE BENEFITS

A conservative estimate places the percentage of the workforce with a drinking problem at 5 percent. Employees who are problem drinkers use on the average three times the health dollars under health insurance as other employees.[1] One classic study found that workers with alcohol problems were absent from work at least 2.5 times more than other workers and had an on-the-job accident rate 3.6 times that of the other employees.[2] Unfortunately, alcohol isn't the only substance-abuse problem. A government-sponsored study in 1977 pegged the bill for on-the-job drug abuse at a staggering $25.8 billion when the estimate is converted into 1983 dollars.[3]

A Brief History

While drug abuse has only recently drawn attention as a health problem, formal recognition of alcoholism as a disease was accorded by the World Health Organization in 1951 and by the American Medical Association in 1956. The first company alcoholism or personal assistance programs, however, date from 1943 (DuPont) and 1944 (Eastman Kodak). By 1959 it was estimated that fifty companies had occupational programs for alcoholism. Meanwhile, alcohol treatment in general hospitals began to be covered to a limited degree as a benefit under regular hospitalization insurance in the 1950s, and it increased as mental health benefits were added to coverage by most employers. In the 1960s, insurers began selling separate coverage for alcoholism services, in part because sanctions were apparently being applied against alcoholics by both insurance carriers and service providers.[4]

As of the end of 1981, thirty-three states had enacted legislation regarding alcoholism benefits under group health insurance policies. Legislation either requires that all policies sold include alcoholism coverage (seventeen states) or that alcoholism coverage be made available to all policy holders (sixteen states). Frequently under these statutes the type of benefit, coverage period and/or treatment to be covered will be specified.[5]

Breadth of Coverage

Few insurance plans, either public or private sector, provide both comprehensive inpatient and outpatient coverage for alcoholism treatment on the same basis as other illnesses. In 1977, based on sixty-one of sixty-nine local plans responding to a national Blue Cross survey, fifty-four covered detoxification and forty covered rehabilitation, while only eight covered both inpatient and outpatient services in both hospital and nonhospital settings.[6] Federally qualified HMOs have been required to provide basic health services to treat alcoholism and other drug addictions. However, many HMOs have provided only a minimum benefit.[7] Where not required by state law, alcoholism benefits frequently have not been readily accepted by employers. For instance, Blue Cross of Maryland started offering an alcoholism benefit to its subscribers in 1972, but only 12 percent had enrolled for this benefit by 1980.[8]

In general, alcoholism benefit premiums have been inexpensive. Blue Shield and Blue Cross, for example, have offered a benefit which provides expanded coverage for residential and outpatient rehabilitation both in hospitals and freestanding facilities at a monthly premium ranging from $.35 to $.82 for an individual and $1.00 to $1.70 for a family.[9] The main reason that alcoholism and other drug treatment benefit riders have been priced low initially and have remained low is a very low rate of reported utilization, usually less than 0.5 percent of the insured adult population annually.[10] The stigma of diagnosing a patient as an alcoholic leads physicians and other health professionals to indicate other reasons for what was in truth treatment for alcoholism. Not only are diagnoses submitted to insurers frequently erroneous, but many insurers do not record sufficient information provided on claims forms to ascertain the reported costs and utilization associated with a particular condition.[11]

A further reason that insurance companies may feel that the marginal cost of alcoholism coverage is low derives from reports that alcoholism treatment reduces use of other covered services. While few, if any, studies of this effect have been sufficiently controlled to provide definite conclusions, the information available to date appears to point uniformly in the direction of lower utilization. One California study reported a 26 percent decline in overall medical care utilization after alcoholism treatment, while another indicated a 48 percent decline in hospital, medical and surgical costs.[12]

Costs of Alcoholism Treatments

Most companies and insurers have not reported on the costs of alcoholism treatment. An exception is a 1979 report from General Motors that employees referred to treatment by their employee assistance program to an unnamed facility cost $3,190 for a fourteen-day stay, while a ten-week outpatient program had an average cost per patient of $400.[13] In a study aided by diagnosis-related group (DRG) analysis, the DRG for alcoholism, when reported from a major company, showed a variation in cost of hospitalization of $1,000 to $8,500.[14]

A review of eleven studies on cost and utilization of alcoholism care covering varying periods from 1973 to 1979 concluded that the average number of inpatient admissions per client per year was between 1.0 and 1.2, with an average length of stay of fifteen days (range 7.8 to 25.0 days). The percentage of the covered population using inpatient alcoholism services, usually in community-based general hospitals, ranged from 0.02 to 0.1 percent per year, and 60 to 69 percent of all users were male. In general, daily inpatient charges were lower in alcoholism treatment centers than community hospitals, but stays tended to be longer in the former.[15]

Aetna's 1979 report on the federal employees it insures indicated that daily charges for alcoholism treatment centers averaged $116 as opposed to $147 for community hospitals. However, total charges per stay were very close, $2,438 and $2,583, respectively. Outpatient visits cost an average of $21 in the specialized treatment center versus $37 in hospital settings.

According to the National Institute of Alcohol Abuse and Alcoholism, in the late 1970s annual costs per client were $740 for solely outpatient treatment, $4,730 for intermediate services, $20,780 for hospital detoxification and $13,730 for hospital alcoholism treatment.[16] Partial hospitalization appears to work as well and at a considerably reduced cost as continued treatment for most patients with alcohol problems.[17] While it would be a mistake to believe that each treatment modality and setting can directly substitute for another, there seems little question that many stays in general hospitals could be avoided without sacrificing the chances of a good outcome.[18] Most workers with alcohol problems can probably benefit from outpatient and partial hospital programs as well as the more popular and well-advertised residential programs. Over time, attendance at Alcoholics Anonymous meetings is probably a more important predictor of abstinence than the type of treatment program initially provided.[19]

Based on these cost differentials and the growing notion that not all alcoholics require inpatient care, some insurers have improved coverage in specialty treatment settings. Kemper Insurance Companies, for example, extended group health coverage for alcohol to both outpatient and inpatient treatment centers in 1973. Aetna extended its coverage to free-standing and outpatient treatment in 1978. Since many of the free-standing alcoholic treatment centers were established and continued primarily on government support, one measure of changes in the private sector is the proportion of total revenues that insurance companies contribute to these organizations. In 1979, total reported funding for over 4,000 alcoholism treatment facilities indicated that only 18 percent derived from private insurance.[20]

Identifying quality nonhospital programs for reimbursement has been a continuing problem for insurers. The Joint Commission of Accreditation of Hospitals (JCAH) now accredits nonhospital treatment programs, and states license such programs. In some cases insurers will reimburse any program deemed of reasonable validity by a company's occupational health director. Still to be developed is a system for granting reimbursement privileges based on outcome rather than facility standards, staffing patterns and types of services provided. The National Association of Insurance Commissioners has recommended model legislation and a model benefit package to include coverage for alcoholism and other drug dependency treatment in free-standing centers, provided that they are either affiliated with a hospital, licensed or otherwise approved by the state or accredited by the JCAH. It also includes coverage for family counseling and care. Coverage for thirty days of inpatient care for alcoholism or other drug dependency is recommended for each benefit period, as are thirty outpatient visits.[21]

Recommended Approaches

Providing appropriate care for employees and their dependents with alcohol and other drug problems can improve productivity and save employers money in reduced health care expenditures over time. However, paying for the initial treatment can also be expensive, often more expensive than necessary. Of special concern are the large number of special alcohol and drug abuse units that have been franchised in hospitals. Based on what has been learned from reports to date, the most cost-effective and humane approach is:

1. Cover inpatient hospital confinement for alcoholism and other serious drug dependency at the same level as any other illness. In most cases this will require some deductible and co-insurance as for other hospitalization.
2. Limit the inpatient coverage to 30 days per benefit period.
3. Cover treatment in free-standing facilities that are licensed, certified, accredited or otherwise affiliated with a licensed hospital. When in doubt about the appropriateness of a facility, have a consultant with specialized knowledge about alcohol treatment help make the decision about coverage.
4. Cover outpatient and partial day treatment, with an incentive to use this option instead of inpatient treatment. Incentives could be to eliminate co-insurance or deductibles for this use.
5. Limit use of these services to 30–40 visits per benefit period.
6. Pay for outpatient treatment by a physician, psychologist, social worker or licensed counselor with demonstrated experience in working with alcohol/drug problems. Reimbursement should be based on usual and customary fees.
7. Collect data over time to estimate the total cost of alternate treatment settings for alcohol and other drug dependency. If possible, and with appropriate concern for privacy and confidentiality, try to relate alternate treatment sites and modalities to differences in outcomes of employees and dependents using them.
8. Suggest that the medical department and employee assistance counselors recommend outpatient and day treatment to substance abusers.
9. Cover family counseling and care for families of alcoholics and other workers with drug dependency problems.

PRESCRIPTION DRUG BENEFITS

Approximately 60 percent of the population uses at least one prescription medication during the year. Among those who use any prescribed medicines, the rate is 7.5 per person per year. Prescribed medicines are used most frequently among the very young and the elderly, with 65.5 percent of those less than six years, 69.1 percent of those in the 55-to-64-year group, and 75.2 percent of the 65-and-over group obtaining at least one prescribed medicine annually (Table 8.1). Increasing age corresponds to increasing patterns of use. The average person 65 and older using at least one prescription drug fills a mean of 14.2 prescriptions for drugs per year, compared to 11.9

TABLE 8.1
USE OF PRESCRIBED MEDICINES BY AGE

	Persons without prescribed medicines	Persons with at least one prescribed medicine	Prescribed medicines per person	Prescribed medicines per person with at least one prescribed medicine
	Percent		Mean	
Total	*41.8*	*58.2*	*4.3*	*7.5*
Age in years				
Less than 6	34.5	65.5	2.8	4.3
6 to 18	55.4	44.6	1.6	3.6
19 to 24	46.8	53.2	2.6	4.9
25 to 54	40.9	59.1	4.2	7.0
55 to 64	30.9	69.1	8.2	11.9
65 or older	24.8	75.2	10.7	14.2

Source: U.S. Department of Health and Human Services, National Center for Health Services Research, "Prescribed Medicines, Use, Expenditures and Source of Payment." Data Preview 9 from the National Health Care Expenditures Study, April 1982. DHHS Publication No. (PHS) 82–3320.

for those 55 to 64 and 7.0 for those 25 to 54. In general, females are more likely to have obtained prescribed drugs than males (64.8 percent as opposed to 51.2 percent) and their rates of use are higher (8.3 prescriptions per person per year as opposed to 6.4).[22]

In 1977, the average drug expense for individuals who used at least one prescribed medication was $46, highest for persons 65 and older ($93), followed by those 55 to 64 ($79) and 25 to 44 ($44). Females had per capita expenses of about 20 percent higher than males. The average charge per prescribed medicine was $6.24, with 74 percent paid by the family, 12 percent by private insurance, 8 percent by Medicaid, and 6 percent from other payment sources. The average per capita cost for the employed population was $39.[23]

Private insurance paid for 14 percent of the prescription drug charges of employees, according to the 1977 survey. Of this group, 6.5 percent paid no out-of-pocket expenses, 41.4 percent paid $1 to 49, 5.3 percent paid $50 to 99, 3 percent paid $100 to 249, and 0.4 percent paid over $250. Over two-fifths of employees had no prescribed medications during the twelve-month period that was studied.[24]

Thus, prescription drugs constitute either no expense or a small expense for most employers. However, a prescription drug benefit has been bargained for or voluntarily provided by employers to an increasing extent, probably because they are visible and moderately priced additions to the benefit package. According to a 1983 survey covering 854 benefit plans, 6 percent of employer plans covered prescription drugs as a separate benefit plan, 89 percent covered them under major medical only, 4 percent covered them under both, and only 1 percent had no coverage. The total monthly unit costs (both employer and employee portions) for the separate prescription drug plan averaged $6.12 per employee plus an additional $11.62 for all dependents. Twenty-two percent of plans had co-insurance, usually 20 percent from employees, and co-payments were required in 66 (85 percent) plans. Of these, 4 required an annual deductible, and of those with co-payments per prescription, 5 used less than $1, 28 each used $1, 22 used $2, and the remainder used between $2 and $5.[25]

Many employers do not feel separate drug coverage is worthwhile. As the employee benefits director of a foods company with 25,000 employees lamented, "Prescription drug plans are an idea whose time should never have come. Too much is spent on administration versus the cost of drugs."[26] Many benefit managers feel that the expenses associated with prescription drugs can best be handled by having them subject to the same co-insurance and deductible provisions as other low-cost health care items.

Probably the greatest value of a prescription drug plan can be to those who have chronic medical problems for which they take medications on a continuing basis. For this group, prescription drug costs constitute a significant expense. The new approach to serving these employees and dependents is to provide mail order drugs, either as a separate benefit or through a prescription drug plan. Instead of receiving a thirty-day supply from the local pharmacy, the enrollee sends to a distribution house for a ninety-day supply, receiving the order back in about one week. This saves money for the enrollee, who has to pay one instead of three deductibles for a ninety-day supply. It saves money for the plan, since distribution houses buy in bulk and substitute generics, which cost less, whenever possible. Dispensing costs are reduced by dealing in large numbers of pills. Plan administration costs go down because there are fewer transactions.[27] Since older people use more prescription drugs on a maintenance basis, the program is particularly appropriate for retiree populations.

VISION CARE BENEFITS

As with prescription drugs, vision care is perceived as a highly conspicuous yet relatively inexpensive benefit. Unlike prescription drug benefits, however, vision care is provided by a relatively small percentage of employers, probably no more than one-quarter of large employers and under 10 percent of smaller employers. However, vision care is likely to grow as a benefit under two circumstances: 1) as companies look for attractive, inexpensive additions to the benefit package, and 2) as companies increase employee cost sharing for other more expensive health care benefits and want to add a new benefit to temper the adverse employee relations effect.

Usually, vision care is covered under a separate plan, although it can be built into a major medical or comprehensive benefit. In most cases, the full premium is paid by the employer for both employees and dependents. Separate deductibles are frequently included in vision plans. Plans almost uniformly cover routine eye checkups and eyeglasses. The majority also cover contact lenses under defined circumstances. Frames are covered separately in most plans. Payment is based either on a percentage of usual and customary charges or, more frequently, on fixed dollar schedules which are periodically updated. Panels of vision specialists, usually optometrists, have been developed by some plan carriers as the primary eye examination and prescribing resource.

Typical of a vision plan provided to employees and their families is that provided by the American Stock Exchange. It provides a comprehensive examination every twenty-four months for employees and spouses and every twelve months for unmarried dependent children up to age 19 (age 23 if a full-time student). The plan pays a maximum of $30 for examinations, $16 for frames, $12 for single-vision lenses, $18 for bifocals, $23 for trifocals and $170 for contact lenses when medically required. Enrollees who choose contact lenses for cosmetic reasons are given a $60 allowance. Not covered are vision training, nonprescription lenses or medical or surgical treatment of the eyes. Employees pay a $10 deductible per claim.[28]

Looking ahead, the overall cost of vision care, lenses and frames can be expected to grow rapidly. As the population ages, a higher percentage will need corrective eyeglasses or contact lenses. Since designer eyeglass frames have become a fashion item, they are much more expensive than comparable standard frames and are purchased with greater frequency. Contact lens technology has moved ahead considerably and, with the advent of soft lenses that breathe, are likely to be preferred by the majority of those who need vision correction. Under most current vision plans, controls on frequency, maximums for each service and requirements that contacts be paid for under the plan only if medically necessary will adequately control costs. However, it is likely that there will be growing pressure for employers to broaden the vision care benefit so that it can pay for a larger percentage of the total costs and substitute payments in tax-deductible dollars for payments currently made in after-tax dollars by employees.

HOSPICE CARE

The combination of a rapidly growing older population and an even more rapid increase in the cost of hospital care has helped to spur interest in alternatives to hospital inpatient services for those with chronic and terminal conditions. One alternative which has come to prominence in recent years is the use of hospice care. A hospice provides care to the terminally ill either in an organized homelike setting or the patient's own home. It can provide the dying patient with humane, dignified care that involves family participation and responsibility. According to Willis Goldbeck, president of the Washington Business Group on Health, "Hospice is a rich concept: rich in humanity, rich in its breadth of understanding about quality of life and dignity of death; rich even in its capacity to save our scarce economic resources for application to the real needs rather than wasted on acute care hospitalization."[29]

Many feel that regardless of whether a hospice saves money, it is the best form of care for the terminally ill patient. Given the high percentage of lifetime health costs utilized in the last year of life, particularly in acute care hospitals, it has frequently been assumed that the use of a hospice as an alternative to time spent in a hospital reduces overall costs for the illness. However, with few exceptions, the evaluations to date of the relative cost-effectiveness of hospice care versus usual care have been incomplete, difficult to generalize and/or subject to a number of serious methodological problems which have affected the credibility of their conclusions. The only findings in which there is agreement among virtually all those that have conducted evaluations are that the per day cost of hospice care is lower than hospitalization in an acute facility and that the average patient spends more days under hospice care than in an acute-care setting when a hospice is not available.

One of the largest of the hospice evaluations is the National Hospice Study which was initiated to determine whether the hospice benefit implemented by Medicare in 1982 can save the program money. Interim findings from this evaluation, which covers 26 hospices and 14 conventional care settings, suggest that hospices can reduce the costs of care for the dying, especially if patients are admitted to the hospice when they are expected to live only for another month or two. Home-based hospices are particularly economical since the cost of home-based hospice care generally doesn't exceed the cost of conventional care for six months.[30]

About 35 percent of the 600 plus respondents to the 1983 Health Research Institute (HRI) health care cost-containment survey indicated that they provide hospice care as part of their plan(s), more than triple the figure two years earlier.[31] Because HRI surveys the 1500 largest employers in the country, this figure probably overstates the true prevalence of this benefit. Nonetheless it is very likely that employers will increasingly add a hospice benefit over the next few years, following the lead of larger employers. From a cost-containment point of view, it is essential to limit the payment for hospice care to some reasonable rate. Medicare, which is beginning to reimburse for hospice care, has developed principles of reimbursement which may be appropriate for private payers. An important caution to companies considering a hospice benefit is to try to develop coverage conditions and reimbursement principles that encourage the growth of hospices which retain a home care/family orientation and are also not burdened with the overhead of acute-care hospitals. If an estimate of how many enrollees will utilize the benefit is an important consideration in deciding whether to offer it, the demographics of the workforce and especially the size and composition of the retiree population should be carefully assessed. Since death rates rise at an increasing rate with age, companies with older workforces and large retiree populations will naturally experience the greatest utilization.

HOME HEALTH CARE

Another benefit which has been marketed as an alternative to hospitalization is home health care services. Home health care services generally provide full coverage for post-hospital home treatment of an illness, injury or related condition. Home health benefits will generally pay usual and customary fees charged by a home care agency up to a dollar maximum per visit (usually $40 to 60) for up to 50 to 150 visits per year. Coverage is usually contingent upon the patient's physician certifying that continued confinement in a hospital or nursing home would be required in the absence of a home care benefit.

As many as one-half of large employers provide some home health care coverage, but the extent of coverage and incentives for use vary considerably. Smaller employers are less likely to offer this benefit. A home health care benefit instituted by Planning Research Corporation, which is typical of what some companies are offering, includes:

- nursing care provided or supervised by a registered nurse;
- services provided by a housekeeper or "homemaker" supplied by an approved home health care agency;
- physical, respiratory, speech, or occupational therapy by a qualified therapist;
- nutrition counseling provided by or under the supervision of a registered dietician;
- medical supplies, laboratory services, and drugs and medications prescribed by a doctor.[32]

As with hospice care, it is unclear whether this type of convalescent care will save money. A key to the answer is the degree to which home care substitutes for inpatient care. Appropriate use of home care can be increased through aggressive concurrent utilization review that identifies all possible opportunities for early discharge and makes an impact on overall length of stay, especially for the very long hospital stays. A good data system is needed to evaluate if this effect is being achieved. As home care programs grow in number and dollars expended, carriers should consider initiating utilization review of the home care programs to prevent abuse. In the longer term, home health care services, and hospice services as well, may be provided under fixed-price contracts that have an incentive to reach the maximum health benefit as efficiently as possible rather than to maximize the number and frequency of services. Fixed payments per hospital stay based on diagnosis-related groups will provide strong incentives for hospitals to discharge patients more quickly and to encourage the use of home health care. However, getting enrollees to use home health services requires, at a minimum, that the cost-sharing provisions be no greater than for hospitalization. Until these benefits are well understood and patients feel confident that they are reliable and of high quality, it may be necessary to provide them with substantially less cost-sharing than hospital-based services. Perhaps the clearest opportunity to achieve cost savings through a home health plan is to provide incentives for early discharge of mother an infant after delivery. For example, in upstate New York, employers subscribing to the Blue Cross of Rochester's home maternity program have reported savings as high as $800 per patient. The cost of inpatient care for mothers and newborns in the Rochester area was reported at $300 per day as opposed to $70 per day for home care for mothers and newborns, and which included nursing visits, homemaker services and any necessary laboratory work. Coverage under home health care requires that mothers and their infants leave the hospital within twenty-four hours of delivery.[33]

NOTES

1. Sarvis, K.S. *Insurance Cost Savings Due to an Adequate Alcoholism Health Benefit.* Prepared for the State of Florida Department of Health and Rehabilitative Services, November 1976. Reproduced by the National Institute on Alcohol Abuse and Alcoholism. RPO 183.
2. Observer and Maxwell, M.A. "A Study of Absenteeism, Accidents and Sickness Payments in Problem Drinkers in One Industry," *Quarterly Journal of Studies on Alcohol,* Vol. 20, No. 2, 1959, pp. 302–312.
3. "Taking Drugs on the Job," *Newsweek,* August 22, 1983, pp. 52–60.
4. Hallan, J.B. "Health Insurance Coverage for Alcoholism: A Review of Costs and Utilization," in Health Insurance Coverage for Alcoholism: Policy Issues, *Alcohol Health and Research World,* Vol. 5, No. 4, 1981, pp. 16–21.
5. Graham, G. "Occupational Progams and Their Relation to Health Insurance Coverage for Alcoholism," in Health Insurance Coverage for Alcoholism: Policy Issues. Ibid., pp. 31–34.
6. Williams, W.G. "Nature and Scope of Benefit Packages in Health Insurance Coverage for Alcoholism Treatment," in Health Insurance Coverage for Alcoholism: Policy Issues. Op. cit., pp. 5–11.
7. Op. cit. (Hallan).
8. Regan, R.W. "The Role of Federal, State and Voluntary Sectors in Expanding Health Insurance Coverage for Alcoholism" in Health Insurance Coverage for Alcoholism: Policy Issues. Op. cit., pp. 22–26.
9. Blue Cross Association. Alcohol and Drug Abuse Treatment Benefit Project Update, May 1982. Distributed by the National Institute on Alcohol Abuse and Alcoholism. RPO 336.
10. Op. cit. (Hallan).
11. Op. cit. (Graham).
12. Op. cit. (Hallan).
13. Op. cit. (Hallan).
14. Presentation made by Willis Goldbeck, President of the Washington Business Group on Health, to the Health Care Coalition of Memphis in Memphis, Tennessee, on April 23, 1983.
15. Op. cit. (Hallan).
16. Op. cit. (Graham).
17. Longabaugh, R. *The Cost-Effectiveness of Partial Hospitalization Vs. Continued Inpatient Treatment in the Treatment of Alcoholism.* December 1980. Prepared for Blue Cross and Blue Shield of Rhode Island. Distributed by the National Institute of Alcohol Abuse and Alcoholism. RPO 381.
18. Op. cit. (Graham).
19. Cator Report on Inpatients Followed at Six and Twelve Months Post-Discharge. April 1982. Distributed by the National Institute on Alcohol Abuse and Alcoholism. RPO 390.
20. Op. cit. (Graham).
21. Report of the National Association of Insurance Commissioners (C–1) Alcoholism, Drug Addiction, and Insurance Task Force. May 26, 1981. Distributed by the National Institute of Alcohol Abuse and Alcoholism. RPO 341.

22. Department of Health and Human Services, National Center for Health Services Research. National Health Care Expenditures Study, Data Preview 9. *Prescribed Medications: Use, Expenditures, and Sources of Payment.* Pub. No. (PHS)82–3320. Washington, D.C.: U.S.G.P.O. April 1982.

23. Ibid.

24. Ibid.

25. Hay Associates. *1983 Noncash Compensation Comparison.*

26. Gawla, L. "Most Employers Don't Plan to Offer Drug Plans," *Business Insurance,* June 21, 1983, p. 1.

27. Drury, S. "Mail-order Drugs Cut the Cost of Prescription Plans," *Business Insurance,* March 21, 1983, pp. 1, 30.

28. "AMEX Employees Get Vision Care Benefit," *Business Insurance,* December 28, 1981, p. 4.

29. Goldbeck, W.B. Medicare Reimbursement for Hospice Care: Testimony on H.R. 5180 to the Health Subcommittee of the Committee on Ways and Means, March 25, 1982.

30. Wallace, C. "Hospices Cost Less Than Hospitals," *Modern Healthcare.* Vol. 13, No. 12, 1983, p. 120.

31. Health Research Institute. *Health Care Cost Containment Third Biennial Survey.* Walnut Creek, California, 1983.

32. Lawson, J.C. "Cutting Costs By Going 'Home,'" *Business Insurance,* February 14, 1983, pp. 3, 30.

33. Lawson, J.C. "Employers Cozy Up to Incentive Plans," *Business Insurance,* February 14, 1983, pp. 3, 31.

9 FROM DATA TO USEFUL MANAGEMENT INFORMATION: HOW TO ANALYZE HEALTH INSURANCE CLAIMS

A midwestern bank paid 4 percent less per insured for health insurance premiums last year than the year before. The officers were extremely pleased but weren't sure whether to pat themselves on the back for having reduced the slope of perennial cost increase or to fear a spectacular increase for the following year.

An eastern service company paid the premiums in Table 9.1 under a self-insured arrangement. It wondered what it could do to stop the apparent trend of accelerating increases.

A western-based publishing company experienced a 25 percent increase in medical claims per employee over a one-year period. Management decided immediate action was required to slow the rise, but what form would that action take?

The answers to these questions and similar ones asked by most employers come only through a very careful look at the components of health benefits utilization.

REASONS TO EXAMINE HEALTH CARE CLAIMS

Buried within the general concern over cost increases are at least four major reasons to delve more deeply into claims reports.

TABLE 9.1
HEALTH INSURANCE COSTS FOR AN EASTERN
SERVICE COMPANY

Year	Premiums paid	Increase from previous year (%)
1	$4,560,000	6.0
2	5,013,000	9.9
3	5,826,000	16.2
4	7,249,000	24.4

1. *To review and monitor the administrative cost control efforts of the carrier(s)* Financial reports are useful not only as a basis for audit of the financial controls but to assess the extent of efforts to limit payments to services that are medically necessary. Financial auditing procedures are routinely employed by carriers and by larger subscriber groups to ascertain that payments claimed have been made and that gross overpayments have not been made on a consistent basis. However, the claims payment systems of different third-party payers vary significantly in terms of sophistication, types of procedures utilized, quality control activities and documentation of what has been done at various stages of the claims payment process. A financial audit will frequently not disclose whether the claims payment system is adequate from a cost control standpoint. In auditing claims, a *financial* auditor can miss many problems that only someone skilled in health care delivery issues would pick up. Questions that should be asked by benefit managers of auditors and of themselves include, for example:
    a) How many large bills were questioned?
    b) With what frequency was justification requested for particular items (e.g., excessive length of hospital stay, questionable treatment, etc.)
    c) Was the denial rate in line with expectations?
    d) Were proper edit checks built into the system to identify inappropriate submissions, such as a dilatation and curettage in an eight-year old girl, or a transurethral resection (partial removal of the prostate) in a woman?
    e) Were adequate treatment screens employed to identify if any treatments provided were not explainable by the diagnoses submitted?
    f) Were eligibility updates current so that former employees were not being reimbursed for services that occurred after their eligibility had expired?
    g) How extensive were the efforts to coordinate benefits to minimize employer costs?

These are but a few of the important questions that frequently require a more specialized audit than is standard practice. The audit team should include one or more health professionals, preferably a physician.

One recent audit of a large insurance company revealed a net 5 percent overpayment rate for one large account. While this may be an exception, even a 1 percent overpayment rate can mean many tens of thousands of dollars per year in unnecessary expense for an employer. It seems reasonable if there are significant dollars involved to have an audit performed by individuals very knowledgable about administrative cost containment opportunities in claims processing.

A particularly helpful report in assessing the aggressiveness of a carrier's administrative cost containment efforts is the absolute amount and percentage of dollars billed that were disallowed due to:

a) coordination of benefits
b) ineligibility of claimant
c) ineligibility of provider
d) uncovered services
e) excess charges
f) deductible
g) coinsurance

These categories of reduction should also be reviewed for each major service category as suggested in Table 9.2.

Changes in benefits, reimbursement principles, practice patterns, general economic conditions, special arrangements with particular providers and other factors can lead to variation in rejected charges from period to period. Nonetheless, large differences suggest that more in-depth analysis may be warranted to try to identify reasons and to assist in making projections for future periods.

2. ***To perform financial planning and allocation*** Review of the claims experience reported by each management unit can permit allocation of costs to each unit based on their actual experience. This provides an incentive for line managers to become concerned about opportunities to limit the cost of ill health of their employees. If each manager only receives a prorated allocation of total costs based on number of employees or number of insured, there is much less incentive for him or her to make this a priority, even if the costs are very high. Frequent review of company-wide experience and experience by management unit permits better monitoring of whether actual expenditures are in line with projections and which locations/units are experiencing greater than average increases. It can also provide base information for refining projections of health benefit costs for future years.

3. ***To estimate impact of changes in benefits/financing arrangements*** Knowing how many dollars are being spent for specific services under existing coverage can be helpful in estimating the dollar impact to the

TABLE 9.2
REASONS FOR CLAIMS REJECTION

	Total billings		Coordination of benefits		Ineligible person		Ineligible provider		Uncovered service		Excess charges		Deductible		Co-insurance	
	($)	(%)	($)	(%)	($)	(%)	($)	(%)	($)	(%)	($)	(%)	($)	(%)	($)	(%)
Period 1	1,000,000	100	80,000	8	22,000	2.2	13,000	1.3	28,000	2.8	81,000	8.1	56,000	5.6	145,000	14.5
Period 2	1,265,000	100	117,645	9.3	31,625	2.5	20,240	1.6	30,360	2.4	110,055	8.7	80,960	6.4	189,750	15.0

Hospital inpatient
Period 1
Period 2

Physician inpatient
Period 1
Period 2

Physician ambulatory
Period 1
Period 2

Dental
Period 1
Period 2

employer of alternative benefits based on a certain set of assumptions. For example, if a $100 deductible per family member were raised to $200, what would be the impact on paid claims assuming a) no change in utilization, or b) a 10 percent decline in utilization? How much could this save compared to reducing the payment for physician services from the 90th percentile based on UCR to the 80th percentile? While these estimates by employers are not a substitute for actuarial help in estimating impacts of possible changes, they do provide a framework to consider alternatives in terms of costs to the company without an elaborate and sometimes expensive analysis by outside experts.

4.   *To assess cost control opportunities*   Assessment of cost control opportunities necessitates a detailed analysis of both the costs and utilization of health care services. It requires trying to uncover those occurrences and related costs in the delivery system which may be altered to contain employer costs without having an adverse impact on the health of the insured. Chapters 5 through 8 describe the major categories of health insurance benefits currently offered by many employers, including the relative costs of each benefit and important considerations for designing the health benefit package. The remaining portion of this chapter elaborates an approach to analyzing where the money goes.

ANALYSIS OF HEALTH INSURANCE CLAIMS

Although there are an infinite number of important questions concerning analysis that can be posed, the three most basic are:

1.   How much is each of the available benefits being used?
2.   Who is using each of the benefits?
3.   What are the characteristics of the increases in the costs of the health benefits?

How Much Is Each Benefit Used?

Each benefit should be considered with respect to utilization and cost. Initially, general categories should be examined, as set up in Table 9.3. For each type of benefit, the key elements to review are total incurred claims, cost per claim and the rate of utilization per insured or per employee.

Next, within each category of service, the major types of service should be broken down, as shown for dental claims in Table 9.4.

A natural question that arises is what are reasonable total costs, cost per insured and utilization rates for each of the elements in Tables 9.3 and 9.4. Unfortunately there are no accepted norms for amounts and costs of

TABLE 9.3
HEALTH CARE BENEFITS COST AND UTILIZATION: CALENDAR YEAR 1982

Benefit category	Total incurred claims*	Cost/claim	Cost/insured**	Utilization measures
Hospital inpatient	$	$/Admission	$	Days/1000; ALOS***
Hospital outpatient	$	$/Visit	$	Visits/insured**
Physician inpatient	$	$/Admission	$	Visits/stay
Physician ambulatory	$	$/Visit	$	Visits/insured**
				Prescriptions/MD visit
Mental health inpatient	$	$/Admission	$	Days/1000; ALOS***
Mental health outpatient	$	$/Visit	$	Visits/insured**
Dental	$	$/Visit	$	Visits/insured**
Vision	$	$/Visit	$	Visits/insured**
Pharmacy	$	$/Prescription		Prescriptions/insured**
Skilled nursing	$	$/Admission	$	Days/1000; ALOS***
Home health	$	$/Visit	$	Visits/insured**

*If incurred claims not available, use paid claims.
**Preferably cost/insured; if unavailable, use cost/covered employee.
***ALOS = average length of stay.

TABLE 9.4
DENTAL PAYMENTS BY PROCEDURE CATEGORY
CALENDAR YEAR 1982*

Procedure category	Total dollars (%)	Dollars per insured
Diagnostic	42,737 (4.9)	$ 5.7
Preventive	44,756 (5.2)	$ 6.0
Restorative	331,325 (38.3)	$ 44.1
Endodontics	82,282 (9.5)	$ 11.0
Periodontics	77,343 (8.9)	$ 10.3
Prosthetics	101,729 (11.8)	$ 13.6
Oral surgery	68,210 (7.9)	$ 9.1
Orthodontics	101,690 (11.8)	$ 13.6
Emergency services	1,127 (0.1)	$ 0.2
Miscellaneous	13,215 (1.5)	$ 1.8
TOTALS	864,415 (100)	$115.3

*Average number of insured—estimated at 7,500.
Source: Control Data Corporation, Management Report, unpublished.

FIGURE 9.1
COMPARISON OF HEALTH CARE UTILIZATION AND
COSTS OF TWO OIL COMPANIES

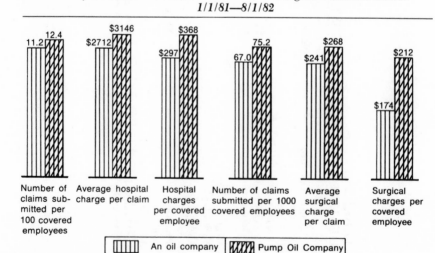

FIGURE 9.2
PERCENTAGE OF TOTAL HOSPITAL CHARGES:
A COMPARISON BY MAJOR DIAGNOSTIC CATEGORY
(1/1/82—8/1/82)

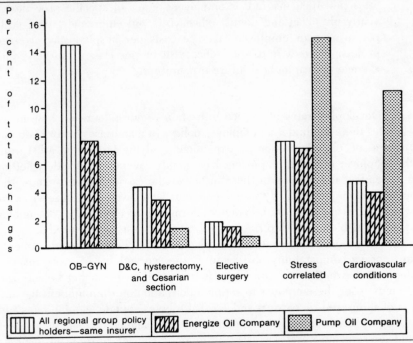

care to which a particular employer's experience can be compared. In their absence common approaches include:

1.   Comparison of current year data with the company's experience in preceding years.

2.   Comparison of a company's experience to that of companies with similar characteristics. For example, an oil company might compare itself to another oil company in the same geographic area, serviced by the same carrier (Figure 9.1).

3.   Comparison of a company's numbers with many other companies of comparable size in the same area (Figure 9.2). A number of employer coalitions to control health care costs are attempting to collect this type of information from all member companies using a common format and common definitions.

4.   Comparisons of area-wide and national data bases collected by insurers, Peer Review Organizations (PROs), the Health Care Financing Administration (HCFA), the National Center for Health Statistics and a num-

ber of nonprofit and for profit data analysis organizations. Among the kinds of information available are average length of stay (ALOS), days of care (for acute hospitals and nursing homes) per 1000 insured or employees (both unadjusted and adjusted for age), hospital admission rates per 1000 insured or employees, rates of specific procedures, yearly physician and dentist office visits and emergency room visits per insured or employee, average costs per hospital stay, average professional fees, e.g., per office visit or per class of surgery, and average per capita health care expenditures.

Developing rates of service utilization requires knowing how many eligibles there are under an employer policy. In many cases insurers do not require specific information on dependents, relying instead on a list from the employer of which employees have family coverage to decide whether a dependent is covered. In these instances the carrier will only know the number of employees and those dependents who have incurred services for which claims are submitted. Without information on the number of eligibles it is impossible to develop *rates* of utilization or costs per insured and only rates per covered employee can be calculated. To perform age adjustments to improve comparability also requires the age and sex of every insured dependent, whether or not a claim is submitted on his or her behalf. The remedy is for the employer to supply information on the number, age and sex of all insureds to carriers and ask that it be used in developing rates and age adjustments. Alternatively, employers can maintain the data themselves and develop the necessary rates by merging the information on insureds with cost and utilization data available from the carrier.

Caution is warranted in making comparisons of covered groups of employees with regional data collected by federal and state agencies directly from health care institutions and from a sample of both individuals and providers of care. Employed individuals are as a group more healthy than the overall population, which includes many persons with health problems that preclude participation in the workforce. Data bases reflective of the general adult population also need to be age-adjusted, since they usually contain a larger percentage of older individuals who use a disproportionate amount of health care services. Finally, governmental data bases generally eliminate the problems of double coverage which can skew data on an employed population (see page 138 for discussion on problems of workers and dependents being covered under two policies and resulting distortions in statistics).

One of the lessons from review of regional data is the significant variation in utilization and costs. Surgical rates, overall hospital days/1000, and average length of stay (ALOS) vary tremendously among different regions of the country, without any indication that persons in areas with lower utilization and/or lower costs are any worse off with respect to either the quality of service or, more important, to their health. Based on lack of differences in health outcomes, a reasonable objective for a company might be to bring its costs and utilization in line with those of regions with the lowest levels that do not appear to have an adverse effect on health.

Advantages of the comparative approach by employers to decide on the appropriateness of their costs and utilization include the ability to objectively assess where an individual company stands with respect to others. The analysis may identify benefits for which a company's utilization are much higher than that of similar employers. It can provide a sense of confidence that a company's health care costs are not unreasonable compared to others. Such evidence can be an especially important exhibit for the benefits, human relations or financial staff if they have to bring a request for a large premium increase to the executive committee or CEO.

Unfortunately, the many differences among employers with respect to structure and level of benefits, cost-control efforts, demographics of the work force, nature of claims review systems and reporting systems and formats often make it very difficult to feel that similar circumstances are being compared and to understand the significance of whatever differences are found. In addition, the fact that others may have the same discouraging experience provides very limited solace for companies that consider health insurance benefits increases as a major source of erosion of profits. Finally, comparative information, even combined with trend data, provides only minimal information on the effectiveness of existing cost-control measures.

In many ways the situation is analogous to grouping hospitals for planning and regulatory purposes to compare and contrast their performance or to set cost or service norms. Average performance is assumed to be an acceptable standard. If every hospital has many unnecessary patient days and/or is grossly overpriced, this approach to promoting efficiency is ineffective. By similar reasoning, having the figures from Table 9.3 and their comparisons to reference groups may not lead to easy conclusions about which numbers are clearly out of line or are priority targets for cost-control efforts.

Limited satisfaction with comparative information has contributed to increased concentration on a more complete dissection of the employer's own costs. This usually includes two main emphases: 1) carefully assessing the trends in utilization and cost experience over time for each service category and subcategory; and 2) looking very hard at each of the cost components in order to identify specific problems and related cost-control opportunities. As an example, a careful analysis of an employer's hospital costs might lead to the conclusion that two of the twenty-nine hospitals commonly used by the employees of the company are out of line in terms of average price per day or per stay.

For larger employers, especially those with a large covered population in a town, small city or region, it may be worthwhile to identify which specific institutions and other health care providers account for the highest charges/costs per service and highest utilization, e.g., ALOS. Of course, this is only a useful exercise if the employer is willing to consider cost-control activities that would provide economic incentives to use the lower cost providers or to pressure higher cost providers to moderate their charges.

Analysis of Hospital Costs

Hospital costs are a preferred target for analysis because:

- hospitals represent the largest single category of health care expenditures;
- hospitals are few in number compared to other health care facilities and provider organizations;
- hospitals provide itemization (although nonstandard) of charges; and
- hospitals provide more information on a routine basis than other providers on the nature of patient health care problems and diagnostic and therapeutic measures employed.

Also, average inpatient hospital charges can frequently be ascertained without use of the employer health insurance benefit data base. Regional planning bodies generally publish room rates and average cost per hospital day and sometimes cost per stay for the institutions in their area. This information may also be available for Medicaid patients from the state agency responsible for the operation of the Medicaid program. Information on the costs by hospital for Medicare patients should be available from the Health Care Financing Administration, the responsible federal agency. In states with established rate setting programs, the state agency responsible for administering the program should be able to provide detailed information on costs per hospitalization for patients covered by payers under their jurisdiction. Peer Review Organizations will have aggregate information on average length of stay, frequently by diagnosis and sometimes adjusted for age, for most publicly financed patients entering each hospital in their jurisdiction. In the few cases where PROs or related physician review organizations review hospital charges, they may also have charge information on these patients as a group and by diagnosis. Where these organizations are conducting hospital utilization review for employers or unions, a data base for these patients will be maintained but may not be generally available.

Knowing the age and sex distribution of the insured population and the average number of hospital days per 1,000 population (adjusted for age and sex) in the area permits an estimate of how many days of hospital care can be expected on an annual basis. The same technique can estimate expected number of admissions for different diagnoses and procedures, and the number of hospital days associated with each. Table 9.5 shows the difference between expected and observed hospital admissions for five major diagnostic categories for insurees of a large Northwestern consumer products manufacturer. This type of analysis can aid in determining whether aggregate use of inpatient hospital facilities is within the expected range and, if not,

TABLE 9.5
EXCESS ADMISSIONS FOR 5 MAJOR DIAGNOSTIC CATEGORIES, ADJUSTED FOR CASE MIX XCT CORPORATION*

Major diagnostic category	Actual number of admissions for XCT	Expected number of admissions for XCT**	Difference in number of admissions for XCT	
			Number	Percent
Diseases and disorders of the kidney and urinary tract	98	56	42	75
Diseases and disorders of the circulatory system	254	149	105	70
Mental diseases and disorders	112	68	44	65
Diseases and disorders of the digestive system	317	194	123	64
Diseases and disorders of the musculoskeletal system and connective tissue	241	167	74	44

*Fictitious name.
**Expected number of admissions was calculated from a major insurer's claims' data base. Expected rates of admissions (or patient days) can also be calculated based on regional or national norms.

121

TABLE 9.6
INPATIENT HOSPITAL CHARGES IN A CITY (6/30/81 to 6/29/82)

Hospital	Total charges	÷	Total days	=	Average cost/day	×	ALOS	=	Average cost/stay
A	$1,215,196		2,636		$461		7.8		$3,597
B	1,390,634		4,352		320		8.6		2,748
C	542,857		2,196		247		6.4		1,581
D	2,156,319		6,071		355		7.4		2,627
E	103,234		404		256		5.6		1,434
F	334,111		1,198		279		5.3		1,479

Source: U.S. Corporate Health Management.

what types of health problems or procedures are more frequent than expected. Such analyses can lead to specific actions to control costs. If, for example, many more knee surgeries are being performed than expected, knee surgery is a strong candidate for inclusion in a mandatory second opinion program.

Employers with insureds concentrated in one or more geographic areas may wish to begin their analysis of hospital use and cost through analyzing their own experience. Table 9.6 suggests a format for initial review of inpatient services.

Significant differences in costs will invariably be identified. The next step should be an attempt to assure that like services are being compared since hospitals vary considerably in the types of services (procedures) performed and in the mix of services. The same chart can be constructed separately for each major category of hospital inpatient admissions, including at a minimum medical, surgical, psychiatric, alcoholism and other drug abuse, and obstetrics/gynecology. Frequently it is helpful to separate orthopedic from the rest of surgical services. It is very desirable that the tables reflect only adult (age 18 and over) admissions.

Another useful way to display inpatient costs is to break them down into the major components of a bill for each hospital (Table 9.7). One third-party administrator provides a report similar to Table 9.8 which gives additional detail both on the distribution of costs by specific procedure and how the claims were paid. Figure 9.3 sets out the uses of Table 9.8 suggested by the carrier.

In order to make a fair comparison between hospitals based on performance measures such as length of stay, mortality rates, and costs, the case mix of the hospitals must be taken into consideration. Case mix refers to the degree of complexity associated with patient care at the hospital, i.e., patient demographics such as age and sex, diagnoses treated and severity of diseases. Naturally, the treatment of more complicated patients requires a greater use of resources, such as more ancillary services, tests, medications and a longer length of stay.

Diagnosis-Related Groups Diagnosis-related groups (DRGs) were introduced by Yale University in 1975 as a way to group patients on the basis of discharge diagnosis, and they are considered by many to be the most advanced method for determining a hospital's case mix. Originally developed both as a management/planning tool, and to facilitate utilization review, case mix defined by DRGs is now being used as the basis for hospital reimbursement in a number of state programs and nationwide for Medicare patients. In 1983 Medicare began the transition from paying hospitals retrospectively on the basis of allowable costs to paying them a prospectively determined amount for each hospital discharge, based in large part on the patient's DRG. A major purpose of this change was to eliminate the incentive for hospitals

TABLE 9.7
AVERAGE COST PER STAY BY HOSPITAL SERVICE FOR AREA HOSPITALS (6/30/81—6/29/82)

Hospital	Average cost/stay ($)	Room & board		Pharmacy		Laboratories		Radiology		Other ancillaries		Indirect*	
		($)	(%)	($)	(%)	($)	(%)	($)	(%)	($)	(%)	($)	(%)
A	3,597	705	19.6	119	3.3	227	6.3	108	3.0	744	20.7	1,694	47.1
B	2,748	590	21.5	162	5.9	107	3.9	112	4.1	568	20.7	1,209	43.9
C	1,581	395	25.0	61	3.8	142	9.0	89	5.6	286	18.1	608	38.6
D	2,627	512	19.5	121	4.6	218	8.3	113	4.3	602	22.9	1,061	40.5
E	1,434	184	12.8	53	3.7	112	7.8	57	4.0	267	18.6	761	52.8
F	1,479	305	20.6	68	4.6	139	9.4	59	4.0	345	23.3	563	38.1

*Indirect costs include expenditures for research, education, general services, fiscal services and administrative services.
Source: U.S. Corporate Health Management.

TABLE 9.8
CLAIMS PROCEDURE FREQUENCY REPORT BY PROCEDURE CATEGORY (1/1/81—12/31/81)

Procedure category	Units billed	Dollars billed	Dollars rejected	Units cutback	Dollars cutback	Co-insurance and deductible	Dollars paid by others	Exceed maximum	Amount paid ($)
Diagnostic radiology	15,723	1,386,304	187,742	29	7,753	171,357	44,504	102	972,845
Therapeutic radiology	841	121,155	15,009	4	887	7,046	5,901	0	92,312
Operating room	6,402	2,337,742	299,211	201	9,450	173,132	94,171	0	1,761,767
Anesthesia	4,199	472,484	58,905	5	1,002	37,818	19,087	0	355,673

Source: Control Data Corporation. Management Report, unpublished.

FIGURE 9.3
USES OF TABLE 9.8

1. Identification of high expenditure/frequently utilized procedure categories and specific procedures for focused cost containment consideration.

2. Identification of specific savings achieved by claims administrator through prevailing fee cutbacks, co-insurance and deductible, coordination of benefits, benefit maximums and ineligible expenses.

3. Identification of those procedures which exceed reasonable and customary as well as the amount by which they exceed it.

4. Identification of scope and impact of prevailing fee cutbacks on employees, e.g., 2,000 cuts at $1 to 3 each.

5. Analysis of actual inflation rates and/or shifts in providers by comparing procedural categories over time, e.g., room and board 1981–1982, obstetric fees 1981–1982.

6. Monitoring shifts in medical practice patterns or case type (for example, ratio of Caesarean sections to normal deliveries 1981–1982, or psychiatric days to total days) can then focus communication efforts to employees.

7. Comparison of utilization and cost patterns by area or division and/or with other employers.

Source: Control Data Corporation. Benefit Services Division.

to maximize services which is inherent in retrospective cost-based reimbursement. Another purpose was to move toward paying a more uniform amount for a defined health problem in institutions with greatly varying historical costs.

Knowing that they will only receive a fixed sum of money for each discharge regardless of their costs provides a strong incentive for hospitals to provide only those services that are necessary and to minimize lengths of stay. While some doctors and hospitals are concerned that the new incentives may lead to less than required care, there is considerable evidence that substantial cost cutbacks could occur without having a detrimental effect on outcomes of health care. In some cases, reductions could lead to improved health outcomes.

Since there is only limited experience in DRG-based hospital payment, it is premature to consider the system a panacea for hospital cost problems. DRGs should not yet be considered either the best method to determine case mix or the most appropriate system for reimbursement under private health insurance. Nonetheless, DRGs are an attractive, innovative concept that can double as a management tool for hospitals and a means for employers to identify variation in costs per hospital for similar problems. When there are a sufficient number of hospital admissions in a defined geographic area, a DRG analysis can assist employees in:

1. Deriving a case mix adjustment index for each hospital utilized. Such an index permits a more accurate and meaningful comparison of the charges and ALOS by individual hospital.
2. Facilitating year-to-year comparisons of an individual hospital's charges to identify relative rates of increase on a case mix-adjusted basis.
3. Deciding which hospitals may be the most appropriate targets for negotiating discounts.
4. Determining if benefit plan incentives to substitute ambulatory care for inpatient care are working. A lower percentage of low-resource consumption DRGs suggests that those incentives may be affecting insurees' use of the health care system.

Analysis of Physician Costs

Since physicians recommend hospitalization, order all medical services in the hospital and decide on discharge dates, it is appropriate to analyze the hospital costs associated with individual physicians.

Tables similar to 9.6, 9.7, and 9.8 can be used to identify physicians who tend to be associated with a high ALOS, high average costs per day and costs per stay; and/or higher charges for physician services per hospital stay. Case-mix adjustment by DRG or other method can increase the usefulness of this information. In developing these profiles it is important to compare physicians within the same specialty, such as general surgery, general internal medicine, cardiology, orthopedic surgery, etc. These analyses are intended to identify individual physicians who might be shown their pattern compared to their peers and those who might be selected as preferred providers. However, some experiences with such analyses in the PSRO program have suggested that it may be difficult to define homogeneous groups of physicians to make valid comparisons. For example, one orthopedic surgeon may specialize in hip replacements, whereas another may concentrate primarily on sports medicine problems and perform mainly arthroscopic knee surgery. The former surgeon will undoubtedly have a much longer length of stay, higher costs per admission, etc., due to more extensive surgery and an older patient population. Likewise, the experience both in and out of the hospital of an internist specializing in endocrinology may differ from that of an internist specializing in gastrointestinal problems.

Despite some obvious limitations in the types of conclusions which can be drawn from these types of analyses, the same general approach can be fruitfully used to develop tables of hospital outpatient services, use of prescription drugs, vision care and dental care. Table 9.9 displays such an analysis of dental services for a medium sized employer.

TABLE 9.9
DENTAL PAYMENTS BY PROCEDURE CATEGORY AND ELIGIBILITY CLASSIFICATION CALENDAR YEAR 1982

Procedure category	Total dollars	Employee				Dependent					
		Male		Female		Male		Female		Children	
		($)	(%)	($)	(%)	($)	(%)	($)	(%)	($)	(%)
Diagnostic	42,737	11,189	26.2	12,944	30.3	2,548	6.0	5,778	13.5	10,276	24.0
Preventive	44,756	10,880	24.3	13,405	30.0	2,097	4.7	5,747	12.8	12,627	28.2
Restorative	331,325	90,793	27.4	111,617	33.7	28,219	8.5	60,252	18.2	40,444	12.2
Endodontics	82,282	26,765	32.5	29,977	36.4	6,510	7.9	12,979	15.8	6,051	7.4
Periodontics	77,343	21,853	28.3	33,576	43.4	5,176	6.7	13,696	17.7	3,041	3.9
Prosthetics	101,729	35,044	34.4	36,084	35.5	5,807	5.7	21,833	21.5	2,961	2.9
Oral surgery	68,210	9,353	13.7	23,114	33.9	4,095	6.0	7,063	10.4	24,585	36.0
Orthodontics	101,690	1,443	1.4	12,433	12.2	122	0.1	4,677	4.6	83,014	81.6
Emergency services	1,127	207	18.4	495	43.9	90	7.9	250	22.2	84	7.5
Miscellaneous services	13,215	2,241	17.0	4,033	30.5	677	5.1	2,062	15.6	4,202	31.8
TOTALS	864,415	209,770	24.3	277,680	32.1	55,341	6.4	134,338	15.5	187,286	21.7

Source: Control Data Corporation, Management Report, unpublished.

128

Who Is Using Each of the Benefits?

A second cut at analyzing costs and identifying cost control opportunities is to focus on who is using the health care benefits. Two steps are basic:

1. *Analyze utilization and costs by employees, spouses, other dependents and retirees for each sex* For each of these groups, start with overall cost and utilization figures and then break them down into smaller service categories, such as those in Tables 9.3 and 9.4. The results may be surprising. In general, 40 to 60 percent of costs are for dependents (spouses and children), but the percentage can differ considerably among service categories. An example of distribution of dental payments from Table 9.4 by employee, dependent adults and dependent children, and by sex is provided in Table 9.9.

2. *For each insured group, determine costs and utilization by age group and sex* Since health problems differ by age it is useful to see how the age distribution of employees and dependents affects overall costs and utilization patterns. Age group breakouts might include:

	Employees	Spouses	Children	Retirees
Age groups				
0–12 months	N/A	N/A		N/A
1–15 years	N/A	N/A		N/A
16–21				N/A
22–30				N/A
31–40			N/A	N/A
41–50			N/A	
51–60			N/A	
61–64			N/A	
65–69			N/A	
70–74	N/A		N/A	
75–79	N/A		N/A	
Over 79	N/A		N/A	

A logical next step is to determine per capita costs and utilization rates associated with each category of insured for each service category listed in Table 9.3. This refinement answers the question, "How much per person is being spent for employees, spouses, other dependents and retirees in each group for each major service category?" Per capita rates are more useful than aggregate rates because the former permit better judgments regarding the reasonableness of expenditures and utilization rates (e.g., days of hospital care per 1,000 insured). It also permits a better comparison with past experience and with the experience of other companies.

Surprisingly, most of the utilization and costs under a majority of health insurance plans can be traced to a small proportion of the covered population. Although figures vary from employer to employer, in several companies approximately 10 percent of insureds account for over 50 percent of total benefit costs. One way to approach this problem is to get a printout of the characteristics of the 10 percent of the insureds with the highest claims payouts, e.g., age, sex, division between categories of insurees (employees, spouses, etc.). At the same time the services accounted for by this 10 percent should be summarized. In some companies there have been some unexpected findings. For example, a medium-sized service company in the northeast discovered that many of the highest cost insureds were premature newborns, whose hospital bills were sometimes $50,000 to $100,000 each, and those with chronic psychiatric problems, both for inpatient and outpatient treatments. Based on these findings they decided to reinforce, through communication, the importance of prenatal care and to institute a utilization review program for psychiatric admissions.

What Are the Characteristics of Health Benefits Cost Increases?

If costs this year were no greater than the last there would not be the same clamor for cost-control initiatives. So, many employers start their claims analysis by seeing how this year differed from the last and the one before that. In analyzing *increases* it is essential to make the comparison per insured. Looking at gross numbers neglects changes in numbers of employees and dependents. All cost increases can be divided into:

1. *Increases in utilization rates*   For example: more hospital days per 1,000 insured, more physician visits per insured, more laboratory tests per hospital day.
2. *Increases in price*   For example: the same electrocardiogram costs more.
3. *Changes in methods or technology*   For example: the ultrasound has been improved and the new procedure carries a higher charge, or a new diagnostic procedure such as nuclear magnetic resonance is employed for the first time.
4. *Changes in covered services*   For example: outpatient treatment for alcoholism is provided for the first time under the benefit package.
5. *Changes in reimbursement provisions*   For example: a stop-loss provision was added so that employees' obligation for payment ends when aggregate charges for a family exceed $5,000 in any one year.

One way to begin this analytic process is shown in Table 9.10. It builds upon basic Table 9.3, adding separate columns for changes over the past two years.

For example, a Southern California consumer products company that

TABLE 9.10
FORMAT FOR INITIAL DISPLAY OF HEALTH CARE BENEFIT COST AND UTILIZATION

Benefit category	Total incurred claims		Cost per claim		Cost per insured		Utilization measures	
	Year 1	Year 2 % change	Year 1	Year 2 % change	Year 1	Year 2 % change	Year 1	Year 2 % change
Hospital inpatient								
Hospital outpatient								
Physician inpatient								
Physician ambulatory								
Mental health inpatient								
Mental health outpatient								
Dental								
Vision								
Pharmacy								
Skilled nursing								
Home health								

had a one-year 24 percent increase in overall costs found that during that period utilization increases occurred in mental health inpatient days/1000 (38 percent), dental visits/insured (21 percent), skilled nursing days/1000 (9 percent) and home health visits/insured (28 percent), while no significant increases were noted in mental health outpatient visits/insured or in total physician visits/insured. Such information is essential in developing well-targeted cost management strategies.

When this company filled in Table 9.10, they naturally found incurred claims had increased drastically for those categories experiencing a large increase in utilization. However, only some categories showed an increase in cost/claim. And some service categories without large increases in utilization still experienced major cost increases, due to increases in price and/or changes in technology/treatment methods. For example, cost per hospital admission rose 18 percent in the face of no change in length of stay or in days/1000 insured. Physician cost per claim was up 16 percent and cost per outpatient mental health claim was up 23 percent.

While trend information is an important part of analysis of health care costs, conclusions frequently are drawn with greater certainty than warranted. Among the factors that can contribute to erroneous conclusions are:

1. *Changes in the sex distribution of the workforce* Women, especially young women, tend to have higher per capita utilization and costs than men.

2. *Changes in the average age and age distribution of employees* In one large manufacturing company the average age of employees increased from 37 to 43 years over a five-year period. Since utilization and costs increase at greater and greater rates for each higher age group, this change alone caused a great growth in the cost of the health insurance benefit.

   To get an idea of the importance of age in determining health care costs, consider that in 1978 persons aged 65 and over spent $2,026 per capita for health care. This is seven times the $286 per capita for persons under age 19 and two and one-half times the $764 per capita for persons aged 19–64, who constitute almost the entire workforce. For physician services in 1978, per capita costs were $75.06 for those under 19, $163.56 for those 19–64 and $365.70 for those 65 and over.[1] Annual prescriptions per capita rise from 4.2 in the 17–24 age group to 5.6 (ages 25–44), 8.7 (45–64) and 14.4 (65 and over).[2] While the 19–64 age group accounts for the same percentage of total health care expenditures as the percentage it comprises of the U.S. population, those over 65 are about 11 percent of the population but use over 30 percent of the total health care resources.[3]

   The increase in costs with age reflects the inevitable increase in frequency of health problems with advancing age, as shown in Figure

FIGURE 9.4
PERCENTAGE OF PERSONS WITH CHRONIC CONDITIONS, BY AGE AND SEX, 1974

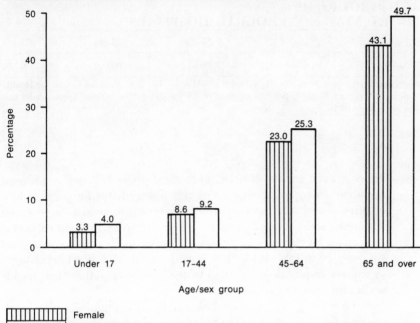

Source: Fisher, C.R. "Differences by Age Groups in Health Care Spending," *Health Care Financing Review*, Vol. 1, No. 4, 1980, pp. 65–90.

9.4, which displays the percentage of persons with chronic conditions, by age and sex. And more illness translates into higher utilization rates. Figure 9.5 shows how age affects days of care in short-stay nonfederal hospitals per 1,000 civilian population in 1978.[4] Of all the categories of professional services, only dental care does not seem to display an accelerating pattern of use and costs with age.

Overall, a change of one year in the average age of insureds can be expected to increase employer health care costs by between three and five percent depending on the original average age, even in the absence of price inflation. The increase could be even greater with a one-year increase in average age if it was primarily attributable to more retirees rather than changes in the average age of the active work force.

3. *Changes in the geographic distribution of the workforce* There is significant variation of medical costs between areas. As shown in Table 9.11, the same operation costs twice as much in New York as in Dallas.

FIGURE 9.5
DAYS OF CARE PER 1000 CIVILIAN NONINSTITUTIONALIZED POPULATION BY SELECTED AGE GROUPS, 1980
SHORT STAY NONFEDERAL HOSPITALS

All ages	Under 15	15–44	45–64	65+
1230.8	316.4	793.1	1607.0	4327.5

Source: United States Department of Health and Human Services, National Center for Health Statistics. Series 13, Number 64, Utilization of Short Stay Hospitals—Annual Summary for the United States, 1980.

There is even greater disparity in medical prices between major metropolitan areas and smaller cities and towns. Rates for surgical procedures also display considerable variation from area to area, even within the same state (Table 9.12). There are still significant variations in average lengths of stay between different parts of the country, as summarized in Table 9.13. Thus, changes in ALOS may reflect changes in the geographical distribution of the workforce rather than trends within the same area.

One soft-goods company believed that its employees were heeding its informational campaign to be more prudent in using health care services because over a two-year period costs per insured had grown only 14 percent. However, during that period they had closed two facilities in the Los Angeles area and expanded three plants in small towns in the southeast. When adjustments were made to provide true comparisons, differences in costs per insured rose about 38 percent

TABLE 9.11
COST OF SELECT OPERATIONS IN SOME MAJOR CITIES*

	New York	Chicago	Los Angeles	Dallas
Triple coronary bypass	$20,740	$19,048	$17,613	$12,715
Total colectomy	15,212	13,602	12,928	8,094
Radical hysterectomy	13,514	11,518	11,459	7,097
Radical mastectomy	8,523	6,943	6,880	4,122
Gallbladder removal	8,368	7,436	7,241	4,350
Appendectomy	5,320	4,357	4,591	2,718
Hemorrhoidectomy	4,390	3,777	3,651	2,183
Inguinal hernia	3,632	2,886	3,187	2,026

*Data: Prudential Insurance Co., *BusinessWeek* estimate.
Source: Reprinted from August 30, 1982 issue of *BusinessWeek* by special permission © 1982 by McGraw-Hill Inc.

TABLE 9.12
1973 AGE-ADJUSTED INCIDENCE OF SURGICAL DISCHARGES AND NINE COMMON PROCEDURES IN MAINE AND MAINE HOSPITAL SERVICES AREAS WITH POPULATIONS OF 20,000 OR GREATER. PROCEDURES PER 10,000 POPULATION

Area	All surgical discharges	Appendectomy	Prostatectomy (males)	Inguinal hernia (males)	Hysterectomy (females)	Varicose veins	Hemorrhoidectomy	Dilation and curettage (females)	Tonsillectomy	Cholecystectomy (females)
State as a whole	689	17	25	45	59	5	7	77	62	35
1	613**	18	18*	40	46**	4	6	58**	36**	37
2	670	11**	22	37	59	6	3*	86	23**	46*
3	742**	13**	35**	47	63	4*	4**	83*	54**	33
4	606**	17	28	45	41**	3	5	49**	47*	36
5	594**	14	18	45	48	6	5	84	35**	31
6	640*	21	26	51	47	6	6	87	60	35
7	688	17	27	49	93**	6	9*	76	59	34
8	738**	17	40**	45	67	4	7	117**	55	50**
9	688	15	20*	53	39**	10**	9	74	62	29
10	864**	19	25	52	58	5	9	114**	105**	55**
11	954**	22**	33**	49	60	8*	19**	86	122**	39
12	579**	19	18**	35**	51**	5	5	67*	77**	
13	764**	19	31	60*	48	7	14**	78	77**	27

*Chi-square significant at the .05 level.

**Chi-square significant at the .01 level.

Source: Wennberg, J.E. and Gittelsohn, A. "Health Care Delivery in Maine I: Patterns of Use of Common Surgical Procedures," *Journal of the Maine Medical Association*, Vol. 66, No. 5, 1975, pp. 123–149.

TABLE 9.13
GEOGRAPHIC VARIATION IN AVERAGE LENGTHS OF STAY—1980

Conditions	All regions	Northeast	North central	South	West
Malignant neoplasm	10.5	12.0	10.7	10.1	8.5
Diabetes mellitus	10.5	12.9	10.3	9.8	8.1
Cataract	3.6	3.5	4.1	3.4	3.2
Acute myocardial infarction	12.6	13.9	13.4	12.3	9.6
Appendicitis	5.5	6.1	5.6	5.5	4.6
Inguinal hernia	4.7	4.5	5.0	5.2	3.8
Sprains and strains of back (including neck)	7.1	8.6	7.0	7.1	5.8

Source: U.S. Department of Health and Human Services. National Center for Health Statistics, Series 13, No. 64. Utilization of Short-Stay Hospitals: Annual Summary for the United States, 1980.

during the two-year period, more in line with the dismaying experience of similar companies. Based on this analysis, the company decided to pursue a more active set of cost control options.

4. ***Changes in benefits*** Few health insurance benefit plans escape some changes over a period of several years. Frequently it is difficult to sort out the effects of benefit changes from unrelated increases in utilization. This is further complicated by the fact that the full impact of benefit changes may occur over a period of years, and not at a constant rate. For example, one company found that during the first two years after adding an outpatient psychiatric benefit, utilization was still low. During the third year utilization more than doubled. While the reasons are not entirely clear, this is not an isolated example and is one of the reasons that predicting claims experience can be a frustrating experience.

Some benefit changes can result in significant substitution of one service for another. For example, a decision to cover home health services or hospice care can lead to a reduction in hospital days. Without taking this benefit change into account, inaccurate conclusions could be reached on trends in hospital utilization and in the total costs associated with treating a specifc health problem.

5. ***Changes in reimbursement policies*** With growing frequency employers are requiring more cost sharing on the part of employees. Deductibles and co-payment provisions are replacing first-dollar coverage. These changes can impact both cost and utilization to the same or to different degrees. For example, increasing co-insurance might lead to a 15 percent decrease in per insured costs of physician visits while having a minimum effect on the number of physician visits per insured.

6. ***Changes in accepted practice approaches*** What doctors consider the best ways to diagnose and treat specific problems change, sometimes very quickly. For example, ulcer surgery is far less common today than several years ago before the development of effective drug therapy. Coronary artery bypass surgery was almost unheard of in the early 1970s and ten years later accounts for several billions of dollars in health care costs. Without knowledge of trends in medical practice it is difficult to analyze the effects of some cost-control efforts. A second opinion program that covered cardiac bypass surgery might be considered a failure if the rate of operations increased by 15 percent during the first year, but compared with national data showing a 25 percent increase, the program might be considered quite successful.

7. ***Inaccuracies in coding*** When looking at categories of health problems that either carry a social stigma or whose definitions are not well standardized it may be difficult to assess the true frequency of the problem. For example, alcoholism is frequently reported as another problem or as "ill-defined signs and symptoms." Schizophrenia may

be reported as a nonspecific nervous disorder. It is not unusual for a company to find the first-rank category of diagnoses associated with hospitalization as "mental and nervous disorders," but inadequate precision in coding usually precludes reliable additional categorization.

8. **Double coverage** As the percentage of family members covered by more than one policy increases, the average utilization and cost experience per policy tends to decrease. For example, if on the average the percentage of spouse hospitalization expenses paid for by an employee's family policy through Employer A declines from 80 to 60 percent due to improved coverage by the spouse's own employee policy under Employer B, dependent utilization and associated costs under Employer A's policy would appear to decrease, or at least rise at a slower rate than employee utilization and costs under A's policy.

    Differences in the scope of benefits, reimbursement provisions and company contribution to premiums between employers can also have effects on *apparent* utilization. Suppose that Company B, the wife's employer, has a much richer benefit package than Company A. Both companies pay the full premium for dependents, encouraging each spouse to cover the other. If the husband, covered now by both A and B, is hospitalized, he may claim reimbursement under Company B's policy, since it means lower out-of-pocket costs to the family. His hospital stay could therefore appear on Company B's report, leaving Company A with an understatement of actual utilization for its covered population. Coordination of benefits provisions usually require that the first payer for the hospitalization be Company A, but insurers may not be aware that the spouse has other coverage.

9. **Changes in when claims are submitted** Cost-sharing features can have an important impact on when claims are submitted and, as a result, on the experience reported to companies by their carriers. For example, when employees are covered by a plan with no co-payment and no deductible, they will tend to submit claims as they are incurred, since they can obtain reimbursement on virtually all of them. By contrast, employees in a company with a $200 per person/$500 per family deductible and 20 percent co-payment have less incentive to submit claims early in the contract year which they know are nonreimbursable due to the deductible, and a slightly reduced incentive to submit even after the deductible is satisfied, since they will receive only 80 percent of allowed charges.

    The impact of deductibles on claims submission is demonstrated in Table 9.14. Note that the percentage of claims submitted within any given number of months increases in later months of the year when the deductible has probably been satisfied. A company that reduces its co-payments and especially its deductibles may get a higher percentage of total incurred claims submitted in the early months of the

TABLE 9.14
COMPANY POLICY WITH $100/250 DEDUCTIBLE DURING EACH CALENDAR YEAR
CUMULATIVE PERCENTAGE OF CLAIMS SUBMITTED WITHIN X MONTHS AFTER SERVICE* (1/1/83–12/31/83)

Month of service	0	1	2	3	4	5	6	7	8	9	10	11	12+
January	0.8	25.4	67.7	76.2	82.8	85.5	89.6	93.3	95.2	96.4	97.6	98.8	100.0
February	0.7	40.6	74.6	87.4	90.8	94.3	95.7	96.7	98.0	99.0	99.7	100.0	
March	3.7	51.8	69.4	81.3	90.9	94.0	95.9	97.0	99.0	99.4	100.0		
April	5.3	36.2	61.3	82.8	92.1	94.7	97.3	98.3	99.3	100.0			
May	2.1	21.7	65.9	89.1	96.4	97.1	97.3	99.3	100.0				
June	0.4	51.1	76.6	86.7	92.2	97.0	98.2	100.0					
July	9.0	52.8	74.0	91.9	96.5	99.3	100.0						
August	11.9	57.4	82.7	94.9	99.0	100.0							
September	13.4	63.7	80.9	95.8	100.0								
October	11.6	63.8	91.8	100.0									
November	6.6	67.6	100.0										
December	20.6	100.0											
TOTALS	6.3	49.3	77.1	89.7	94.7	96.9	97.6	98.6	99.2	99.5	99.7	99.9	100.0

*Insurance policy year January 1—December 31.

contract year than before the change, and year-to-year comparisons of those months would lead to overestimates of the rate of increase in claims. Conversely, if the deductible is increased, a higher proportion of total incurred claims for the contract year can be expected to be submitted to the carrier in the latter months. In a company that had recently instituted more cost sharing, employees would tend to be slower in submitting claims for services early in the contract year. Early paid claims reports could therefore overestimate any reductions in services incurred due to increased cost sharing. In this case, month-to-month trend analysis early in the contract year could underestimate the rate of rise in claims during the year and lead to anticipating smaller increases for the next year than should be expected.

10. *Changes in when care is sought for health problems* Changing economic circumstances also create differences in the pattern of claims incurred and paid. In recessionary times employees and their families concerned about possible layoffs tend to have elective procedures done if they are well-covered by a company-sponsored health insurance plan. Insurers remark that in layoff situations you can expect to see a large increase in costs during the period between the announcement of layoffs and when health insurance coverage terminates. In addition, in bad economic times claims are likely to be submitted more quickly after incurral dates because families need the money. Comparing the experience of the same company during periods of relative prosperity and periods of recession is therefore difficult.

Use of Incurred Claims

For assessing trends and comparing information from several periods it is very helpful to have reports based on *incurred* claims. Currently most data provided by insurers is for what claims were paid during a period. This preference is understandable since it is the dollars paid out that are the primary contractual concern. However, as discussed above, many factors can affect payment rates independent of the incurrence of services and related costs. For example, in times of recession and high unemployment, when claims submittals are being accelerated, claims paid will rise at a faster rate than incurred claims. The number of days a claim is held before payment can also affect the payment rate. A fluctuation of five working days in the period between receipt of claims and payment by the carrier from month to month can artificially increase the claims paid experience one month and decrease it the next, obscuring important trends during that period.

A telling example comes out of the experience of a diversified service company in the southwest. Its cost report for 1982 indicated a 9.3 percent increase (from $43 to $47) in average per capita monthly expenditures under its health insurance, a major difference from the anticipated 20 percent per capita increase. Trying to explain this phenomenon, the company was surprised to see the smaller than expected increase reflected in all major cost categories. It turned out that the problem was in the number of insured used to figure per capita costs. The number of employees and dependents used by the carrier to compute per capita figures was a weighted average of the monthly totals they received from the employer. But during that year an increase in business led to gradual hiring of about 25 percent of the workforce, and therein lay a common problem. The claims paid during any month reflected the usual time lag from service to payment. Therefore, during any month fewer individuals were submitting claims than were incurring costs. The weighted average of insureds divided into the total claims for the year seriously understated the obligations that a growing workforce had incurred. When the number of active employees for each month of the year divided into the claims *incurred* during that month, a more accurate measure, the average monthly per capita cost was found to have risen 18.6 percent (from $43 to $51) over the prior year (Table 9.15). This experience underlines the necessity of using reliable and appropriate numbers of employees and dependents in the calculation of cost and utilization trends and of looking at per capita costs on a monthly basis. It also argues for the importance of having information based on incurred rather than paid claims, especially during periods of fluctuation in the strength of the workforce.

Problems of Multiple Sites

Large employers have the inherent difficulties in analysis of health insurance claims compounded by having to combine information from a number of different carriers. For example, one large company estimated that to compute total health care costs they must combine information from fifty-six different sources, including some Blue Cross/Blue Schield plans, a spate of commercial carriers and nineteen HMOs. Contract years may be different, and due to differing financial arrangements the net figures for premiums and even claims paid almost invariably reflect different accounting systems and conventions.

For companies with these multiple sites, analysis that only examines the aggregate expenditure pattern of all sites together is frequently insufficient to decide what should be the priority targets for cost control efforts. As Table 9.16 demonstrates, cost per employee can vary by as much as 100 percent from the lowest cost to the highest cost sites. This selected data, covering about one-half of a large employer's twenty major sites, reflects the variation usually found in cost per employee or cost per insured. Equally striking are variations found in rates of per capita increases. Some of the highest cost cities in Table 9.16 are showing slow growth while some of the lowest cost cities have expenditures growing at among the fastest rates. The annual rate of increase of cities rank ordered from 1 to 3 was more than

TABLE 9.15
1982 CLAIMS REPORT*

Month	Number insured	Incurred claims ($)		Paid claims ($)	
		Total	Per capita	Total	Per capita
January	4,783	224,801	47	205,669	43
February	4,926	231,522	47	216,744	44
March	5,024	241,152	48	224,801	45
April	5,123	251,027	49	231,522	45
May	5,225	271,700	52	241,152	46
June	5,329	266,450	50	251,027	47
July	5,435	277,185	51	271,700	50
August	5,543	288,236	52	266,450	48
September	5,659	282,950	50	277,185	49
October	5,772	305,916	53	288,236	50
November	5,891	318,114	54	282,950	48
December	5,986	329,230	55	305,916	51
Average	5,391	274,024	51	255,279	47

*1981 Average monthly claims cost per capita = $43.

TABLE 9.16
COMPARATIVE COMPANY HEALTH CARE COSTS BY CITY

City	Region of country	Health care cost per employee, 1979		Increase/employee 1975–1979	
		$/employee	Rank	Percentage	Rank
A	Mid-Atlantic	1,240	1	77.4	1
B	West	1,066	2	38.8	6
C	Southeast	979	3	45.6	3
D	Mid-Atlantic	889	4	40.1	5
E	Mid-Atlantic	883	5	32.5	7
F	Mid-Atlantic	849	6	30.9	9
G	Southeast	814	7	48.1	2
H	Mid-Atlantic	729	8	17.6	10
I	West	715	9	42.6	4
J	Northeast	603	10	32.3	8

Source: Unpublished company data.

143

TABLE 9.17
HOSPITAL, SURGICAL AND MAJOR MEDICAL PLANS—COST PER EMPLOYEE—1979

City	Total cost/employee ($)	Hospital ($)	Rank	Surgical ($)	Rank	Major medical ($)	Rank
A	1,240	580	1	270	2	390	1
B	1,066	465	3	292	1	309	2
C	979	487	2	218	6	274	3
D	889	431	4	233	3	225	5
E	883	426	5	232	5	225	5
F	849	376	7	233	3	240	4
G	814	422	6	193	8	199	7
H	729	371	8	198	7	160	9
I	715	360	9	172	9	183	8
J	603	329	10	129	10	145	10

Source: Unpublished company data.

144

double the rate of growth of the three lowest ranked cities. Having only information on the *average* annual rate of increase would obscure the fact that for cities ranked 1 to 3 it was 14.26 percent and for cities ranked 8 to 10 it was only 6.7 percent. Looking ahead one could argue for focusing considerable efforts on those areas with the fastest rate of increase since it is easier to affect increases than shave away any of the base costs.

Another way of looking at costs is to determine which services account for the major cost increases. Were basic hospital services a major problem? Major medical expenses? Surgical fees? Table 9.17 displays dollar costs for one year for each of these categories by principal sites for the same employer.

As expected, for each service type the highest cost areas have expenditures over twice the level of the lowest cost areas. While this level of analysis is helpful in singling out both sites and services for priority attention, it invariably yields more questions than answers. Finer-grain analysis is usually needed to decide what should be the specific cost-control initiatives.

An Example of Finer-Grain Analysis: Mental Health Benefits in a Large Manufacturer

At a large midwestern manufacturer, mental health costs had grown over a four-period at a compound rate of 18.9 percent per year. Among the steps undertaken to better understand the nature of the problem were:

1. *Determining the relative contribution to total costs of employees, spouses and children* The results were unexpected. Almost 45 percent of inpatient psychiatric benefits were for children, a quarter for spouses and slightly less than one-third for employees (Table 9.18). Thus a mental health promotion program aimed primarily at employees would not be directly targeted at those responsible for over two-thirds of the inpatient psychiatric costs. And the effectiveness of a program to reduce the psychiatric admission rates for employees and the attendant cost of their psychiatric inpatient services could be obscured by increases in dependent utilization, which account for about 70 percent of dollars.

   Additional investigation found that of 140 insured who utilized $15,000 or more in psychiatric benefits, 117 (84 percent) were employees rather than spouses or children. Further, employees associated with this high cost tended to be in the central office (38 percent), although only about 11 percent of total employees were based there (Table 9.19). Therefore, efforts to reduce the frequency of high cost psychiatric benefit use might aim at finding lower cost treatment options for central office employees.

2. *Determining the distribution of expenditures by service component over time* For example, Table 9.20 provides a summary of the distribution of outpatient psychiatric benefits (which totalled 80 percent of the cost of inpatient benefits). Despite a stable employee count, the

TABLE 9.18
INPATIENT PSYCHIATRIC BENEFITS PAID—1979

	Number of cases	Percentage of cases	Average cost/case	Percentage of total dollars
Employee	1,083	34.0	$3,334	31.2
Spouse	1,024	32.1	2,730	24.1
Children	1,082	33.9	4,792	44.7
Totals	3,189	100.0	$3,635	100.0

Source: Unpublished company data.

TABLE 9.19
EMPLOYEES AND DEPENDENTS: $15,000 AND OVER,
PSYCHIATRIC BENEFITS

Location	Number of individuals*
Headquarters	44
A	17
B	13
C	11
D	7
E	7
F	6
G	5
H	5
I	5
J	5
K	4
L	4
M	4
N	3
	140

*Employees, 117; Spouses, 20; Children, 3.
Source: Unpublished company data.

numbers of individuals using the services increased about 45 percent over a three-year period, while payments increased 60 percent. Further analysis showed that for 74 percent of families utilizing the benefit, the total payment was less than $3,000 in one year (Table 9.21). Thus, the majority of utilization was not for prolonged intensive psychotherapy as has sometimes been observed in other companies.

3. *Determining the difference in total mental health benefit costs per site (Table 9.22)* Unlike the two-fold variation in overall health benefit costs per site, mental health cost variations were almost ten-fold. The three highest cost sites averaged 341 percent of the cost of the three lowest cost sites, and this pattern of extreme variation also prevailed for both inpatient and outpatient annual costs per employee when considered separately.

To further zero in on aspects of the problems that would be amenable to correction, additional analyses might be performed on:

a) the admission rates and average length of stay for hospitals where employees in high cost cities were admitted;

b) utilization and gross billing profiles of psychiatrists, mental health centers and other clinics associated with high outpatient mental health costs;

TABLE 9.20
DISTRIBUTION OF OUTPATIENT PSYCHIATRIC BENEFITS PAID

	1976	1977	1978	1979
Benefits paid	$620,000	$720,000	$870,000	$990,000
Increase (%)	—	16	21	13
Major medical (%)	21	20	22	22
Numbers of employees/ spouses/children receiving benefits	1,250*	1,406	1,620	1,818

*Estimated.
Source: Unpublished company data.

TABLE 9.21
OUTPATIENT PSYCHIATRIC BENEFIT PAYMENTS
RECEIVED—1979

Dollar range	Number of individuals	Benefits paid (family)
$ 1– 999	13,702 (84%)	$307,014 (31%)
1,000– 1,999	1,856 (11%)	262,264 (27%)
2,000– 2,999	508 (3%)	154,950 (16%)
3,000– 3,999	178	109,968 (11%)
4,000– 4,999	68	62,524 (6%)
5,000–24,999	51	91,886 (9%)
Totals	16,363*	$988,606 (100%)

*Employee, 45%; Spouses, 32%; Children, 23%.
Source: Unpublished company data.

c) distribution of specific health problem categories, e.g., alcoholism, other drug problems, specific categories of psychiatric disorders, etc.

Being armed with this level of information permits a much more targeted cost management approach than with what is usually available

TABLE 9.22
DIFFERENCE IN TOTAL MENTAL HEALTH BENEFIT
COST PER SITE—1979

Location	Average cost/employee	Rank
Headquarters	$182.3	2
A	130.7	5
B	148.5	4
C	102.3	9
D	314.9	1
E	93.1	10
F	124.4	7
G	68.5	12
H	81.6	11
I	157.8	3
J	54.6	13
K	31.5	15
L	118.6	8
M	130.7	5
N	51.9	14

Source: Unpublished company data.

from routine reports of claims activity. Based on the analysis above, for example, the following steps could be considered:

1) Change the benefit structure to increase cost sharing of outpatient psychiatric benefits, especially after the first five visits (see Chapter 7).

2) Place a ceiling on annual family mental health outpatient benefits, possibly $5,000.

3) Conduct frank discussions with facilities charging high rates or having long lengths of stay.

4) Subject bills for psychiatric inpatient services to retrospective prepayment peer review.

5) If patterns of overutilization or overcharging are identified, institute preadmission and concurrent utilization review.

6) Suggest that facilities with high total daily costs consider giving a discount for individuals covered by that employer.

7) Provide incentives to employees and dependents for being hospitalized in a lower cost facility nearby. Consider eliminating deductibles and/or reducing co-payments for hospitalizations at the lower cost facility and widely publicize those incentives at repeated intervals to employees and covered dependents.

Naturally, which options are pursued depends upon the company's constraints, such as union contracts, concern about effects on employee relations, other priorities and opportunities for health care cost management, characteristics of the health care community in high cost cities and company expertise and commitment to invest in improved monitoring and control systems, among other factors.

A decision to undertake this level of analysis already suggests a willingness to address uncovered problems. However, in some cases considerable resources have been expended in data analysis and strategizing about potential initiatives to affect costs and utilization when top management is not prepared for aggressive actions and their potential repercussions. It is therefore advisable to have the signals clear before embarking on extensive data analysis.

FROM DATA ANALYSIS TO COST-CONTROL STRATEGIES

The purpose of all this analysis is to determine two things:

1. What cost-control approaches are likely to work?
2. Over time, do these cost-control efforts meet their objectives?

A high payoff from a particular cost control strategy is much more likely if the strategy is chosen to respond to the specific problems of a particular employer. Problems should be selected and priorities initially assigned based on possible savings over a defined period of time. Strategies selected as the highest priorities will differ if immediate savings are the objective rather than if a longer time horizon is selected.

It is usually helpful to think of the cost-control options according to which components of overall costs are to be affected:

- utilization of services
- prices of services
- nature of services

The following examples (with changed company names) show how this approach can be operationalized.

A. The Hudwig Company, as the result of careful analysis of its health care benefit costs, concluded that its major problem was hospital costs. These costs per capita were growing at 23 percent per covered individual per year. Admission rates for about twelve surgical procedures and average length of stay for all hospital admissions were the major sources of the increase. Based on this problem they decided to pursue three strategies:

1. Requirement of mandatory second opinion for the twelve surgical pro-
    cedures found in published studies to be associated with high non-
    concurrence rates plus six others for which their rates of occurrence
    were significantly above the average for their area (Chapter 11).
2. Contracting with a local physician organization for utilization review
    of all inpatient stays (Chapter 12).
3. Active consideration of contracting with an emerging preferred pro-
    vider network of hospitals (Chapter 14).

B. Protex Enterprises, which has a benefit plan providing first dollar coverage, found their major cost problems to be high utilization of physician services, although their cost per claim was in line with other subscribers of the same carrier. A $200 deductible was introduced. At the same time, to mitigate negative employee reaction, a prescription drug benefit, costing about one-third of the anticipated savings, was also introduced (Chapter 8).

C. The cost problems of Omni Services were somewhat more complex. Hospital admission rates were 25 percent above the average for their area, even after adjusting for age and other possibly confounding factors. Admission rates were high for virtually all categories. Length of stay was in line with the regional average, but 20 percent of hospital admissions were to the maternity service. In addition, outpatient emergency services per capita had increased over 100 percent during the past three years. Strategies selected to deal with these problems included:

1. Restructuring the benefit to increase the coinsurance from 20 to 40 percent for those with elective hospital admissions who did not participate in an elective preadmission certification program developed with the local professional review organization (Chapter 12).

2. Increasing the coinsurance from 20 to 40 percent for emergency visits to the hospital, unless a visit led to an inpatient admission.

3. Paying $150 in cash to any woman who left the hospital within twenty-four hours of delivery of her baby.

4. Actively promoting membership in the two local HMOs whose premiums were about 10 percent lower than the self-insured plan administered through a commercial carrier (Chapter 13).

D. Stansens, a regional department store chain, discovered that they had two major problems. First, hospitalization costs were high, primarily due to a long average length of stay. Further investigation revealed that the median length of stay was close to average for the area but that there were many very long stays, which had a large cost impact. A second problem was high inpatient costs for poorly defined mental health problems, which were most likely due to alcohol and drug abuse.

Since employees had foregone a scheduled pay increase due to the company's economic problems three months before, it was decided that benefit cuts would not be made. Instead Stansen's decided to add benefits which could substitute for hospitalization at lower cost and couple these with utilization review to prevent overutilization. Among the changes were:

1. Improvement in skilled nursing facility benefits, which had been limited to fifteen days after hospitalization.

2. Initiation of a hospice benefit, with preadmission certification to the hospice by a team of local oncologists.

3. Experimentation in one area with expanded outpatient mental health benefits and with day treatment benefits for selected alcohol and drug treatment programs.

E. Seconia County, employing 2,560 unionized workers, had a broad health benefit package that had been in place for six years. Per insured costs in virtually all categories were 20 to 40 percent above average for private employers in the county. Analysis of the types of health problems revealed a higher than average incidence of heart disease, stroke, lung cancer and accidents, which also contributed to a very high disability retirement rate.

The union had stated its strong opposition to any reductions in health insurance benefits. In view of these circumstances Seconia decided to:

1.   Strongly support the development of a closed-panel HMO for the area.
2.   Make clear to the union that the desired addition of dental coverage would only be considered if the union would accept some medical benefit changes, including penalties for not receiving a second opinion, utilization review for all hospital admissions and some coinsurance feature.
3.   Enlist the union's cooperation to jointly sponsor a health promotion program for all county employees, covering hypertension control, smoking cessation, nutrition and off-site accident prevention (Chapter 18).

F. Corinth-Multiflex, a large manufacturer with 6,000 employees in and around Clayton, Ohio, found that two of the six local hospitals had a cost per day and cost per admission 27 percent above that of the other four. Analysis of the case mix using diagnostic related groupings (DRGs) revealed no significant differences among hospital case mixes. Surgical admission rates for hysterectomy, cholecystectomy (gallbladder removal) and herniorrhaphy (hernia repair) were almost 50 percent above national and regional averages. Profile analysis showed that five surgeons, two gynecologists and three general surgeons, were associated with most of these operations.

Armed with this information the company resolved to:

1.   Negotiate with the high-cost hospitals to get them to provide a substantial discount for Corinth-Multiflex employees. If these costs could not be brought in line with those of the other hospitals, the company would agree only to pay 60 percent rather than their usual 80 percent share towards those hospital costs and would mount a concerted informational campaign to inform employees of the cost implications of choosing to be hospitalized in those facilities.
2.   Review the information on differential costs with the two company executives who serve on the boards of those hospitals and suggest to them the company's strong interest in moderating future cost increases to bring these hospital costs in line with other area hospitals.
3.   Engage in frank discussions with the five surgeons, reviewing with them what the company analysis revealed, indicating that the company planned to a) carefully monitor rates in the future, and b) institute a mandatory second opinion program for those procedures and as rationale provide the aggregate profile information to employees if the rates did not decline.

Much more complex is the problem for a company with a myriad of different locations, many different benefit plans, several or more carriers and a host of different unions. For these organizations it is important to decide which sites have the greatest problem and which local circumstances are most propitious to effect substantial improvement. The characteristics of the medical services, the interest and sophistication in health benefit issues of senior management at that site, industrial relations considerations and other factors will help to identify priorities. Over time the best ways to assure that the relevant issues can be effectively addressed are to:

1. Charge each operating unit for their health benefit experience.
2. Permit each operating unit to modify the benefits (within defined limits) to respond to local circumstances.
3. Provide each operating manager with the basic analyses of how the benefit was used.
4. Provide training for a senior manager from each operating area to become more knowledgable and skilled in management of the health benefit.
5. Provide technical assistance to those managers in their efforts to make appropriate changes in benefit design, in developing communications programs to employees and their families and in how to deal effectively with providers and carriers.

In all of these cases the assumption has been made that the employer is at risk for the health benefit costs. Given the growth of self-insurance, minimum premium plans and other shared risk arrangements, and the need for a third-party carrier, if at risk, to translate this year's adverse experience into next year's premium increases, the assumption has validity.

As the largest purchaser of health care, employers can make a major difference in the cost and delivery of health care services. However, employers must be willing to become actively involved and to devote the necessary resources to protect their own interests. In the absence of such commitment employers are unfortunately relegated to their former role, contributing more to the problem than the solution.

NOTES

1. Fisher, C.R. "Differences by Age Groups in Health Care Spending," *Health Care Financing Review,* Vol. 1, No. 4, 1980, pp. 65–90.
2. Ibid.
3. Ibid.
4. Unger, W.J. "Challenges in the New Era of Competition," *Issues in Health Care.* Vol. III, No. 1, 1982, pp. 8–14.

SELF-INSURANCE *10*

*T*wenty years ago most companies financed their health plans by allowing their carriers to assume all the risk, handle all the claims and problems with providers and collect the premiums without much argument about requested increases. In a 1980 survey of 100 corporate benefits managers of large employers, the extent to which traditional arrangements had changed was underscored by the finding that only 8 continued to finance and administer their health plans in the traditional way. Although only 5 of these 100 companies were entirely self-insured and self-administered, a variety of different financing arrangements were reported. Seventy-five of the health plans surveyed had claims paid by the insurer or by a service plan under a shared risk or administrative services only arrangement; 15 paid claims themselves; 10 were totally self-insured, primarily through a 501(c) (9) trust. Stop-loss contracts were incorporated into 22 plans. Thirty-two companies had retrospective rating, 20 had minimum premium plans and 20 had delay of premium payment arrangements.[1]

A 1983 survey of a large number of major employers throughout the country reported that, based on responses from 773 companies, only 35 percent of non-HMO plans were fully insured. Ten percent of companies completely administered their self-insured plans, 29 percent utilized the administrative services of a private insurer while assuming the full risk of health benefits and 22 percent protected themselves from a major financial loss due to excessive benefits through a minimum premium plan.[2]

Employee benefit surveys uniformly confirm a swift growth in self-insured medical plans. For example, the 1981 Conference Board Profile of Employee Benefits found that one plan in four was at least partially self-insured, compared with only 7 percent in 1973.[3] And according to the 1980 Employee Benefits in Industry survey by the Department of Labor, whose sample includes a much broader representation of companies than most other surveys, 15 percent of health insurance participants had some portion of their plan (usually major medical benefits) funded through self-insurance.[4]

This welter of new financing and payment options has developed from a growing perception that employers can reduce the rate of health care cost increases by cutting the "gap"—the amount between the employer's costs and aggregate payments to the providers.

A BRIEF HISTORY

Spurred by a desire to eliminate premium taxes (typically in the range of 2 percent), Metropolitan Life Insurance Company and Caterpillar Tractor Company agreed in 1962 that Caterpillar would pay the first 90 percent of expected claims and Metropolitan would pay the rest. In 1966 Metropolitan offered a similar plan whereby the employer would pay premiums only for claims in excess of 90 percent but the insurer would pay all the claims, the first 90 percent of expected dollars from employer funds and the rest from premiums. This permitted employers to save premium taxes on the first 90 percent without having to enter into the claims-paying business. A second milestone agreement was written in 1970 whereby Equitable Life Assurance Society would pay claims and provide other administrative services for the 3M Company's own self-funded noncontributory health plan. This type of arrangement became known as an administrative services only (ASO) contract. A 1980 estimate of the percent of health plans using an ASO contract was 10 to 20 percent.[5]

EFFECTS OF SELF-INSURING

Movement towards self-insuring can impact on each of the various components of retention (the part of premiums not going to claims):

1. **Premium taxes**   Averaging 2 percent, this cost can be a considerable portion of the 6 to 15 percent net retention. In most states there are several ways to avoid or reduce these taxes. One way is to have a minimum premium plan. A second is to set up a 501(c)(9) trust from which health benefits are paid. Blue Cross/Blue Shield plans, operating as nonprofit entities, are exempt from premium taxes.

2. **Risk charges** In the quotation received for any insurance is some amount of money that represents the risk of insurance. It is the amount the insurance company wants to be willing to take the risk of guaranteeing payment even if the premiums don't cover the full cost. Any arrangement which shares the risk reduces this expense to employers. As employers move to minimum premium plans and other shared cost arrangements, their risk increases while their risk-related payment to the carrier decreases.

3. **Reserve requirements** In a health insurance contract, reserves are needed to pay for claims that were incurred during the covered period but for which claims were not submitted until later. In traditional insurance, this was essential to prevent the carrier, who has a legal responsibility to pay these claims, from being stuck with these costs if the contract was not renewed. Reserves can be a very substantial 20 to 40 percent of yearly premiums which, from the employer's point of view, both creates an additional cash drain and robs the employer of the return on those dollars were they put to work in any other manner.

   Ability to eliminate a reserve fund is one of the reasons for self-insurance. Insurers have countered employer concerns over the size of reserves by combining minimum premium plans with return of most of the reserves to the employer. To have this arrangement the employer must accept liability for claims incurred but unreported during the contract period and for any other unpaid claims. This approach, which permits the employer to have use of previously reserved money, is termed "retrospective rating." To minimize the time lag between when the employer pays the insurer and the insurer pays the claims for that period, many insurers have introduced arrangements in which employers delay premium payments by as long as ninety days. When employers do pay reserves, insurers credit a certain amount to the employer's account as compensation for the loss of those funds during the period the reserve was in place. Because so many variables need to be considered in deciding how much can actually be saved by eliminating reserves, there is not even a good rule of thumb for savings. However, savings on the order of 2 to 4 percent of premiums appear possible in cases where the insurance carrier has been crediting the employer based on interest rates well below those achievable in short-term, high-quality debt instruments.

4. **Administrative costs** Competition is very keen for health insurance business, and contracts are won or lost based on a fraction of one percent difference in the administrative overhead involved in processing claims. Over the past decade a number of new organizations have sprouted up which only administer a claims system and don't

accept the financial risk for level of payment. A number of them are quite efficient in claims processing and have invested in automated claims processing systems with high efficiency potential. Benefitting from self-insuring employers' desire for efficient ASO contracts, these plan administrators, also called third-party administrators (TPAs), have been able to secure a significant share of the market, especially for medium-sized employers.

Administrative services only contracts generally cost from 4 to 6 percent of claims, although many of these contracts are negotiated for a flat fee or are based on number of employees or number of claims processed.[6] Among insurance companies, a significant percentage of total dollar volume is in ASO contracts, alone or in combination with minimum premium plans (MPPs). It was reported as early as 1979 that the combination of ASO contracts and MPP business made up more than half of Metropolitan Life's group health business, with about 10 percent of it coming strictly from ASO contracts.[7] In 1975 ASO/MPP contracts accounted for less than 5 percent of total insurance company group coverage. Approximately 30 percent of the 1981 group insurance business of commercial carriers was represented by ASO/MPP arrangements.[8] Insurers catering to smaller companies tend to have a smaller share of these types of arrangements. However, according to one benefits consultant, more and more medium-sized employers are getting into self-insurance or shared risk arrangements.[9]

5. *Profit* The best way for an employer to minimize the profit portion of retention is to invite competitive bids on some periodic basis. Ironically, however, the increased use of a competitive bidding process also tends over time to increase health care insurance costs. Responding to a large number of requests for proposals is costly, as in trying to convince existing accounts that there is no need to entertain competitive bids. In addition, there is a significant cost in assuming an account from another company, as there is in terminating an account. Frequent changes in carrier has been most common among medium-sized and smaller companies.

6. *Brokers commission* Use of self-insurance coupled with third-party administration can significantly reduce the cost of brokerage commissions which are buried in health insurance premiums. Some employers buy health insurance claims administration directly, without use of a broker. However, performing this function requires an investment in internal personnel or outside consultants to help in the selection process and contract negotiations.

SOME MAJOR SELF-INSURANCE ISSUES

1. *Estimating cost savings* Accurately projecting savings from self-insurance or from increased risk sharing can be difficult. How can an employer factor in the possibility that claims experience during a contract period will be worse than expected? If the insurer bore the excess cost, the company might save considerable dollars during that period. And if the carrier tried to recoup the losses by requesting a very large rate increase for the next round, the employer can rebid the contract and possibly find a hungry competitor willing to give a significantly lower rate.

    Reducing the reserve usually provides a net savings. But how much of a savings depends on the difference between what the employer could get investing the money elsewhere and what the insurer would credit to its account for having the use of those dollars over the contract period. This difference can vary considerably over time and is sensitive to interest rate fluctuations and the marginal return on investment that the employer can attain. In some cases the employer might be better off "investing" the money with the insurer.

2. *Deciding what plan size warrants self-insurance* Some argue that any employer with over 100 employees should seriously consider self-insurance. Others feel that below 1,000 employees self-insurance is generally a poor option. Again, the issue is risk. A smaller organization has a much higher chance that its experience will be more than any specified percentage above predicted levels. For example, a consulting actuarial firm suggests that the probability that annual claims will be 120 percent or more of predicted claims levels is 25 percent for a plan size of 100 to 250 lives but only 6 percent if the plan has 2,500 or more lives (Table 10.1).[10] A plan covering 100 to 250 lives can expect considerable swings in cost experience from year to year, in some years exceeding 200 percent of predicted levels. For some companies this degree of variation could create cash flow difficulties and/or have a major impact on the bottom line.

3. *Limiting yearly risk* To prevent wide fluctuations in costs from year to year, employers who self-insure for health increasingly buy "stop-loss" insurance, which limits the top margin for claims fluctuation. Under stop-loss the insurance company agrees to pay claims which exceed a specified percentage of aggregate predicted claims, such as 125 percent. The ceiling can alternatively be applied to an individual policy and, for example, pick up the excess over $25,000 per year for any claimant. Some carriers provide stop-loss contracts in connection

TABLE 10.1
SIZE OF COVERED GROUP AND SELF-INSURANCE RISK

Number of covered lives	Probability that annual claims will be at least 120 percent of predicted levels
100–250	25%
250–500	23
500–1,000	22
1,000–2,500	12
2,500 or more	6

Source: *Employee Benefit Plan Review.* Vol. 34, No. 12, 1980, p. 24.

with ASO contracts. Others sell stop-loss only in conjunction with life insurance. Safeco Life Insurance Company only will sell stop-loss coverage if the plan is administered by a third-party administrator acceptable to Safeco.[11]

4. *Choosing a plan administrator* For a self-insured company wishing to select a plan administrator, cost should not be the only criterion. More important should be systematically going through a careful checklist, both independently and by asking these questions of former and current clients:

   a. What is the reputation of the organization? Has it had considerable experience? What do companies who have been using it for several years say about it?

   b. If a TPA, is it large enough not to be dependent on one or two individuals for efficient operation?

   c. How good is its system, both in design and execution, for containing costs in the claims payment process?

   d. What historically have been its savings from coordination of benefits? The differences between efficient COB and less efficient COB can be several percent per year on total costs, much more than the likely difference in price among ASO bidders.

   e. How effective is its claims processing system in identifying inappropriate procedures (e.g., a hysterectomy in a male, a dilatation and curettage in an eight-year-old female), overcharges, uncovered services? What percentage of claims are kicked out of the automated claims processing system and require individual review?

   f. Does it have an audit from a well-recognized impartial auditor showing a very low error rate, especially in overpayments?

   g. Who performs this review, based on what guidelines, and with what results in the past? Can the claims processing and payment system be easily programmed to provide for new and innovative approaches, such as paying a higher percentage of the total when the claimant has received a second opinion?

h. Has it interfaced well with organizations doing utilization review of hospitalization?

i. What kind of information can the plan administrator provide? Is there the flexibility to provide virtually any report on costs, utilization and their linkage? Does the administrator routinely code diagnoses and procedures? Can these be provided so that you can track the specific health problems accounting for the majority of your costs?

j. How timely are the reports? Are the reports of sufficient accuracy and detail that different sites and periods of time can be readily compared?

k. Is there an efficient system for answering inquiries by the employer and by insured?

Some experts feel that only plan administrators that can sell stop-loss coverage should be considered and that the cost for such coverage and convenience of having the same source for both should be important factors in the decision. However, some excellent administrators may not offer such coverage. Tying the insurance and administration together risks making a good fiscal decision but a poor systems decision, which could in fact cost more in the long run. Increasingly carriers are willing to unbundle their services. It is possible, for example, to buy claims handling and administration, loss-control engineering, data processing and actuarial services either alone or in combination from insurers but without insurance policies.

A most interesting development has been the large brokerage houses going into competition with the carriers and independent plan administrators in selling claims processing, actuarial and other risk management services. Among those brokerage firms who have established subsidiaries for this purpose are Marsh and McLennan Inc., Alexander and Alexander, Frank B. Hall and Co. and Fred S. James.

5. *Self-administration* A number of larger companies have decided to take on the full burden of administering their health insurance plan. They point to the increased control of having the function in-house and feel that it provides the best method to get a handle on how to minimize costs. Some indicate that it has saved them substantial dollars compared with outside administration.

A frequently cited example of a self-administered plan is that of Deere and Company. For more than a decade Deere and Company has been self-insuring and self-administering its employee health benefits. The primary motivators of Deere's decision to self-insure were financial considerations, including a desire to avoid premium taxes and the ability to earn interest income on reserves maintained on its own books. The corporation chose to self-administer its plan rather than contracting with an insurer for ASO because:

a) It was felt that more coordinated claims management would result from a claims function centralized in one office (as opposed to an insurance company's decentralized claims management).

b) Insurance company overhead charges could be avoided.

c) Internally administered claims would enhance the corporation's ability to work with local providers on health care problems. (Deere is the largest employer in areas where it has its largest plants.)

Self-administration of Deere's plan has been facilitated by the geographic concentration of employees and the company's offering one consistent benefit package to all beneficiaries, simplifying claims administration and management. According to one company representative, Deere has processed more than 18,000 claims per week and total administrative costs have not exceeded 3 percent of claims paid. In addition to significant savings from avoidance of premium taxes and interest credits on reserves, advantages of self-administration cited by Deere include the ability to pay claims in a consistent and timely fashion and achievement of much tighter coordination of benefits.[12]

Before embarking on a course of self-administered medical benefits, a number of caveats are in order. Becoming self-administered is like starting a new business, with all the attendant problems. Whether entering this new line of business makes sense in terms of the other opportunities within an organization is usually a difficult question. Although improved minicomputer-based claims processing data systems are available, they still require a significant initial investment both for purchase and initial programming and data entry. Without experience in this field it is unlikely that those running the system will know which claims should be questioned. In addition, getting into the claims paying business forces the employer to confront directly providers without the insulation of a buffer organization such as an insurance company or plan administrator. It also means employees with problems regarding their payments will be coming to someone within the organization rather than to a third party. To the extent that there are problems, the company bears the brunt of them.

It is also important to look at the true costs of self-administration compared to outside contracting. In some instances analysis has not included indirect costs of bringing the operation inside such as prorated facility and related costs, extra costs to legal and other departments and more senior managerial time from within the human resources and finance departments. Can the company do its own actuarial experience studies, or are outside consultants needed? Who will prepare

liability estimates? What about the preparation of contracts and amendments? Who will audit the claims payment? The answers to these
questions have discouraged some employers who were initially enthusiastic about self-administration.

6.   *Legal considerations*   Prior to the passage of the Revenue Act of 1978
which added Section 105(h) to the Internal Revenue Code, primarily
through self-insurance employers could provide certain employees with
medical reimbursement for items not covered under insured health
plans. This reimbursement was tax deductible for employers as a business expense and not considered a part of the employee's gross income.
Beginning January 1, 1980 Section 105(h) of the IRC prohibited employees from excluding from income medical reimbursement received
through a self-insured plan* if the plan discriminated in favor of "highly
compensated" individuals** with respect to plan eligibility or benefits.[13]

Under the eligibility requirements of the act a plan will not be
considered discriminatory if it benefits at least 70 percent of all employees or at least 80 percent of all eligible employees if at least 70
percent of all employees are eligible for benefits. A plan will also be
considered nondiscriminatory if it benefits a classification of employees
not among the "highly compensated."[14] For purposes of eligibility, the
following categories of employees may be excluded from consideration:
a)   employees with less than three years of service
b)   employees under age 25 at the initiation of the plan year
c)   part-time or seasonal employees
d)   union members covered by a collective bargaining agreement ne-
     gotiated in good faith
e)   nonresident alien employees who do not receive income from the
     employer in the United States[15,16]

In order to meet the benefits test of the act, the same type and
amount of benefits must be available to all plan participants irrespective
of the participant's age or length of employment. Benefits available to
dependents of highly compensated employees must be available to
dependents of all plan participants. Benefits available cannot be established at a level proportional to compensation, since this would
discriminate against less compensated employees.[17]

If a medical plan is discriminatory, any excess reimbursement to
the highly paid employee must be considered part of his or her gross

---

*A self-insured plan is considered to be any portion of a medical plan for which the risk is not
transferred to a third party.
**A highly compensated individual is defined as: 1) one of the five highest paid officers; 2) a
shareholder owning more than 10 percent of stock; or 3) one of the highest paid 25 percent of
all employees.

income. However, there are certain benefits which may be excluded from the discrimination rules. Diagnostic procedures including routine physical examinations, blood tests and x-rays can be provided to certain employees without regard to the regulations. The procedures, however, must be offered at a facility which only provides medical services. Reimbursement for the testing, treatment or cure of a known condition is considered part of the regular medical plan and therefore may not be provided selectively to key employees.[18]

THE FUTURE

Most large companies have already moved in the direction of self-insurance, and smaller organizations are starting to follow. The number of possible arrangements currently available is so large that a smaller company is likely to rely on a broker/agent to help make the decision. Unfortunately, even for larger employers, answering some of the tougher questions about cost-effectiveness is very time consuming and may require expertise not available internally. Despite the growing complexity and insecurity many employers feel in moving away from more traditional insurance relationships, cost savings from reducing carrier risk-sharing are achievable. Often companies benefit from using outside experts to help them choose among a number of confusing financial and administrative alternatives.

NOTES

1. "Economics and Flexibility Persuade Companies to Change Funding Methods for Their Health Plans," *Employee Benefit Plan Review*, Vol. 34, No. 12, 1980, pp. 18–19.
2. Hay Associates. 1983 Noncash Compensation Comparison.
3. The Conference Board. *Profile of Employee Benefits: 1981 Edition.*
4. U.S. Department of Labor. Bureau of Labor Statistics. *Employee Benefits in Industry, 1980.*
5. "Who Takes the Risk? A Variety of Methods Are Available for Financing the Health Care Plan," *Employee Benefit Plan Review*, Vol. 34, No. 12, 1980, pp. 12–14.
6. "Self-insurance Spurt Slashes Cost of ASO Contracts: Poll," *Business Insurance.* December 10, 1979, p. 40.
7. Ibid.
8. Health Insurance Association of America. *Source Book of Health Insurance Data 1982–1983.* Washington, D.C., 1983.
9. Op. cit. (*Business Insurance*, 12/10/79).
10. Determining Present and Future Health Claim Costs," *Employee Benefit Plan Review*, Vol. 34, No. 6, 1980, pp. 20, 24, 116.
11. Op. cit. (*Employee Benefit Plan Review*, Vol. 34, No. 12, pp. 12–14).
12. Stokehold, K. "Deere and Company," in Egdahl, R.H. and Walsh, D.C.

(Editors), *Containing Health Benefit Cost: The Self-Insurance Option.* New York: Springer-Verlag, 1979.

13. Arbulu-C., A.V. "Self-insured Medical Reimbursement Plans Revisited," *Taxes*, Vol. 59, No. 7, 1981, pp. 482–486.
14. Ibid.
15. Ibid.
16. "Self-insured Medical Expense Reimbursement Plans," *Federal Taxation*, Vol. 1, No. 3, 1980, pp. 47–48.
17. Op. cit. (Arbulu-C.).
18. Op. cit. (Arbulu-C.).

11 SECOND OPINION SURGERY PROGRAMS

THE IMPACT OF UNNECESSARY SURGERY

Changing financing and administrative arrangements can save money, but about 95 percent of employer health insurance costs goes for health services. A number of cost management approaches have been developed to reduce the number of specific services. One of the biggest cost categories is for surgery.

About 45 percent of all hospital stays involves significant surgery. Every year about one in ten Americans has an operation. And in 1978 the cost of surgery and accompanying hospitalization was estimated at $65 billion.[1] Based on an average annual increase of 15 percent, these costs rose to $150 billion in 1984. These statistics alone should be sufficient to direct considerable attention to possible overutilization of surgical techniques. However, more grist for the mill comes from statistics displaying a clear rise in the rate of elective surgical procedures. From 1971 to 1978 the rate of elective surgical procedures increased 24.3 percent, from 7,805 to 9,703 per 100,000 population.

In general, there were greater increases in surgical utilization among females, in part due to increased gynecological surgery, including sterilization. Different surgical procedures, however, grew at different rates. During this period when the U.S. population grew about 5 percent, the number of appendectomies, generally considered a good index of *emergency* surgical procedures, declined 6 percent. Most *elective* surgeries, however, increased much more rapidly than the population. Cholecystectomies (gallbladder removals) rose 15.8 percent, hysterectomies 13.0 percent, back surgery 21.1 percent, cataract removal 46.9 percent and knee surgery 70.3 percent. The only two common elective surgical procedures that rose more slowly than the population increase were hernia repair, rising only 1.6 percent, and tonsillectomy and/or adenoidectomy, which fell 43.4 percent in the wake of growing media reports that most of these procedures were unnecessary and could possibly endanger the ability of the body to withstand infections in the future.[2]

Some of the recent increases in surgery are due to the aging of the population since older people have higher surgery rates. A study using data from the National Health Interview Survey estimated that when adjusted for age and sex, the elective operation rate declined 3.5 percent and the rate of all surgery increased only about 0.5 percent.[3] Nonetheless, small changes in overall rates tend to obscure the significant increases in rates for some elective procedures.

Contrary to what common sense might suggest, there is great regional variation in rates for specific surgical procedures. A study conducted by the American College of Surgeons and the American Surgical Association reported that overall surgical rates varied by geographic areas from a low of 58.5 procedures per 1,000 population annually to a high of 91.5.[4] Even within the same state, rates differ considerably. Within Maine, for example, cholecystectomy rates ranged from 27 to 55 per 100,000 population per year. In nearby Vermont, hysterectomy rates ranged from 20 to 60 per 100,000 population. Similar but independent studies in Kansas, New Hampshire and Wisconsin indicated similar patterns of variation.[5] These differences exist in the absence of clear differences among the populations or mortality rates. Discrepancies in rates are particularly pronounced with respect to newer procedures. For example, cardiac bypass surgery rates in 1978 for males varied from 46.1 per 100,000 in the Northeast to 117.5 per 100,000 in the North Central region (Table 11.1).[6]

Rapidly rising surgical rates for new procedures, as typified by the rate of increase in cardiac bypass procedures in Table 11.1 between 1971 and 1978, are another cause for concern. Most of these procedures occurred before there were any carefully controlled studies of the efficacy of this type of surgery. By 1981 sufficient evaluation had been done to suggest a clear benefit for one or two types of specific patterns of coronary artery blockage. However, the majority of the annual 100,000-plus operations of this type were probably being performed on individuals whose problems did not fit into these categories. Many other procedures have become widespread before any study has been done either on their effectiveness or, equally im-

TABLE 11.1
CARDIAC BYPASS RATES (per 100,000 population)

Region	Males			Females		
	1971	1978	Change (%)	1971	1978	Change (%)
Northeast	9.6	46.1	+380.2	2.2	8.6	+290.9
North Central	33.2	117.5	+253.9	5.4	20.0	+270.4
South	15.0	65.8	+338.7	4.3	11.7	+172.1
West	24.9	76.7	+208.0	1.3	11.9	+815.4
Total U.S.	20.6	77.4	+275.7	3.6	13.2	+266.7

Source: National Center for Health Statistics, Hospital Discharge Surgery, reprinted by permission of the publisher from Eugene G. McCarthy, Madelon L. Finkel and Hirsh S. Ruchlin, *Second Opinion Elective Surgery* (Boston: Auburn House Publishing Company, 1981).

portant, on their complications. An excellent example of this is gastric wrapping, whereby an inert plastic polymer sheet is wrapped around the entire stomach and permanently left there to reduce its functional size. This operation is being performed as a cure for obesity. It can, however, lead to serious complications, such as infections and loss of blood supply to the stomach or other organs, and in some cases the result can be death. Growth in popularity of this procedure has preceded any careful, controlled studies.

Disturbingly, the rate of surgery appears to correlate in significant measure with the concentration of surgeons and with the availability of specialized facilities for some types of procedures. It appears to be no coincidence that the United States has both the highest surgery rate in the world and the highest concentration of surgeons. As of 1980 the United States had a surplus of surgeons that some experts estimate to be 30,000 to 40,000.[7] Projections released in the Summary Report of the Graduate Medical Education National Advisory Committee (GMENAC) indicate that by 1990 there will be 50 percent more general surgeons than estimated to be required, as well as substantial surpluses in almost every major surgical specialty.[8] And the surplus will probably continue to grow until training programs in surgery are reduced and/or surgeons finishing training are unable to find work. As the concentration of surgeons continues to increase, more surgeons will not have a sufficient volume of patients requiring surgery to maintain their skills. Although the correlation between concentration of surgeons and rates of elective surgery augurs for continuing increases in surgical rates in the foreseeable future, it is unlikely that elasticity of demand for surgical services will grow as rapidly as the number of surgeons.

There exist many justifiable reasons for concern about the appropriateness and necessity of many operations currently being performed. A 1976 report by the Congressional Subcommittee on Oversight and Investigations of the Committee on Interstate and Foreign Commerce estimated that 2.4 million unnecessary operations were performed in 1974, at an estimated cost of $3.9 billion. It further estimated that 11,900 deaths could be expected from those unnecessary operations based on an average surgical mortality of 0.5 percent.[9] Estimates in that congressional report and a similar report reiterating its conclusions in 1978 were based on a nationwide extrapolation of the findings of a 17 percent nonconfirmation rate for elective operations reviewed in a second surgical opinion program of a New York City union.[10] Looking at two operations thought to be overperformed, one researcher estimated that at least one-third of the hysterectomies and two-thirds of the tonsillectomies were unnecessary.[11] Yet, despite many claims of excess surgery, developing reasonable and widely accepted criteria for what surgery is necessary has been a frustrating task. Concensus appears easier to achieve in some surgical subspecialties than in others and for some common procedures than others.

A BRIEF HISTORY

Because of these problems, second opinion programs, whose recent history goes back only to 1972, have considerable intuitive appeal. They are a response to rapidly rising health care costs without threatening the elimination of necessary services or requiring the payer of insurance premiums (the employer, the government or, more rarely, an individual) to pay a greater proportion of the costs of care. Second opinion programs are likewise a response to concerns about the quality of care. Operations carry an appreciable risk, both related to the procedure and to general anesthesia, the latter thought to be responsible for two deaths per 10,000 operations.[12] In addition, second opinion programs empower the consumer to take a more active, informed part in making important decisions about the care he or she receives, and to weigh more intelligently the potential benefits against the risks and monetary costs.

In 1972 the Storeworkers Union in New York asked Cornell University to establish a mandatory second opinion program for elective surgery. The same year a companion voluntary program was initiated for some New York City municipal employees through their union. Blue Cross and Blue Shield sponsored some pilot programs in 1976, and a major federal initiative followed in 1977. Of all the current efforts to impact positively the cost and quality of health care by altering some aspect of the delivery system, second opinion programs are the best studied.

Most complete and longest term results emanate from the original New York programs. In these, as in many that have been spawned by these models, not only a second but also a third opinion is paid as a regular benefit. The final decision regarding surgery rests with the employee, and if the patient decides to have the surgery, the consultant surgeon cannot perform it. These programs utilize a "closed" panel of surgical specialists whose resumes are reviewed by the Cornell group before being placed on the consultant panel. The panel considers such information as specialty board certification, hospital affiliation and training.

During an eight-year study of the two Cornell programs, approximately 12,000 individuals, slightly more than one-half of whom were in the mandatory program, received a second opinion. In the voluntary group, 33.4 percent of consultations led to nonconfirmation of the need for the proposed surgery, compared with 18.7 percent in the mandatory group. Nonconfirmation rates for various diagnostic categories in each of the two populations is presented in Table 11.2. Particularly high nonconfirmation rates were seen for knee surgery, gynecological procedures and prostatectomy. However, in the voluntary program group, all the selected categories of surgery had nonconfirmation rates of over 10 percent, and only two categories in the mandatory group were below this level.[13]

Upon questioning six months after receiving a second opinion, the

TABLE 11.2
SELECTED DIAGNOSTIC CATEGORIES OF INDIVIDUALS NOT CONFIRMED FOR SURGERY, BY VOLUNTARY AND MANDATORY GROUPS, 1972–1980

Selected diagnostic category	Voluntary		Mandatory	
	Total	Not confirmed	Total	Not confirmed
Bunionectomy	47	22 46.8%	147	48 32.7%
Cholecystectomy	215	25 11.6	270	22 8.1
Dilatation and curettage	144	45 31.3	465	82 17.6
Deviated septum surgery	94	32 34.0	185	47 25.4
Hernia repair	302	44 14.6	599	35 5.8
Hysterectomy	545	225 41.3	397	122 30.7
Knee surgery	190	97 51.1	124	29 23.4
Mastectomy and other operations of breast	299	84 28.1	353	61 17.3
Prostatectomy	161	66 41.0	90	26 28.9
Removal of cataract	209	63 30.1	270	41 15.2
Tonsillectomy and/or adenoidectomy	117	35 29.9	315	32 10.2
Varicose vein excision and ligation	104	34 32.7	215	28 13.0
Total	2,427	772 31.8%	3,430	573 16.7%

Source: Reprinted by permission of the publisher from Eugene G. McCarthy, Madelon L. Finkel and Hirsch S. Ruchlin, *Second Opinion Elective Surgery* (Boston: Auburn House Publishing Company, 1981).

majority of individuals had followed the advice of the consultant. In the voluntary group, 70 percent of those confirmed for an operation had proceeded with it, compared with only 17 percent of those who had received a negative consultant recommendation. In the mandatory group, 86 percent of those with confirmatory opinions had the surgery versus 32 percent of the nonconfirmed group. An important observation is that a substantial number who had received two opinions that surgery was necessary decided not to have it within six months. Follow-ups at one and two years after the second opinion was given showed about the same results. If a person had not proceeded to surgery within ninety days of the second opinion, it was unlikely that he or she would have the operation. Perhaps less surprising are the persons who went ahead with the operation in the face of a negative second opinion.

Rates of no surgery in individuals confirmed and nonconfirmed for surgery are presented in Table 11.3 for selected diagnostic categories. That almost one-third of the mandatory second opinion group proceeded despite negative advice from the consultant may reflect their negative reaction to the requirement that they get a second opinion and their personal assessment of its value. On the other hand, it may reflect their confidence in the original physician or surgeon who recommended the operation. (These individuals would not have been likely to seek a second opinion in a voluntary program.) Reasons most frequently cited by patients not undergoing surgery despite the need being confirmed by a panel consultant were: the patient's judgment that the surgery could be postponed (39.5 percent); the advice of another doctor was followed (17.0 percent); the condition was tolerable (12.6 percent); and the fear of surgery (10.1%).[14]

Computation of the savings from the second opinion surgery program provided for the largest mandatory group in the New York study involved adding the savings of health resources and the savings from avoided work loss. These savings were then compared to the program expenditures. Program costs were estimated to be $203,000, including $78,700 for administration, $101,520 for consultation and testing fees and $23,080 in out-of-pocket costs to patients to obtain the consultation. With savings of $534,791, slightly over two-thirds of which derived from health resource savings, every dollar of investment in the program returned $2.63.[15]

From a significant number of studies of voluntary second opinion programs in a number of states, nonconfirmation rates are usually 25 to 35 percent. Results from the smaller number of mandatory programs show greater variation, from 5 to 7 percent in a Wisconsin program, to 19 percent in the Cornell program. With reasonable consistency, the highest nonconfirmation rates are found for orthopedic, gynecologic, podiatric and ophthalmic procedures.

Nonconfirmation rates may underestimate the savings, especially for mandatory programs, as illustrated by a mandatory program for Medicaid

TABLE 11.3
RATES OF CONFIRMATORY AND NONCONFIRMATORY SECOND OPINIONS FOR SELECTED PROCEDURES

	First follow-up data: individuals not confirmed for surgery who had not had surgery performed, by selected diagnostic category, 1972–1980						First follow-up data: individuals confirmed for surgery who had not had surgery performed, by selected diagnostic category, 1972–1980					
	Voluntary			Mandatory			Voluntary			Mandatory		
Selected diagnostic category	Total not confirmed	Total NCNS*	Percentage	Total not confirmed	Total NCNS*	Percentage	Total confirmed	Total CNS**	Percentage	Total confirmed	Total CNS**	Percentage
Breast surgery	71	61	85.9	49	40	81.6	97	19	19.6	55	9	16.4
Cataract surgery	53	43	81.1	34	22	64.7	69	23	33.3	45	4	8.9
Foot surgery (excluding bunionectomies)	74	69	93.2	80	56	70.0	40	18	45.0	15	4	26.7
Hernia repair	39	33	84.6	30	20	66.7	113	24	21.2	127	15	11.8
Hysterectomy	183	139	76.0	95	45	47.4	125	26	20.8	58	7	12.1
Knee surgery	86	71	82.6	24	22	91.7	37	13	35.1	21	0	0.0
Prostate surgery	59	47	79.7	26	16	61.5	48	21	43.8	12	2	16.7

*NCNS: not confirmed, no surgery.

**CNS: confirmed, no surgery.

Source: Reprinted by permission of the publisher from Eugene G. McCarthy, Madelon L. Finkel and Hirsch S. Ruchlin, *Second Opinion Elective Surgery* (Boston: Auburn House Publishing Company, 1981).

recipients in Massachusetts. Initiated by the state legislature in 1976, the program required a second opinion for eight elective operations: tonsillectomy and/or adenoidectomy, hysterectomy, hemorrhoidectomy, cholecystectomy, meniscectomy (removal of cartilage in knee), nasal surgery, excision of varicose veins and disc surgery/spinal fusion. Reports covering the initial period of program implementation indicated a nonconfirmation rate of 14.5 percent, with small variations among different regions of the state.[16,17] Of greater interest, however, was the finding of an important "sentinel effect" whereby the mere existence of a mandatory second opinion program appeared to reduce initial recommendations for surgery, presumably surgery of questionable value. Comparing changes in surgical rates for Medicaid recipients in Massachusetts and in neighboring Connecticut, it was felt that the sentinel effect had reduced the incidence of the targeted elective surgical procedures by 20 percent. Looking at the total number of surgeries avoided due to the program, 71 percent were attributed to the sentinel effect and 29 percent to decisions made as a result of patients receiving a nonconfirmation.[18] When the estimated sentinel effects and savings from deferred operations were combined, the cost:benefit ratio improved to 1:3.9.[19]

GROWTH OF PROGRAMS

Third-party carriers have rapidly expanded their offerings of second surgical opinion programs. By 1980, fifty-three of sixty-eight Blue Cross and Blue Shield plans offered this as a benefit, twenty-five plans had referral centers and fifteen limited their consultants to board certified surgeons. All the programs were voluntary with the exception of an experimental mandatory program offered by Weirton Steel in Pennsylvania. One voluntary program, available to 350,000 U.S. Steel employees starting in 1981, was limited to eight procedures: hysterectomy, tonsillectomy and/or adenoidectomy, cataract extraction, spinal and vertebral surgery, prostatectomy, cholecystectomy, mastectomy and excision of breast cysts, tumors or other lesions of the breast.[20]

Many commercial carriers also now offer second opinion program coverage to their insured accounts and suggest it to their ASO (administrative services only) accounts. To date no systematic assessment of availability of this feature or its results has been reported. Increasingly second opinion has been coupled with plan design changes which introduce economic incentives for its use. For example under a plan introduced by Prudential, if the subscriber does not obtain a second opinion or if the original recommendation for operation is not confirmed by the consulting physician, the patient's surgical benefit for that surgery is cut to 80 percent of the scheduled benefit or half the reasonable and customary benefit. Subscribers are offered a 6 percent premium deduction if they choose this option.[21] Many other carriers have similar plans. While these plans are not mandatory in the sense that those covered receive some payments in the absence of second opinions, the reduction in benefits is usually sufficient to encourage very high rates of participation.

Surveys of large employers confirm the growth of second opinion programs. From 1979 to 1983 the percentage of the largest U.S. employers reporting to a survey that they made second opinions available to employees increased from 52 to 73 percent. Over the same period those firms reported to be utilizing a mandatory program increased from none to 3.6 percent.[22] In 1981 almost one-half of employers (47 percent) reported actively encouraging employees to use the program, up from 22 percent in 1979. The importance of active promotion of the benefit is underscored by the uniformly low participation experience in voluntary programs, that is programs without economic incentives for use. Low participation rates in the voluntary programs probably account for the estimated savings of 0.7 percent reported by employers responding to the survey in 1981.[23]

Eighty-four percent of 213 employers (all among the Fortune 1,000) responding to a survey conducted in late 1981 by Towers, Perrin, Foster and Crosby reported providing coverage for surgical second opinion as a health care cost-containment measure. However, only two percent of companies with programs required second opinion be obtained from among a company-approved group of physicians.[24]

Nonetheless, there are many individual companies taking the initiative with regard to surgical second opinion, and some report substantial savings as a result. For example, Connecticut General began a pilot second opinion program in 1977 for its own employees and dependents. In 1980 this program was offered to selected markets and large customers. Reported 1980 savings were $174,000 and 237 disability weeks for 162 patients using the program. Of all those seeking a second consultation, regardless of confirmation, 43 percent did not have the originally suggested operation.[25]

FMC Corporation initiated an innovative incentive second opinion program in 1982 under which the plan will pay the full reasonable and customary fee of the second physician, as well as any diagnostic tests required by the second opinion, for the following procedures:

- bunionectomy
- cataract removal
- cholecystectomy (removal of gallbladder)
- coronary bypass
- dilatation and curettage (D&C)
- hemorrhoidectomy
- hysterectomy
- knee surgery
- laminectomy (removal of part of vertebra)

- ligation and stripping of varicose veins
- mastectomy and other breast surgery
- prostatectomy
- submucous resection (repair of deviated septum)
- tonsillectomy and/or adenoidectomy

Employees may select any board certified specialist for the second opinion as long as the specialist is not financially or otherwise professionally connected to the physician initially recommending surgery. If the two opinions differ, the plan will fully pay for a third opinion. Failure to obtain a second opinion results in the plan paying only 50 percent of full plan benefits related to the surgery.[26]

Consumer Education

Convinced of the merit of second opinion programs, the federal government initiated a national consumer education effort and two mechanisms to facilitate referrals of interested individuals to participating physicians. A national hotline, which as of mid-1984 was 800–638–6833, refers callers to 294 local referral centers which include Peer Review Organizations (see Chapter 12), Medicare carriers (primarily Blue Shield plans), local medical societies and federal regional offices. All centers maintain lists of physicians willing to give second opinions. Calls per month to the hotline in 1984 averaged about 1,200.[27] A media campaign has also been launched to educate the American consumer about the benefits of second opinion. In 1980 press kits and public service announcements were distributed to 700 television stations and 6,000 radio stations, and most of the calls to the hotline have been the result of television coverage, public service announcements and newspaper and magazine reports. However, the impact of the program on the number of surgeries being performed or second opinions obtained is unknown.[28]

Why Voluntary Programs Frequently Fail

A major problem with voluntary programs (those without incentives for use) is poor participation. A number of employers who have initiated these programs have found that only 1 to 3 percent of eligible beneficiaries avail themselves of it. As a result, the overall impact on utilization and costs of health care services under insurance is minimal. When the number of patients using a program is low, the administrative costs per second opinion can be very high. In a Medicare demonstration project in Michigan, for example, the average cost per consultation for the first two years, including start-up costs, was $733.[29]

One of the prime reasons for low usage of second opinion programs is patient confidence in the primary physician. A survey by Pennsylvania Blue

Shield found that of patients who chose not to seek a second opinion, 76 percent did so because they had great faith in the first physician or referring physician. Many patients, however, decide not to seek a second opinion out of concern that their initial physician will not like it. In a Michigan Medicare second opinion demonstration project, 43 percent of patients using the program asked that their physician not be informed that they had requested a second opinion.[30] One of the arguments for building incentives into a second opinion program is that it removes the patient from the awkward position of requesting a second opinion based solely on personal interest.

Reinforcement of the importance of confidentiality comes from a failed program undertaken by the Cumberland County Medical Society in Portland, Maine. In 1978 the society's surgeons offered free second opinions to patients recommended for elective surgery. Part of the program was discussion between the referring and the consulting surgeons. Only 41 of 8,500 surgical patients requested a second opinion in the initial period, and utilization dwindled to almost zero.[31] Patient concern regarding lack of confidentiality may have contributed to this low use. Another problem with this type of approach is that it may be much more difficult for a peer to recommend against the surgery suggested by his colleague after they have discussed the proposed surgery.

DESIGNING A WORKABLE PROGRAM

What's the best strategy for a second opinion program? While experience is still rather limited, it appears that a program without economic incentives is hard to justify if the major criteria is a favorable cost-to-savings ratio. However, from a point of view of quality, any unnecessary operation prevented is a significant benefit. An elective program might have higher utilization if a strong continuing promotion was undertaken to make employees feel much more comfortable about elective surgery *if* they had a confirmatory second opinion.

Programs with economic incentives appear to be quite effective, although they can engender some employee dissatisfaction. The sentinel effect can be even more important in reducing surgery than nonconfirmations leading to a decision not to have the surgery. The greatest payoff in all programs comes from covering only those procedures associated with a reasonably high percentage of nonconfirmation. A good starting point might be those with high rates of nonconfirmation from the Cornell studies or the FMC corporation list, modified over time on the basis of local experience and adding procedures with great rate increases. Local rates can be compared to regional and national norms to determine whether they are out of line with what would be expected. A major advantage of an incentive program is that it removes the onus from employees, who can tell their doctor that a second opinion is required to obtain maximum reimbursement under the company health insurance plan.

Selecting a Physician Panel

Although many programs permit any physician board certified in the specialty area in question to produce a second opinion, a better result can probably be achieved by limiting those giving second opinions to a defined panel. However, choosing a panel is a touchy problem. In many programs with defined panels only surgeons are eligible to participate. In others medical specialists are part of the panel. Although some of the defined panels are limited to carefully chosen physicians respected by their peers, most others are broad, such as any member of a local foundation for medical care are open to all physicians who are board certified in a relevant specialty. Best results can be anticipated when the panel is carefully chosen as opposed to opening membership to all those who meet minimum criteria. In some locales, however, this may not be a feasible alternative and reliance must be placed on a medical society or area peer review organization that is reluctant to play a strong gatekeeper function. In a few instances small proprietary organizations have been established to sell the second opinion services of a well-screened, usually small panel. A national panel has been developed by Dr. Eugene McCarthy and a national second opinion service is being marketed to employers.

From the employee perspective, having a larger panel is desirable since panel members are almost always close by and readily accessible. A program in a smaller community is more difficult to make work since the physicians may all know each other, be on the staff of the same hospital and be reluctant to contradict the opinion of the first surgeon. It should come as no surprise that nonconfirmation rates are lower when there has been discussion between the first and second physician.[32]

While most panels are exclusively or primarily composed of surgeons, in many cases the best second opinion can probably be rendered by a nonsurgical specialist. A surgeon has been educated to believe in the value of surgery. Given this, it is surprising that surgeons render nonconfirmatory opinions as often as they do. Nonetheless, the alternative to surgery is some form of medical treatment, often managed by a medical specialist. For example, the alternative to surgical treatment for an ulcer is medical therapy. A gastroenterologist is thus frequently in a better position to provide an alternative viewpoint than another general surgeon. In similar manner, it seems most appropriate that a cardiologist rather than a cardiovascular surgeon provide a second opinion regarding bypass surgery. In support of this approach is the experience of HMOs, which have been able to reduce hospital utilization in part by having the primary care physician control referrals to surgical subspecialists. A panel of medical specialists should be chosen with the same care as surgical specialists.

THE FUTURE

Second opinion programs can serve the dual objectives of quality assurance and cost containment. If properly presented they can aid employees to take a more participatory role in their own health care. But although elective programs are currently available to the majority of employees, economic incentives are needed to increase utilization. Active promotion of second opinions as the best way to avoid having an unnecessary operation should be undertaken on a continuing basis. Nonconfirmation rates will tend to be higher if scheduling of consultations is done by program staff, if discussion between the first and second physicians is not encountered, if confidentiality is assured and if the panel of consultants is carefully chosen and includes medical specialists. Acceptable cost/benefit ratios require economic incentives for use, although avoidance of unnecessary surgery may be considered sufficiently worthwhile to justify a program with no incentives.

Second opinion programs with economic incentives are rational responses to financial incentives and professional orientations that encourage surgery. It is best to avoid the term "mandatory," and to offer elective second opinion programs where there are reimbursement incentives for having a second opinion (e.g., 100 percent coverage instead of 80 percent if the second opinion confirms the need for surgery, or elimination of the deductible, etc.) and/or disincentives for not obtaining a second opinion or going through with surgery after a nonconfirmatory second opinion. Disincentives can be a larger deductible and/or a higher co-insurance payment.

Whether the 6 percent premium discount offered for participation in Prudential's program will be cost justified from an underwriting standpoint remains to be assessed. However, mandatory programs should probably be able to reduce premium costs in the range of 1.5 to 5 percent. Despite the considerable number of studies to date, net effects of the range of second opinion programs will be difficult to discern for a long time. Most analyses have not taken into account the additional operations performed because an individual received a confirmatory second opinion. Without rates of surgery based on the initial recommendation in the absence of a second opinion program, conclusions about the magnitude of this effect must remain speculative.[33]

NOTES

1. McCarthy, E.G., Finkel, M.L. and Ruchlin, H.S. *Second Opinion Surgery.* Boston: Auburn House, 1981.
2. Ibid.
3. Mitchell, J.B. and Cromwell, J. 1980. *Physician-Induced Demand for Surgical Operations: Final Report.* Boston University School of Medicine, as cited in *Second Surgical Opinion Programs: A Review and Progress Report,* American College of Surgeons. March, 1982.

4. American College of Surgeons and the American Surgical Association. 1975. *Surgery in the United States: A Summary Report of the Study on Surgical Services for the U.S.* (SOSSUS Report).

5. Op. cit. (McCarthy, et al.).

6. Op. cit. (McCarthy, et al.).

7. Op. cit. (McCarthy, et al.).

8. U.S. Department of Health and Human Services. Public Health Service. *1980 Summary Report of the Graduate Medical Education National Advisory Committee.* Washington, D.C.: U.S.G.P.O., pp. 3–37.

9. *Cost and Quality of Health Care: Unnecessary Surgery.* Report by the Subcommittee on Oversight and Investigations of the Committee on Interstate and Foreign Commerce. Chairman John E. Moss. January 1976.

10. American College of Surgeons. Surgical Practice Department. *Second Surgical Opinion Programs: A Review and Progress Report.* March, 1982.

11. Roos, N.P. "Who Should Do the Surgery? Tonsillectomy-Adenoidectomy in One Canadian Province," *Inquiry,* Vol. 16, No. 1, 1979, pp. 73–83.

12. Op. cit. (McCarthy, et al.).

13. Op. cit. (McCarthy, et al.).

14. Op. cit. (McCarthy, et al.).

15. Op. cit. (McCarthy, et al.).

16. Op. cit. (ACS).

17. Gertman, P.M., Stackpole, D.A., Levenson, D.K., Manuel, B.M., Brennan, R.J. and Janko, G.M. "Second Opinions for Elective Surgery: The Mandatory Medicaid Program in Massachusetts," *New England Journal of Medicine,* Vol. 302, No. 21, 1980, pp. 1169–1174.

18. Martin, S.G. "The Sentinel Effect in Second Opinion Programs," *Employee Benefit Plan Review,* Vol. 36, No. 8, 1982, pp. 24, 26.

19. Ibid.

20. Op. cit. (ACS).

21. "Insurers Promote Second Opinion; Offer Discount," *Modern Healthcare,* Vol. 9, No. 4, 1979, p. 12.

22. Health Research Institute. *Health Care Cost Containment Third Biennial Survey.* Walnut Creek, California, 1983.

23. Health Research Institute. *Health Care Cost Containment Survey.* Walnut Creek, California, 1981.

24. Lawson, J.C. "Second Surgical Opinion Plans Popular: Study," *Business Insurance.* April 5, 1982.

25. Op. cit. (ACS).

26. "FMC Amends Plan to Include Second Opinion," *Employee Benefit Plan Review,* Vol. 37, No. 5, 1982, p. 16.

27. Personal communication with Ms. Ann Verano, Project Manager for the National Second Opinion Program Hotline. April 11, 1984.

28. Op. cit. (ACS).

29. Op. cit. (ACS).

30. Op. cit. (ACS).

31. Op. cit. (ACS).

32. Op. cit. (ACS).

33. Op. cit. (ACS).

UTILIZATION *12* REVIEW

U tilization review (UR) is the general title for a variety of mechanisms employed to review the appropriateness, necessity and quality of care provided to a specific group of patients. UR has been applied to ambulatory care, long-term care, and most frequently, hospital care. It has been performed under various auspices, both public and private, medical and nonmedical.

Insurance carriers have for many years reviewed claims where the services appeared to be inconsistent with the diagnoses and, if the services could not be justified to their satisfaction, they have refused to pay the bill or some portion thereof. A number of surgical procedures whose efficacy is questionable have been considered nonreimbursable expenses by Blue Cross and Blue Shield plans and a number of commercial carriers. While in many of these cases decisions have been made only after review by the medical department of the insurer, once policies have been established for what constitutes inappropriate modes of care, administrative procedures have been set in place to preclude payment for these services on a systematic basis. Public financing programs likewise frequently make administrative decisions regarding the necessity of care provided to an eligible person.

From the standpoint of employers, the most important type of UR is hospital review, since this is where 60 to 75 percent of health plan benefit dollars are spent. Physicians, frequently aided by specially trained nurses, review the care under a specific insurance plan to patients who are scheduled for hospital care or are already hospitalized. This type of review is a form of peer review, in contrast to review by nonphysicians generally performed prior to payment by claims analysts employed by the carrier.

A BRIEF HISTORY

Utilization review began in the 1950s when several unions asked medical societies to help them conserve dollars in their health and welfare funds by auditing the health care services provided to their members. In 1959 Blue Cross of Western Pennsylvania, the Allegheny County Medical Society Foundation and the Hospital Council of Western Pennsylvania initiated a program of "utilization review" involving retrospective analyses of individual hospital records and summary hospital utilization data. The remedy for aberrant patterns was physician education and recommendations to hospital administration for any needed administrative changes.[1]

Quantifiable benefits of a UR system go back at least to the early 1960s. For example, a program to control the length of the hospital stay was developed by John Hancock Mutual Life Insurance Company at the request of the New Jersey Teamsters Welfare Fund. It used criteria established by the carrier, was coordinated by the fund, and used the hospital's UR committee to resolve disputes between treating physicians and the fund coordinator. Length of stay for admissions controlled in this way fell from 6.73 days in 1963 to 6.02 in 1966 while the length of stay for noncontrolled admissions increased from 7.72 to 9.10. No information, however, was provided on admission rates or total hospital days per 1,000 covered workers and dependents.[2]

With the advent of Medicaid and Medicare, increasing concern about possible overuse of publicly funded services led a number of states to contract with medical-society-spawned review organizations, called Foundations for Medical Care (FMCs). Under these contracts, the foundations agreed to police the care, especially hospital care, provided to public program beneficiaries with respect to appropriateness, necessity and quality. When the foundations found that care did not meet acceptable professional standards, based on the subjective clinical judgment of the reviewing physician, payment would be denied. A number of states published analyses indicating that savings outweighed review costs. For example, both a preadmission certification program run by the state of California and a concurrent review program run by the Sacramento Medical Care Foundation in 1970 for Medi-Cal recipients were reported to show days of hospital care reduced by about 15 percent below predicted levels. The precertification program showed a denial rate of 3 percent, with the remaining decrease attributed to the sentinel effect (i.e., the discouraging effect of merely having a program in place). However, subsequent analyses of the original study design and data have raised serious questions about the magnitude of an effect from the review systems.[3] Based on reported positive experiences available at the time, however, an impressed Congress passed, as part of the 1972 Social Security Amendments, a fairly detailed law which encouraged the creation and bore the costs of Professional Standards Review Organizations (PSROs). PSRO-enabling legislation required the review of all federally financed patient care in acute hospitals, but contained provisions for review to extend to long-term care facilities and to ambulatory care services as proficiency increased.

As a result, about two hundred PSROs were created around the country. Although precluded by law from being creatures of the medical societies, many of them had leadership with significant overlap to those societies. Physicians viewed PSROs with various degrees of enthusiasm, from total aversion to acceptance of a concept that promised to keep quality decisions out of the hands of bureaucrats. How PSROs work, their problems and evidence of their cost-effectiveness are summarized in Chapter 16. In brief, the overall savings-to-cost ratio appeared to be marginally positive. Detailed due process requirements and other organizational problems related to the public nature of the program appear to have been important obstacles to greater cost-effectiveness. Businesses, however, seeing an opportunity to get some help with their rapidly escalating health care costs, especially for hospital care, have increasingly turned to PSROs, recently renamed Peer Review Organizations (PROs), to perform similar review functions. With declining Federal support of PSROs and private payors and carriers concerned about hospital costs and utilization, many PSROs/PROs have not only been receptive to requests from employers but have actively marketed their services to businesses and insurance carriers, with growing success. A number of entrepreneurial review organizations have been developed around the country by physicians and by nonphysicians. In some areas there is considerable competition to sell UR to employers and carriers. Recently several large commercial insurance companies and some Blue Cross/Blue Shield plans have formed their own utilization review operations, while others have purchased these services from the PROs or other review organizations. The American Medical Peer Review Association (AMPRA), a private national trade organization for PROs, has begun a new project to sell utilization review services to national employers. National employers will be able to contract with AMPRA to do company-wide review and AMPRA will subcontract with local PROs to conduct the reviews.

TYPES OF REVIEW

Review may take a number of forms. Preadmission review requires that the outside group agree with the medical necessity and appropriateness of hospitalization prior to admission for the hospital bills to be paid at the maximum rate under the benefit plan. It is generally considered the most effective form of utilization review in reducing unnecessary hospitalizations and associated procedures. For example, by offering an incentive for covered persons to receive prior authorization for inpatient care, Health Maintenance Life Insurance Company has experienced a substantial decline in inpatient days. Under its program, which began in 1978, if a patient receives prior authorization for hospitalization, the admission is fully (100 percent) reimbursed, whereas unauthorized inpatient days are reimbursed at 80 percent even if retrospective review fully justifies the admission. For every 1,000 individuals insured in 1978, 714 days of hospitalization were requested, 492 were authorized and 510 were used. In 1979, inpatient days per 1,000 insurees decreased to 450 (compared with an industry average of 734), and in 1980 were approximately 400.[4]

Concurrent review, the most frequent form of UR applied, and the one with which the physician organizations have had the greatest experience, monitors the care of patients while they are in the hospital. While the scope of review varies from program to program, it usually audits the necessity of admission, the length of stay, and less frequently reviews the specific procedures and other services provided during the hospitalization.

Retrospective review involves analysis of data relating to the hospital stay after the patient is discharged. It may look at individual hospitalizations or patterns of hospital care. For example, retrospective analysis frequently includes reviews of profiles of care by hospital, physician, diagnosis or procedure. Table 12.1 depicts information received by one review organization on length of stay by hospital for a specific diagnosis. It appears that two of the nine hospitals have a longer stay for similar procedures. Based on this kind of information, concurrent review might be focused on the two hospitals, with efforts made to assure discharge as early as possible. Table 12.2 demonstrates a different problem. Of the six general surgeons who do most of their operations at the same hospital, two have by far the longest length of stay for three common procedures. As a result of this analysis, concurrent review might be focused on the patients of these doctors. If a particular surgeon appears to be performing a disproportionate number of certain operations, precertification might be required of his patients. For example, a general surgeon whose appendectomy cases represent twice the percentage of practice of other general surgeons might be singled out for such review. Retrospective review may also include reviewing whether a certain portion of the hospital care was not appropriate to the problem and should not be reimbursed.

At least two major approaches exist for performing review. One approach relies heavily on having nurses as on-site reviewers in all hospitals within an area. Often review groups which adopt this approach rely on delegating the function to hospitals' own internal utilization review proce-

TABLE 12.1
LENGTH OF STAY AMONG HOSPITALS

Diagnosis					Hospitals				
	A	B	C	D	E	F	G	H	I
Myocardial infarction	12.5	13.2	13.1	11.8	16.3	12.3	15.7	13.0	12.6
Cataract	4.0	4.0	3.6	4.2	6.4	3.4	6.0	3.6	4.1
Inguinal herniorraphy	4.9	5.1	3.9	4.3	7.6	5.2	6.8	4.7	4.7

TABLE 12.2
LENGTH OF STAY AMONG SURGEONS*

| | Surgeons at one hospital | | | | | |
Diagnosis	A	B	C	D	E	F
Gastric resection	10.2	14.8	11.7	10.5	9.4	13.7
Cholecystectomy	5.7	8.2	6.3	5.9	6.2	8.8
Inguinal herniorraphy	3.2	5.1	2.8	3.6	1.9	4.3

*Surgeons with a minimum of 20 patients with a listed diagnosis during the twelve months covered by this data.

dures and personnel. The major strength of the on-site approach is the ability to verify what the physician has provided as reasons for continuing stay and to assess the adequacy of discharge planning efforts. In some cases attempts may be made to intervene before a questionable test or other procedure has been carried out. However, most concurrent review focuses primarily on length of stay. Disadvantages of on-site review can include high cost, delegation where it is not indicated, a close relationship between the UR and hospital personnel which over time affects the judgment of UR personnel, and the difficulty of ensuring uniformity in approach when many different hospitals are involved. In addition, this approach does not prevent unnecessary hospital admissions.

A second approach is to perform review primarily by telephone. Advantages of this approach include greater appropriateness for preadmission review, greater control and uniformity in the review system, efficiency, and ease of use for multiple sites via a toll-free 800 telephone number. Disadvantages are lack of on-site review to verify what the patient's physician has submitted verbally or in written form, and potentially greater physician resistance, based on the absence of local physician involvement. However, many telephone UR programs have overcome these disadvantages by verifying reviewed information against the discharge summary submitted to the carrier and by having a reviewing physician talk to any physician disgruntled about the review process.

Regardless of which of these approaches is used, review is generally coordinated by a nurse with special training in performing UR. Situations which require a decision that could affect payment are then referred to a physician reviewer. Frequently the nurse reviewer is employed by the hospital, which has been "delegated" review responsibility from the PSRO/PRO. In other cases, the nurse is employed by the review organization. Review activities are usually contracted for based on a fee paid to the physician organization for each patient reviewed. However, some organizations bill based on a fee per employee per month, regardless of whether they are hospitalized. This fee may be paid through the carrier as a regular part of a negotiated premium or as a supplement to the regular administrative services contract.

Figure 12.1 provides a partial list of surgical procedures which, under normal circumstances, can be performed on an outpatient basis. Nurse reviewers typically use similar lists to decide if a proposed surgery requires a hospital admission. However, procedures on this and similar lists might warrant hospitalization if the patient has certain other medical problems such as congestive heart failure or a cardiac arrhythmia. In any case where the nurse is uncertain as to the appropriateness of an admission, he or she contacts the reviewing physician, who has ultimate responsibility for such decisions.

The cost-containing effects of UR can be both direct and indirect. A careful, concurrent UR program may help to get patients home from the hospital earlier, with the threat that bills related to unnecessary days of hospital care will not be reimbursed. A preadmission certification program may prevent unnecessary hospitalization, such as a stay for tests that could be conducted on an outpatient basis or for an elective procedure without clear justification. Retrospective review may lead to decisions not to pay part of the hospital or physician bill. However, denial of payment by the employer or insurer may leave the patient fighting with the provider over the bill, and money saved to the insurance plan may come out of the employee's pocket. Some employers, aware of this problem, have taken the position that if the denial results from problems not within the control of the employee, the company will intercede on the employee's behalf. For example, a provision in the contract between John Deere and the United Auto Workers allows the company to cease reimbursement within twenty-four hours of a PSRO utilization review determination that medical service is no longer required. Under this provision, however, the patient is "held harmless" and backed by the company if out-of-pocket payment for the denied portion is demanded by the provider.[5]

Private Review

A large number of insurance carriers, including most of the large commercial carriers offering health insurance and the Blue Cross and Blue Shield plans, have their own internal UR systems, or contract for review with PSROs/PROs or private review organizations. Many individual employers contract for services directly or indirectly. A 1983 survey by AMPRA of 137 of its member organizations found that 89 out of 111 responding PROs had signed contracts for private review and an additional 8 were in the process of negotiating such contracts. A few of the larger employers mentioned in that survey as having signed contracts with one or more PROs were Levi Strauss & Company, Honeywell, 3M, Pillsbury, DuPont, Eaton, Kellogg, Kaiser Aluminum, Quaker Oats, Caterpillar and Boeing.[6]

FIGURE 12.1
FOUNDATION FOR MEDICAL CARE OUTPATIENT/ SAME-DAY SURGERY LIST*

Otolaryngology/auditory system

21310—Treatment of closed or open nasal fracture without manipulation

69420—Myringotomy including aspiration and/or eustachian tube inflation

69433—Tympanostomy (requiring insertion of ventilating tube), local or topical anesthesia; unilateral

69434—Tympanostomy (requiring insertion of ventilating tube), local or topical anesthesia; bilateral

69436—Tympanostomy (requiring insertion of ventilating tube), general anesthesia; unilateral

69437—Tympanostomy (requiring insertion of ventilating tube), general anesthesia; bilateral

General surgery/integumentary system

11100—Biopsy of skin, subcutaneous tissue and/or mucous membrane (including simple closure), unless otherwise listed (separate procedure); one lesion

11101—Biopsy of skin, each additional lesion

11750—Excision of nail and nail matrix, partial or complete (e.g., ingrown or deformed nail), for permanent removal

12000—Repair of superficial wounds
Series

19100—Biopsy of breast; needle (separate procedure)

19101—Biopsy of breast; incisional

General surgery/digestive system

43235—Esophagogastroduodenoscopy, diagnostic

43251—Esophagogastroduodenoscopy; with removal of polyp(s)

45300—Proctosigmoidoscopy; diagnostic (separate procedure)

45310—Proctosigmoidoscopy; with removal of polyp or papilloma

45330—Sigmoidoscopy, flexible fiberoptic; diagnostic

45333—Sigmoidoscopy, flexible fiberoptic; with removal of polyp(s)
Many—Surgical removal of impacted teeth
codes

49500—Repair inguinal hernia, under age 5 years, with or without hydrocelectomy; unilateral

49501—Repair inguinal hernia, under age 5 years, with or without hydrocelectomy; bilateral

Gynecology/female genital system

57500—Biopsy, single or multiple, or local excision of lesion, with or without fulguration (separate procedure)

57520—Biopsy of cervix, with or without D&C

58120—Dilatation and curettage, diagnostic and/or therapeutic

(Partial list)

*This list includes surgeries that should usually be performed on an outpatient basis when they are performed as a routine, primary independent procedure. (Codes are generally adapted from CPT–4 system.)

Source: Midwest Business Group on Health. *Containing Health Benefit Costs: A Business Guide to Utilization Review*, 2nd ed., 1981, App. VI.

The 1981 Hay–Huggins survey of noncash compensation practices among major companies found that 42 of 519 respondents to their inquiry (8 percent) contracted with an outside claims review service such as a PSRO/PRO, and another 62 (12 percent) were considering the initiation of such an arrangement.[7] Assessing the cost-effectiveness of these review activities is difficult. Several companies, particularly in the Midwest, however, have reported improvements in utilization and costs which they attribute to UR programs. Caterpillar Tractor in Peoria, Illinois reported reducing the average length of hospital stay of employees and dependents from 6.5 to 5.5 days after hiring the Mid-State Foundation for Medical Care in April 1978. The PSRO/PRO there monitored the hospital stays of a covered group totaling 110,000 in a fifteen-county area.[8] Caterpillar indicated that the review had saved the company $4 for every $1 spent on UR.[9] John Deere reported that after two years of monitoring by the review organization, its Moline, Illinois plant saved $2.5 million on reductions in medical-benefit payouts.[10] Admissions per thousand for Deere in the Illinois area were down 21 percent after thirty-six months of review.[11]

Although companies generally feel that the private review for which they have contracted is yielding a good return on investment, careful studies are needed to document this result. There is likely to be considerable variation in terms of effects depending on the aggressiveness of the PSRO/PRO, the characteristics of the medical marketplace in the area (physicians per capita, average length of stay, hospital beds per 100,000 population, etc.) and the availability and cost of outpatient surgery. Unfortunately some review organizations report success based on inappropriate measures of effectiveness. For example, days of care requested by the provider but denied by the reviewing group is an example of an inappropriate measure. A much better index is a reduction in days/1000 insured for hospital inpatient care.

A number of other considerations in reviewing the effects of a UR program include:

1. Reductions in average length of stay and/or admission rates may be due to factors other than UR. Changes in age, sex and case mix have effects on these parameters as may changes in reimbursement rates such as DRG-based hospital payments which favor shorter lengths of stay; changes in insurance policies of large employers to cover outpatient surgery on more favorable terms; changes in incidence and prevalence of diseases, such as flu; growth of alternative facilities such as hospices and birthing centers; and even local publicity about unnecessary surgery.

2. Analyses may not include the cost of alternative treatment such as additional outpatient care, care in hospices, and services provided in long-term care facilities associated with reduced use of hospital inpatient services. Some of these costs may be paid by the patient in

addition to those which would show up under different portions of the benefit package.

3. Reductions in hospital utilization do not automatically translate into cost savings. In the absence of reduction in overall costs, reduced demand may simply be translated into higher prices. It is difficult to tease out this effect from the other inherently inflationary forces that cause higher prices in the health care reimbursement system. Also, hospital costs for the last days of a hospital stay, those most affected by concurrent review, are less costly than the early days of stay. Therefore cost savings cannot be accurately computed by multiplying the reduced number of days by the average daily hospital cost.

4. Costs due to denial of payment may be paid by the individual employee or dependent, and thus not represent a net cost savings, but rather only a transfer of payment responsibility from the employer or insurer to the individual.[12]

Since review organizations vary in their ability to achieve cost savings, any employer considering this option should request data from the review organization on the effect of review it has performed for other employers. Employers may want outside help in reviewing this data and the review process proposed by competing organizations, since considerable experience and sophistication in data collection methods and statistical analysis are necessary to judge the validity of most claims by review organizations. Companies with longer lengths of stay for common procedures than other area employers, high costs and long lengths of stay for inpatient mental health treatment or high numbers of hospital days per 1,000 eligibles (adjusted for age and sex) should have a particularly strong motivation to consider contracting with private review organizations.

High quality review organizations will in general maintain a reasonably sophisticated data system that can show trends over time in length of stay for specific procedures and overall changes in practice patterns by physicians associated with patterns of particularly long lengths of stay and high admission rates. Some of the more sophisticated approaches involve targeting review to those situations thought to provide the highest payoff. In some cases the review organization may concentrate on an educational process for physicians. For example, the Delmarva PSRO in Maryland tallied for each physician the number of days on which his or her patients were hospitalized but received no acute care services, and allowed each physician to compare his or her nonacute profile score with those of other physicians in the area. Based on this approach, the PSRO reports dramatic reductions in days of care.[13]

A good resource for companies considering adopting a UR program is "Containing Health Benefit Costs: A Business Guide to Utilization Review," Second Edition, by Elizabeth Ehrhart Blaine, Midwest Business Group on

Health, August, 1981. Suite 6757, 200 East Randolph Drive, Chicago, Illinois 60601.

THE FUTURE

PROs or other review organizations have been built into the Medicare reimbursement system using DRGs. Their role is to monitor hospitalization to Medicare recipients to assure that hospital admission is required, appropriate care is provided, and the diagnostic information on which payment is based reasonably reflects the patient's problems. In addition, the trend toward private review is likely to grow.

Changing incentives for hospitals will quickly lead to growing administrative monitoring for all services ordered by physicians, initially for Medicare recipients but increasingly for all patients. To operate in a more competitive environment, hospitals must control utilization and costs rather than maximize reimbursement through billing for every possible service.

As physicians join with hospitals in risk-sharing under preferred payment arrangements or HMOs, physicians will have a growing interest in assuring that both they and their colleagues are being efficient in ordering hospital services. Internal professional review will be an important mechanism to maximize physician revenue under cost-sharing arrangements.

Clinical data systems used by physicians in hospitals to order tests and treatments offer a rich data base for concurrent review. If a test is ordered which, based on pre-established criteria, is not clearly related to the patient's reported problems, an immediate review may be triggered. This review can take place before the service is provided, potentially reducing costs as well as payment.

NOTES

1. Gertman, P.M., Monheit, A.C., Anderson, J.J., Eagle, J.B. and Levenson, D.K. "Utilization Review in the United States: Results from a 1976–1977 National Survey of Hospitals," *Medical Care*, Supplement to Vol. 17, No. 8, 1979.
2. Chassin, M. "Utilization Review." *Medical Care*, Supplement to Vol. 16, No. 10, 1978, pp. 27–35.
3. Ibid.
4. "Incentives Offered for Preauthorized Hospitalizations," *Employee Benefit Plan Review*, Vol. 35, No. 9, 1981, p. 88.
5. Washington Business Group on Health. *A Private Sector Perspective on the Problems of Health Care Costs*. A working paper prepared for the Department of Health, Education, and Welfare, 1977, as cited in Walsh, D.C. and Egdahl, R.H. *Payer, Provider, Consumer: Industry Confronts Health Care Costs*. New York: Springer-Verlag, 1977.
6. American Medical Peer Review Association. *Private Review Activity*, Washington, D.C., 1983.

7. Hay Associates. *1981 Noncash Compensation Comparison.*
8. LaViolette, S. "Employers Cut Hospital Costs with Utilization Review Plans," *Business Insurance,* April 2, 1979, p. 1.
9. "PSROs Covet Corporate Work," *Medical World News,* September 14, 1981.
10. Ibid.
11. Blaine, E.E. *Containing Health Benefit Costs: A Business Guide to Utilization Review,* 2nd ed., Chicago, Illinois: Midwest Business Group on Health, 1981.
12. *How Business Can Use Specific Techniques to Control Health Care Costs.* Prepared by Interstudy. Washington, D.C.: National Chamber Foundation, 1978, pp. 11–12.
13. "The Politics of Peer Review," *Washington Report on Medicine and Health Perspectives,* May 24, 1982.

HEALTH *13*
MAINTENANCE
ORGANIZATIONS
(HMOs)

DEFINING THE HMO

Many different kinds of organizational and financial arrangements are subsumed under these three letters that some people seem to revere, others condemn and most view with confusion and ambivalence. HMOs can be defined based on a set of behavioral characteristics:

1. Contractual responsibilities to provide some health care services, usually at least physician and hospital services;
2. Provision of service to an enrolled, defined population;
3. Voluntary enrollment of subscribers;
4. Fixed periodic payments independent of how many services an individual uses;
5. Assumption of some financial risk in the provision of services.[1]

In general, HMOs are categorized in two ways: 1) a group/staff model where physicians are paid on a salary or capitation basis and work in an office or hospital owned or leased by the HMO; and 2) an independent practice association (IPA), composed of physicians in private offices who bill the HMO on a fee-for-service basis. However, there is tremendous diversity among HMOs with respect to all the characteristics of interest to employers. They may have as few as several thousand subscribers or more than one million;

their annual turnover rates may be as little as under 5 percent or as great as 75 percent. Their risk of failure can be great or small and is divided differently among the plan, the physicians and the hospitals (if not wholly owned). They may be federally qualified and therefore subject to a significant number of constraints with respect to financial reserves, pricing and scope of services required to be provided, but also empowered to require area employers to offer an HMO as an alternative to their regular insurance coverage.* Or they may be primarily regulated by the state in which they operate, with requirements of varying stringency. Sponsorship is even more varied, with HMO sponsors including commercial insurers, Blue Cross and Blue Shield plans, unions, employers, multispecialty physician groups, hospitals, municipalities, universities and health care management firms.

HMO Enrollment

The 1983 National HMO Census, conducted by InterStudy, a firm very active in the promotion and assessment of prepaid health plans, identified 280 plans with 12,490,780 members. Total enrollment from 1978 to 1982 grew at an average annual rate of approximately 12 percent. During the same period the average family premium grew about 10 to 12 percent per year. As of 1983 there were 181 group models whose membership was 85 percent of total HMO subscribers, and 99 IPA model plans. Average plan size was almost 45,000 members. In 1983 HMOs served 41 states, Guam, and the District of Columbia, with California having the most plans (32) and the largest enrollment (4.9 million). The greatest recent growth in enrollment has been in those plans with 50,000–100,000 members, followed by those with 25,000 to 50,000. In 1983 the 21 plans with greater than 100,000 enrollees represented 60 percent of the total.[2] Based on the 1983 HMO Census, over one-half of the HMOs in the country were federally qualified and comprised 76 percent of the total membership. While many of the HMOs only serve employed populations and their families, the majority of plans had members over age 65, and 80 plans had Medicare contracts.

Costs of HMO Care

There is relatively little dispute that HMOs provide a given range of benefits at a lower aggregate cost than under usual insurance reimbursed care. That is, the total cost of premiums plus out-of-pocket expenditures for HMO enrollees is lower than for comparable people with conventional health insurance coverage. In group models total costs appear to be in the range of 10 to 35 percent less than conventional insurance.[3] Evidence for lower overall costs in IPAs is less clear, with no definitive information having been pre-

---

*Only employers of twenty-five or more employees who offer health insurance as a benefit are subject to this requirement, known as the "dual choice" mandate.

sented. The cost reduction in prepaid group practices does not stem from lower costs per unit of service. Rather it derives from hospitalization rates that are 20 to 35 percent lower than for comparable populations with conventional insurance and hospital lengths of stay that are marginally shorter than for fee-for-service patients in the same geographic area.

Admission rates in HMOs appear to be lower for both surgical and medical cases, and do not appear to be achieved primarily by reduction in so-called discretionary procedures. Admission rates for elective operations such as hernia repair or hysterectomy are reduced but not substantially more than nondiscretionary surgical procedures.[4] For the 193 HMOs that reported utilization rates in 1983, the average plan used 444 hospital days per 1,000 members, and overall the weighted average for all HMO enrollees was 409 days per 1,000 enrollees, with group (network) models averaging 368 days and IPAs 425 days.[5] When adjusted to look at only utilization for enrollees under age 65, in 1981 the average plan rate was 386 days per 1,000 enrollees. By comparison, the average Blue Cross/Blue Shield rate for members under age 65 was 725 per 1,000 insured.[6] The Government Accounting Office (GAO) came to similar conclusions in a study of twelve successful HMOs around the country. They found hospital utilization rates for those HMOs of 451 days per 1,000, compared with 722 days for Blue Cross members and 1,099 days for the general population. When admission rates were adjusted to account for differences in the age and sex distribution of HMO and Blue Cross/Blue Shield members, all the HMOs still showed lower utilization rates, in some cases as much as 70 percent less.[7]

Many reasons for these differences between HMOs and indemnity insurance-financed care have been advanced, but most are hard to prove. However, there is no evidence to suggest that utilization differences reflect any difference in quality of care. Since the financial incentives for both the plan and the physician (who usually receives a bonus based on the plan's experience) are in the direction of ambulatory care, the lower hospitalization rates are not surprising.

Assessing the overall costs of care in HMOs versus private insurance is easy compared to assessing the relative cost of these two options to the employer. Most HMOs offer a broader scope of services than private insurance but traditionally required a higher premium. In general, however, HMO premiums have been reduced relative to private insurance premiums over time so that striking disparities are rare. With increasing frequency HMO premiums are slightly lower than Blue Cross/Blue Shield or commercial carrier coverage. In 1983 the average monthly family premium for the 221 reporting plans was $171.73, with IPAs the most expensive component of this at an average of $179.45. As shown in Figure 13.1 below, over recent years the cost of HMO coverage has increased at a slightly faster rate than the medical care component of the CPI and at a slightly lower rate than per capita personal health expenditures.[8]

FIGURE 13.1
1980–1983

	Increase (%)	Average annual increase (%)
Average family premium for HMO coverage	37.8	12.60
Medical care component of the CPI	31.1	10.4
Per capita personal health care expenditures	26.6*	13.3*

*December 1980 through December 1982.

Sources: The National HMO Census, 1981, 1982 and 1983. Conducte'l by InterStudy and The Economic Report of the President, February, 1984. Council of Economic Advisors. Washington, D.C.: U.S.G.P.O.

Direct comparisons to private insurance increases are difficult in an era where significant changes in benefits have occurred as well as cost shifting from public to private sector. In addition, there are lingering questions about whether HMO enrollees tend to be healthier than those opting for private insurance, invalidating direct comparisons of costs and utilization. Adverse selection effects, if they exist, can be assessed through careful analysis of the characteristics of those in HMOs versus those in indemnity plans and the characteristics of enrollees' utilization before switching to the HMO.

Although HMO premiums have increased, on the average, less rapidly than costs of indemnity insurance, whether the long-run rate of increase will remain lower is still open to conjecture and will depend in part on the degree of competition from other combinations of providers. The presence of several HMOs in a particular area appears to have a moderating effect on the rate of rise in health costs across all providers in the area, suggesting that their presence stimulates competition.[9]

Many employers and most large employers, especially if decentralized in several areas of the country, have considerable experience with HMOs. Out of 660 companies responding to a 1983 employee benefits survey of major employers, 4 offered an HMO as their sole plan while another 418 (50 percent) offered at least one HMO option for medical coverage.[10] In most of the country, however, enrollment in HMOs was found to be low in comparison to typical fee-for-service coverage. Almost half of survey participants offering HMOs indicated that enrollment in them constituted less than 7.5 percent of eligibles and 80 percent indicated that HMO enrollment was less than 25 percent, regardless of whether coverage was contributory or noncontributory.[11] In some areas, however, such as parts of California, HMO enrollees are 25 to 60 percent of the employee populations offered HMO options.

ATTITUDES ABOUT HMOs

Employer Attitudes

A survey on HMOs sent by Charles D. Spencer and Associates to subscribers to its *Research Reports* elicited 170 responses. In this 1980 survey 51 companies had fewer than 5,000 employees, 98 between 5000 and 50,000 and 21 more than 50,000. Of the 170 corporations only 23 had not offered HMOs. Nationwide median enrollment in HMOs by eligible employees of responding companies was 5 percent. Table 13.1 shows the nine items mentioned most frequently by respondent companies as the most positive aspects of HMOs and Table 13.2 lists the most commonly reported negative aspects.[12] Observed cost containment was not frequently cited as a positive aspect of HMOs, but the "potential" for cost containment appears an important consideration. Individual comments on the cost impact of HMOs ranged from "HMOs, based on our experience, truly are controlling the medical inflation rate and are a very favorable alternative to national health insurance" to "our overall company cost of medical care has been increased due to our offering HMO options, because of adverse selection . . . We pay the full rate to the HMOs for employees who formerly cost us much less because they are young and healthy on average."[13]

By far the most comprehensive and scientific survey of how employers view HMOs was conducted by Louis Harris and Associates for the Henry J. Kaiser Family Foundation in 1980. Interviews were conducted with 984 executives in 634 companies nationwide, with the sample equally divided between first-line decision makers (frequently benefits officers) and chief executive officers or their designates. When asked about the argument that HMOs offer strong cost containment incentives to doctors and hospitals, especially compared to traditional fee-for-service plans, 64 percent of benefits officers and 54 percent of CEOs felt that HMOs were at least somewhat effective in containing costs. However, about one-half of CEOs indicated agreement that the emphasis on cost containment by HMOs could lead to sacrifice in the quality of care offered by them. Table 13.3 summarizes responses to questions on the relative cost containment and quality of care characteristics of IPA and closed panel HMOs.[14]

Of all executives with experience in offering dual choice (HMO as well as private insurance), 63 percent felt the HMO option had no effect on costs of health care to their organization, 14 percent believed that it increased costs and 13 percent that it decreased costs. Only 7 percent felt it was too early to tell, and 3 percent were not sure.[15]

TABLE 13.1
POSITIVE ASPECTS OF DEALING WITH HMOs

Items Mentioned by Respondents	Number of companies*
1. HMOs provide competition and are an alternative health care source to employees.	29
2. The administrative burden on the employer is cut, especially as concerns claim forms.	25
3. HMO employees are cooperative.	19
4. The HMO offers at least the potential for cost containment.	19
5. The employees are happy with the HMO.	16
6. Quality care is offered by HMOs.	12
7. The benefits are comprehensive.	9
8. HMOs are responsive to employer/employee needs.	8
9. HMO enrollment increases employee awareness of coverage and costs.	6

*Total number of companies is 170.
Source: *Employee Benefit Plan Review Research Report,* May 1981, pp. 314.1–314.9.

Overall, in the Louis Harris study, HMOs were perceived by executives to be better than regular health insurance on five points:

• the availability of adequate preventive health services
• the availability of health education
• the lower out-of-pocket cost of health care services to employees
• the availability of doctors and medical service twenty-four hours a day, seven days a week
• the total cost (whether met by insurance or not) of health care services.

TABLE 13.2
NEGATIVE ASPECTS OF DEALING WITH HMOs

Items Mentioned by Respondents	Number of companies*
1. Administrative problems, including:	
—additional paperwork for the employer	34
—lack of uniform billing procedures between HMOs	16
—late reports or no reports	14
2. Dissatisfaction with HMO personnel, including:	
—turnover and lack of experience of marketing staff	20
—HMO representatives "force" their product on employers	19
3. Fear of HMO bankruptcy is the major concern.	18

*Total number of companies is 170.
Source: *Employee Benefit Plan Review Research Report,* May 1981, pp. 314.1–314.9.

TABLE 13.3
EMPLOYER PERCEPTIONS OF COST CONTAINMENT AND QUALITY OF CARE IN TWO TYPES
OF HMOs

	Which type, in general, do you feel offers the greatest potential for controlling health care costs—the closed panel or the IPA type of HMO?		Which do you feel generally offers higher quality medical care—the closed panel or IPA?	
	Benefits officers	Chief executives	Benefits officers	Chief executives
Base (number)	276	204	276	204
	(%)	(%)	(%)	(%)
Closed panel	70	68	22	20
IPA	16	16	40	45
Both equally	5	6	25	21
Neither	3	4	3	3
Not sure	6	6	10	11

Source: Survey prepared for the Henry J. Kaiser Family Foundation, Menlo Park, California, by Louis Harris and Associates, 1980.

On four points the fee-for-service system was preferred over HMOs by these executives:

- overall employee satisfaction with services
- convenient location of physician offices
- convenient location of hospitals
- quality of physicians

Consumer Attitudes

In a companion study of HMO members, eligible nonmembers and the general public, HMO members more frequently indicated that they were "very satisfied" with their health care services (57 percent) than nonmembers (53 percent) or the general public (52 percent), although the differences are small. Only 7 percent of HMO members said that the performance of their HMO was worse than expected, compared to 31 percent who felt it was better than anticipated. Only 3 percent of members said that they did not plan to renew their membership. In general, members were more satisfied than nonmembers with the total cost of their health care, the around-the-clock availability of doctors and medical services, the coverage of preventive health services, the wide range of services in one place and the convenient location of their HMO. However, many members voiced criticism of their HMO, particularly lack of choice of physicians and hospitals, quality of physicians, impersonal service and negative attitudes of some doctors and other staff and the length of time necessary to get an appointment with an HMO doctor.[16]

The fact that an employer offers HMO options and a positive attitude of management toward enrollment are the important contributors to HMO membership. In the Louis Harris survey of HMO members, 58 percent had first heard about the HMO from their employer, 15 percent from friends or relatives and 13 percent from husband or wife.[17] Of members, 59 percent (almost three out of five) joined through their employer, 22 percent through the employer of other family members and 8 percent through their union. Only 3 percent of those joining did so outside employer or union groups.

Across the country the majority of people may still not be familiar with the HMO concept and how it works. Seventy-nine percent of the public and a smaller majority (59 percent of eligible nonmembers said in the 1980 poll that they are either not very familiar or not at all familiar with the term "health maintenance organization" or its abbreviation, while only 15 percent of eligible nonmembers and 5 percent of the public indicated that they are very familiar with the term.[18] Therefore, significant efforts by the local HMOs and the employer are likely to be necessary to educate employees about this option and the benefits of membership.

FINANCIAL VIABILITY

One of the major concerns of employers in pushing HMO enrollment is the financial status of many plans. In the 1980 survey of 170 employers, 18 reported that fear of HMO bankruptcy was their major concern.[19] Concerns have grown with the reduction in federal funding for HMOs. According to one report, by the end of 1981 at least 19 federally qualified HMOs had gone broke, and estimates of the number of HMOs that might not survive in the absence of additional federal funds ranged from 20 to 65.[20] Reasons most frequently cited for failure include undercapitalization, bad management and inability to attract top financial talents. Looking at the history of federal funding, Alan Vinick, director of HMOs for Connecticut General Life Insurance Company, stated a commonly held view that, "Funds were given to people who didn't have the breadth of experience to manage a complex business."[21] Even companies who are generally supportive of HMOs, such as Hewlett-Packard, are telling their employees during enrollment campaigns that they are "at risk" if they sign up for HMOs. Hewlett-Packard's caution was borne out of five bad HMO experiences over a three-year period.[22] While there is no absolute remedy for this potential problem, the following checklist will help companies limit their problems:

1.   Carefully evaluating the HMO before it is offered;
2.   If a number of HMOs are available, only choosing to offer those that are financially stable or offer adequate protection to both company and employees in the event of insolvency; and
3.   If the only federally qualified HMO in the area is financially shaky, insisting on signing a contract with it that will help the company and its employees avoid any direct responsibility for medical expenses covered under the HMO's program.

Under the Omnibus Budget Reconciliation Act of 1981, federally qualified HMOs are required to adopt one of four methods to handle outstanding liabilities:

1.   Contract with each regularly used hospital to prohibit hospitals from charging participants for any services that the HMO is obligated to pay;
2.   Purchase insurance;
3.   Hold adequate financial reserves; or
4.   Use any other liability arrangement(s) acceptable to the secretary of Health and Human Services.

HMO EVALUATION

For some companies, even large ones with specialized benefits staff, evaluating HMOs is a new and foreign enterprise. However, such investigation is necessary because HMOs are not homogeneous in any respect except prepayment. In the Spencer and Associates 1980 survey of employer-sponsored HMO coverage, however, 107 out of 133 companies offering HMO benefits reported conducting an "in-depth" independent evaluation of an HMO prior to offering it to employees. Three additional companies accepted the evaluation of the Federal Office of Health Maintenance Organizations. Of those companies completing their own evaluations:

- 100 made site visits
- 93 conducted a legal evaluation
- 91 evaluated financial records
- 80 contacted other employers already offering the HMO
- 78 investigated the quality* of medical services offered
- 65 evaluated the HMO's management[23]

To the extent that it is practically feasible, all of the above factors plus some others should be incorporated into any evaluation whose results are to form the basis for a decision regarding the offering of an HMO. Figure 13.2 elaborates on the areas that should be investigated.

In the future the HMO industry is likely to have the majority of new plans developed by a small number of large nonprofit and investor-owned institutions. A survey by InterStudy of Kaiser Foundation Health Plan, Blue Cross/Blue Shield Associations, CNA Financial Corporation, Charter Medical Corporation, INA Corporation and Prudential Insurance Company suggested that these organizations would invest over $250 million in HMOs over the 1981–86 period. Figure 13.3 indicates the private investment in HMOs in 1981 and the anticipated investment by existing HMO sponsors. While noninsurance employers have in a few well-known instances developed their own HMOs (John Deere and R.J. Reynolds are two commonly cited examples), in most cases employers do not have a sufficient number of employees in an area to organize a viable company-based HMO. However, in an increasing number of instances, coalitions and other groupings of companies in an area are encouraging the formation of HMOs, sometimes providing planning funds and occasionally even lending executives to help assure that the HMO runs on sound business principles.

*Quality was not defined.

FIGURE 13.2
EVALUATING HMOs

HMO feature	Important considerations
Track record	Look at community reputation, reaction of other employers, marketing objectives and whether they were met, financial statements and services offered.
Management expertise	Does management have the experience and skills to plan, organize, staff, direct and control the operation of the HMO? This is particularly important in new HMOs, where there is no track record to evaluate.
Marketing expertise	Especially for new HMOs, the quality and size of the marketing staff and the quality of the marketing plan will be critical factors in whether the break-even point can be achieved before capital and loans are exhausted.
Service area	Is the area large enough to sustain a reasonable HMO membership, given other health care resources, including competing HMOs? Is the employment base from which the vast majority of membership will be drawn stable?
Pricing	Is the current pricing structure competitive? If not, are there realistic financial plans that include moving quickly to pricing which is competitive in the local marketplace?
Financial condition	Especially for new and/or small HMOs, is there an adequate capitalization or loan commitment? Special attention should be paid to sources of funds, projections for break-even points, ratio of assets to liabilities, with analysis of whether these appear realistic and whether contingency plans have been developed if the need for funds is underestimated.[24]
Size	The HMO should be large enough to have all major medical specialties and support equipment available.
Physicians	Questions should be asked about physician qualification, accessibility, time to arrange appointments, grievance procedures and preventive care provided.[25] A good physician-patient ratio is considered to be about one full-time equivalent doctor per 1,000 enrollees.
Facilities	In addition to providing after-hours emergency services, are evening and Saturday appointments available? How difficult is it to get authorization for outside services or special treatments not generally available?[26] A visit to one or more of the sites where care is provided can be an invaluable adjunct to these questions. On-site one can get a good sense of degree of organization, attitudes of staff and types of efforts to maintain a satisfied membership.

FIGURE 13.3
PRIVATE INVESTMENT IN HMOs

Name	Active* HMOs	(Type)**	Number acquired	Number started	Membership	Growth projection
Blue Cross & Blue Shield Chicago	44	(21 G) (9 IPA) (14 Net)	3	41	903,944	25 new plans in development
INA Health Plans, Inc. Dallas	8	(5 G) (2 IPA) (1 Net)	6	2	475,000	Acquisitions and 1 new start per year
Prudential Insurance Newark, New Jersey	7	(7 G)	1	6	116,000	2 new plans/year
Charter Medical Corporation Macon, Georgia	4	(4 IPA)	4	0	50,000	Seeking additional mgmt cont
CNA (Intergroup) Chicago	4	(4 IPA)	0	4	61,500	4–5 new states in development
Connecticut General Hartford, Connecticut	2	(2 G)	0	2	92,000	No plans
Wausau (WI) Insurance Co.	8	(8 IPA)	0	8	75,000	No plans

Company		Type**			Enrollment	Plans
Safeco Seattle	4	(4 IPA)	0	4	41,000	No plans
Medserco St. Louis	2	(2 IPA)	0	2	55,000	3 in development/2 per year
Metropolitan Insurance New York City	1	(1 G)	0	1	20,769	No plans
Health Plans Inc. Nashville	2	(2 G)	2	0	38,000	Mgmt contract with option/2 under agreement
John Deere Moline, Illinois	1	(1 IPA)	0	1	11,000	2 plans in development
R.J. Reynolds Winston–Salem, North Carolina	1	(1 G)	0	1	34,000	No plans
Total	85	(39 G) (34 IPA) (15 Net)	16	72	1,973,213	

*Nationwide General Insurance Co., Columbus, Ohio, and American Medical International, Beverly Hills, California, have minimal investments in HMOs. National Medical Enterprises Inc., Los Angeles, has expressed interest in investing in HMOs. Humana Inc., Louisville, Kentucky, looked at but rejected the idea of operating HMOs.

**Key to type of HMO: G = group practice association, IPA = independent practice association, Net = network.

Source: Ernst and Whinney. *Modern Healthcare*, September, 1981.

THE FUTURE

The basic question for employers is which HMOs to offer and the degree of encouragement that employees should be given to join them. In general, it appears that the support for the HMO principle is reasonably founded. HMOs have economic incentives to contain costs and must provide services to the satisfaction of members if they wish to prosper. The preponderance of the limited data to date suggests that HMOs do provide a wide range of health care less expensively than in the fee-for-service system. However, cost conscious employers are increasingly asking whether their average cost per employee for HMOs is higher than it appears due to adverse selection into the indemnity plan. There currently exist reasonable methodologies to estimate the extent, if any, of adverse selection in comparing plans and the related financial impact. At least one study has recently reported that during the year before enrolling in an HMO, those joining averaged 53 percent fewer inpatient days per 1,000 population and had lower hospital and professional medical expenditures than those remaining in a fee-for-service plan. Lower utilization and costs persisted even after age-adjusting the figures. However, the study covered employee groups in only one major metropolitan area (Minneapolis) and a single year of data may not be sufficient to base conclusions regarding adverse selection. Nonetheless, this study indicates that self-selection may be an important influence on the difference in utilization and costs between HMO and fee-for-service insurees.[27]

Where a number of HMOs have competed in a defined geographical area, they appear to have collectively exerted a moderating effect on the rate of rise of health care costs. However, the HMO industry continues to suffer from financial and organizational problems experienced by some of its members. The industry is not mature and considerable shake-out is likely before this point is reached, especially in an increasingly competitive environment. During this period, careful evaluation of each HMO by individual employers or employer health care coalitions is probably indicated. In general, closed panel or group model plans are likely to have a better rate of success than IPAs, but no particular arrangement is a guarantor of longevity.

As the major payers of care become increasingly aggressive in trying to contain costs, providers will develop a variety of organization and financing arrangements competitive with HMOs. They will compete based on convenience, the ability to see a doctor whenever a patient feels the need without having to first see a nurse, nurse-practitioner or other health professional and the ability to maintain relationships with traditional sources of care. These organizations will resemble IPAs except that payment will be based on services provided rather than a prepaid capitation arrangement. HMOs will still, however, be attractive and continue to grow. They will benefit from a softer job market for almost all types of physicians and the reduced regulatory problems that have prevented some HMOs from buying or building their own hospitals. To provide truly comprehensive care, HMOs will increasingly offer dental care, expanded alcohol and drug treatment and rehabilitation programs, hospice care, home health services and health promotion programs. Often these will require supplemental payments.

HMOs will make concerted efforts to offer a range of services to employers, including occupational health services, executive health assessments, employee assistance programs and health education/health promotion programs. In this market they will compete directly with hospitals and health professional groups.

Over time, HMOs will increasingly rate each subscriber group, with current limitations on such rating by federally qualified HMOs likely to be reduced or entirely removed. Employers who have favorable experience due to both the nature of the workforce (and their dependents) and successful efforts to improve and maintain employee health will benefit from this change. Other employers will see their HMO premiums rise faster than average and reduce any level of cost savings which they felt was obtained from HMO enrollment. HMOs are likely to offer more than one plan, with differences in co-payment and some coverage features. The HMO industry will shift from a preponderance of nonprofit plans to the majority being for-profit. Nonprofits will have continuing difficulty obtaining adequate debt financing, and some nonprofits that are operating profitably will convert to for-profits, opening up capital markets.

More hospital-sponsored HMOs will appear as a defense against loss of market share to other HMOs. While some of these will probably be competitive, current experience suggests that they will have more days of hospitalization per 1,000 enrollees than non-hospital-sponsored HMOs.

NOTES

1. Luft, H.S. "Assessing the Evidence on HMO Performance," *Milbank Memorial Fund Quarterly/Health and Society*, Vol. 58, No. 4, 1980, pp. 501–535.
2. *The National HMO Census, 1983.* Conducted by InterStudy. Excelsior, Minnesota, 1984.
3. Luft, H.S. "How Do HMOs Seem To Provide More Health Maintenance Services?" *Milbank Memorial Fund Quarterly/Health and Society*, Vol. 56, No. 2, 1978, pp. 140–168.
4. Op. cit. (Luft. 1980).
5. Op. cit. (InterStudy).
6. *The National HMO Census, 1981.* Conducted by InterStudy, 1982. Released through The Group Health Association of America.
7. "HMOs Cut Hospital Use in Half: GAO," *Modern Healthcare*, Vol. 12, No. 3, 1980, p. 59.
8. Op. cit. (HMO Census 1981 and 1982).
9. Goldberg, L. and Greenberg, W. *The Health Maintenance Organization and Its Effect on Competition.* Federal Trade Commission. Washington, D.C.: Bureau of Economics, 1977.

10. Hay Associates. *1983 Noncash Compensation Comparison.*
11. Ibid.
12. Charles D. Spencer and Associates, Inc. "Survey Details Many Positive Employer Efforts with HMOs," *Employee Benefit Plan Review Research Reports,* May 1981, pp. 314.1–314.9.
13. Ibid.
14. Louis Harris and Associates. *Employers and Health Maintenance Organizations.* Prepared for the Henry J. Kaiser Family Foundation, New York, 1980.
15. Ibid.
16. Louis Harris and Associates. *American Attitudes Toward Health Maintenance Organizations.* Prepared for the Henry J. Kaiser Family Foundation, New York, 1980.
17. Ibid.
18. Ibid.
19. Op. cit. (Charles D. Spencer).
20. "HMOs: The Financial Picture," *Personnel,* Vol. 58, No. 6, 1981, pp. 56–59.
21. Giesel, J., "HMO Funding Drought Could Be Fatal," *Business Insurance,* May 4, 1981, p. 1.
22. Taylor, R.L. "HMOs Discussed at Group Health Meeting," *Employee Benefit Plan Review,* Vol. 36, No. 2, 1981, pp. 16, 20.
23. Ibbs, P. "Employers Work Hard To Promote HMOs But Get Poor Employee Response, Administrative Problems," *Employee Benefit Plan Review,* Vol. 35, No. 6, 1981, pp. 8–9, 12–13, 75–77.
24. Goran, M.J., quoted in Matlock, M.A., "Check HMO's Health Before Signing Up," *Business Insurance,* March 16, 1981, pp. 1, 75.
25. Ibid.
26. Goran, M.J. as quoted in "Healthy Competition," *Personnel Journal,* Vol. 60, No. 7, 1981, pp. 519–522.
27. Jackson-Beeck, M. and Kleinman, J.H. "Evidence For Self-selection Among Health Maintenance Organization Enrollees," *Journal of the American Medical Association,* Vol. 250, No. 20, 1983, pp. 2826–2829.

PREFERRED 14 ARRANGEMENTS WITH PROVIDERS

DEFINING PREFERRED ARRANGEMENTS

Of all the abbreviations bandied about in discussions of health care cost containment today, none is more common than PPOs (Preferred Provider Organizations). PPOs are groups of hospitals and/or physicians who, directly or through a third party, develop contractual arrangements with payers to provide a specified set of health care services under defined financial arrangements. While the essence of the concept is a contract with providers, employers may have preferred relationships without contracting directly with any provider organization. The organization with which employers contract for services can be an insurance company, a third-party administrator, an insurance broker or a health care management firm. Because of the number of variations, perhaps a better way to describe this class of activities would be "preferred arrangements." In addition to employers, other entities with a responsibility to finance health care services for a defined group, including insurance companies, the government and union trust funds, may all buy health care services through preferred arrangements.

Preferred arrangements are so diverse that common characteristics are difficult to identify. However, to date these arrangements have generally included:

1. A defined set of providers who contract with a single organization;
2. An agreement by the providers to render some or all of their usual

range of services for a predetermined price or for a discount off their usual and customary charges to a defined group;

3. The ability of the patient to choose between seeking care from the providers under the contract or from other standard sources of care not covered under such arrangement;

4. Economic benefits to both payers and patients from the use of the contract providers in preference to other providers of similar services;

5. Some mechanism designed to monitor the utilization and quality of care provided by contract providers; and

6. A data system for efficient payment of claims under the contract(s) and to provide reports on utilization and costs under the contract(s).

The most common preferred arrangement to date is for a group of hospitals and/or physicians to agree to waive some or all patient co-payment requirements under the benefit plan. They also provide a discount to payers or agree to a rate schedule in advance which represents savings to payers.

This concept of preferred arrangements is far from new. In many ways the arrangements now being negotiated with physicians or hospitals are similar to those commonly in place under some dental insurance plans. Under such plans, a covered employee has the option of either seeing a nonplan dentist, with reimbursement limited to a fee schedule rate, or seeing a participating dentist who accepts what the plan provides as payment in full. While employees are not limited to participating dentists, the cost differential is usually sufficient enough to shift a considerable amount of business to the plan dentists. The plan is attractive to dentists because most areas are oversupplied with dentists and many would like much more business than they presently have.

Blue Shield and, to only a slightly lesser degree, Blue Cross plans represent a type of preferred arrangement. Under most of its plans, Blue Shield traditionally has gotten physicians to sign a contract with it that a) limits their fees and b) stipulates that they accept what Blue Shield will pay for the services as payment in full. If patients see other physicians there is no guarantee that the charges will be fully covered under the plan. Blue Cross has negotiated contracts with hospitals that include receiving a discount off usual charges billed to and paid by commercial insurers in return for prompt payment of the hospital bill.

The best way of describing many of the evolving preferred arrangements is as intermediate forms between HMOs and private insurance. Providers under these arrangements constitute an integrated alternative delivery system. For the most part, incentives are provided for efficient delivery of services. Patient out-of-pocket costs are lower for the use of the contract services. However, as with private insurance plans, subscribers are free to use any provider for covered services. Financing is based on payment for services received by individual subscribers, not upon a prepaid capitation fee that covers whatever services are required by the subscriber.

REASONS FOR PARTICIPATION IN PREFERRED ARRANGEMENTS

Employers

Preferred arrangements offer a potential opportunity for employers to reduce how much they are spending without cutting benefits. They also allow the employer, who frequently encounters adverse employee reaction to increasing co-insurance and deductibles, to offer employees a way to save money by offering them, in essence, first-dollar coverage.

Hospitals

Preferred arrangements are a mechanism for hospitals to increase or retain their market share from their best customers, the charge payers. From the administrators' point of view, these arrangements give them leverage to effect institutional changes in the direction of efficiency, including holding physicians accountable for the cost of care they generate. These incentives for efficiency are in line with the diagnosis-related group (DRG) based payment by Medicare that pays a flat rate based on patient problems rather than services utilized or charges generated. Another reason for hospital interest is to increase occupancy rates, and some hospitals or groups of hospitals covered under a single master contract are getting a guarantee from the payers of a certain number of patient days per month or per year. As a condition of participation, hospitals are also frequently demanding more rapid payment for services to improve their cash flow.

In a 1983 California Hospital Association study, over one-third of the responding hospitals were already in the process of developing PPOs, and an additional 39 percent were interested in doing so. Of those hospitals already involved in PPOs, 95 percent reported that an important reason for joining was to increase market share, 88 percent to increase occupancy rates and 54 percent to respond to strong market demand.[1]

Physicians

Physicians are willing to participate in preferred groups because they are concerned about the future. In many areas, HMOs are taking an increasingly large market share. Many physicians are having difficulty maintaining their patient volume. New practices are increasingly difficult to build. Larger and larger numbers of hospitals are no longer offering staff privileges to new physicians. Physicians are therefore anxious to develop arrangements which will increase their patient base, even if at some discount. They also prefer working under contractual arrangements that have been developed with the participation of physicians to working under those dictated by the government through Medicare and Medicaid.

Commercial Insurers

For commercial insurers, preferred arrangements represent a new opportunity to stay competitive. These arrangements allow them an edge in competing for fully and partially insured business, because they can better predict and control costs for some portion of the covered services. For the first time, commercial carriers can also obtain the same types of discounts that Blue Cross and Blue Shield have had for many years. Only very recently have states begun granting insurers the right to enter into contractual agreements with hospitals and physicians. In California, for example, insurance companies were only able to win approval to contract with health care providers in 1982, in conjunction with the state legislature's decision that the best way to help make the Medi-Cal program affordable to the state was to allow it to pay only for care in hospitals with which it could conclude favorable contracts.

Blue Cross and Blue Shield, building upon longstanding contractual relationships with both physicians and hospitals, are working to strengthen those relationships and obtain even better terms. They see the new environment as favorable because they have a natural base for preferred arrangements with a large number of providers, and they are also being allowed to offer the same type of coverage as any other insurers and to price their services to different subscribers based upon difference in costs.

Third-party administrators and other health care service and management organizations see the opportunity to get into a new business that can be synergistic with their existing businesses. In some cases they have excellent data systems to monitor utilization, costs and performance.

Cost-Saving Experience

Given the interest from all sides, the remarkable growth of these arrangements is not surprising. Nor is it surprising that much of the discussion about what preferred arrangements can and cannot mean in terms of dollar savings and changes in patterns of care is based on opinion rather than actual experience. Nonetheless, there have been some reports by employers and others who have developed preferred arrangements, which suggest that they hold promise for quick cost savings to employers.

- Rohr Industries, Inc. of Chula Vista, California was able to obtain discounts ranging from 4 to 25 percent on typical fees at six hospitals in the area through self-initiated contracts which guarantee payment within ten days to participating hospitals. According to Doris J. Ramsey, director of risk management for Rohr, the company saved $300,000 during its first year of PPO operation and is planning to expand the arrangement to dentists and physicians.[2]

- By negotiating agreements with local hospitals and physicians, Martin E. Segal, a health consulting and actuarial firm, reports a saving of $330,000 on $4.5 million in hospital-based claims (7.3 percent) in 1981 for clients representing an aggregate subscriber base of 250,000.[3] According to one estimate, three of Denver's provider organizations, which have contracts with five major hospitals and 800 physicians, are saving the city $200 on every $1,000 for medical services.[4]

- Samuel J. Tibbitts, president of Lutheran Hospital Society of Southern California, reports that the society is saving between $300,000 and $400,000 a year on an in-house PPO which was established to create incentives for employees to utilize Lutheran's seven hospitals and 1,300 affiliated physicians.[5]

- Through a combination of self-insurance initiated two and a half years prior, and contracts with three local hospitals which provide discounts ranging from 3 to 7 percent that began seven months earlier, Chatauqua County, New York, a rural county employing about 1,200 people, reported saving $500,000 over a two-and-a-half-year period.[6]

- Aetna Life and Casualty has been a leader among the commercial carriers in the formation of preferred provider arrangements through its new health plan called CHOICE. Under CHOICE, each enrollee must designate a personal physician through whom all nonemergency care will be initiated. Any licensed physician is acceptable for this role. If specialty services are required by the patient, he or she must be referred to physicians and surgeons with whom Aetna has contracted. Any surgery that is necessary must be performed in a recognized referral hospital. By providing a comprehensive benefit package, including the coverage of preventive services, and eliminating deductibles, the plan is designed to encourage the use of primary health care. By allowing enrollees to select their own personal physician to provide primary care and act as the gatekeeper to more extensive medical care, and by offering a broad assortment of carefully selected specialty referral physicians, Aetna believes it is maintaining the element of personal choice for health care which is so highly valued by the American public. Whether this type of arrangement will save employers money depends primarily upon the ability of the specialists to keep patients out of the hospital.

Unfortunately, experiences with preferred arrangements have not been all good. Some third-party administrators have put together multiple employer trusts, associations of small employers who pool reserves to share their risk of adverse claims experience, and in turn the administrators have contracted with some providers to get discounts for covered employees. Often these trusts have not been subject to regulation by the state insurance department, and some have had poor management, have asked for unrealistically low premiums from the employer, and have inadequately monitored utilization and costs under the plan. As a result, a number of these plans, such as COMPETE, run by Far-West Administrators, have gone bankrupt, leaving millions in unpaid bills and leaving the participating employers with a large potential liability.[7]

TYPES OF PREFERRED ARRANGEMENTS

The diversity of preferred arrangements is limited only by the imagination of their developers. One type of arrangement may simply be a discount compared with usual and customary fees charged by physicians or compared with usual charges for services rendered by one or more hospitals. Some preferred arrangements pay based upon the Relative Value Scale system, which assigns each service a relative weight in relative value units and pays a certain dollar amount per unit. Increasingly, however, there is a demand for more sophisticated arrangements with controls that increase the chances of significant cost savings to payers. Since physicians order most health care services, the reduction in prices to a payer can be quickly lost if there is an increase in the number of services ordered or provided by a physician. Therefore, utilization review and a detailed data system are combined to assure that the volume of services does not grow to compensate for lower per service payments. Another method used to keep utilization down is to offer physicians some extra payment at year's end, based on the experience generated by the entire participating physician group. Unfortunately, as the experience of many independent practice associations (IPAs) has shown, this incentive is quite weak compared to the incentive for each physician to maximize his or her own billings under the plan.

Of course, the greatest savings to employers can be achieved from reducing hospital payments. Many hospital preferred provider networks are forming and offering services to employers and other payers. While some are simply offering discounts that are attractive to payers and subscribers, these groups are usually demanding rapid payment as a quid pro quo. Moreover, they are negotiating increasingly for a guarantee of a certain level of utilization or aggregate payment level. Alternatively, they are tying the degree of discount to the volume generated from the contracting employers or payers. These types of arrangements provide additional incentives for employers to encourage strongly the use of the preferred facilities. Both physician- and hospital-preferred arrangements may only be open to payers maintaining a benefit package and set of reimbursement principles that meet provider-developed criteria. Otherwise providers may not receive reimbursement for a number of their services, or will have to collect a significant portion of the bill from the patient, with attendant collection problems and some expected bad debt.

Discounting of regular hospital charges is only one mechanism of providing price advantages and/or greater predictability of payment levels for contracting payers. Payment of a negotiated, all-inclusive per diem or a flat per stay charge is gaining in popularity. Payment based on DRGs could become an even more common mechanism. It has the advantage from the hospital's standpoint of safeguarding it against having a covered population under a specific contract which requires more than the average amount of hospital services per day or per stay.

Each payment mechanism has consequences for economic incentives and disincentives. A discount on charges provides a disincentive to limit the number or type of services provided. Per diem payment provides no incentive to minimize hospital stay. Most importantly, none of these arrangements provide any incentive to eliminate unnecessary or inappropriate admissions, since payment is strictly related to hospital utilization.

From the employer viewpoint, the best opportunity for promoting cost effective care via preferred arrangements is to contract with organizations which represent an integrated system of physicians and hospitals. Hospitals and their physician staffs are starting to go into business together, through a joint venture or other arrangement. Assuming per stay or per DRG-based payment for inpatient hospital services, physicians can benefit by helping the hospital keep costs down. They can be judicious in ordering tests and efficient in scheduling tests and treatments in a way that minimizes hospital stay. They can recommend that patients leave the hospital as early as medically prudent. A data system can demonstrate which physicians are responsible for the highest use of services for each diagnostic and procedure category (e.g., myocardial infarction or hysterectomy). With this approach, both hospitals and physicians have incentives to perform careful and tough utilization review. A longer than necessary hospital stay costs all the physicians money.

A new type of preferred arrangement may further complicate the already confusing welter of options. Some providers are soliciting contracts to provide a narrow, specialized range of services at an attractive price. For example, at least one hospital with a large open-heart surgery unit is proposing to provide open-heart surgery at a lower price than most hospitals in the same area charge for those services. Cardiac surgeons associated with the hospital are also willing to reduce their fees below the usual and customary in the community. However, the hospital is not willing to provide similar arrangements to cover their other usual inpatient services. A group of urologists has built a large, well-equipped surgical suite and staffed it with high-quality anesthesiologists and all necessary ancillary personnel and equipment. They are offering to perform most routine urological procedures at a substantial discount compared with what the same procedures would cost in the hospital, even as outpatient procedures. The potential cost benefits are clear, but the potential logistic complications and employee education and employee relations problems associated with having a large number of different types of arrangements for different types of care in the same area are formidable.

EMPLOYER OPPORTUNITIES

There is no single best approach that all employers can take toward preferred arrangements. The needs of employers differ, based on their workforce, their claims experience, the variation in costs of providers in the company's major locations, their benefit plan, and pre-existing arrangements. In some companies the health benefit plan is so rich that it will be difficult to find ways to save employees money through the use of preferred providers. Therefore, the providers could not anticipate any benefit from the arrangement. However, in most cases at least some short potential benefits for both company and providers can be predicted if the right kind of contract can be negotiated.

Employer Questions and Answers

Faced with an inherently attractive set of new potential cost containment opportunities presented by preferred arrangements, how should an employer react? While each employer has somewhat different circumstances that will in part dictate its posture, there are basic questions which employers should ask in deciding how to approach preferred arrangements.

1. *Can preferred arrangements yield a significant savings to an employer?*
   In these days of aggressive cost-containment efforts, this is universally the first question asked. The short, direct answer to this all-important question is yes. How much money can be saved depends upon the characteristics of the workforce, its usual patterns of obtaining health care, and the types of preferred arrangements that are available. Companies with low HMO enrollment, a geographically concentrated workforce, strong competition among providers and considerable difference in the charges for equivalent services by various providers are the best candidates for preferred arrangements that can achieve significant savings. By contrast, employers with a very dispersed workforce, locations in areas with limited competition, high HMO enrollment and limited differential between charges in any one area for the same services are unlikely to derive substantial benefit from preferred arrangements. How much can be saved will vary considerably. In some cases an employer may somewhat slow the rate of health care benefit costs. In a small number of situations it is conceivable that absolute dollars spent could remain the same over a year or two, or even decline.

2. *Will preferred arrangements continue to save money for employers over time?* Although less frequently asked, this question is probably more important as companies lengthen their planning horizons. One reason for this question's importance is that significant investment is required to develop preferred arrangements, educate employees, change benefit materials and monitor the program to judge compliance with the agreements and to judge their net effects on costs and utilization. A second reason is that if employers concentrate time and resources on preferred arrangements, they devote less attention to other cost-containment opportunities.

   If the rate of increase in costs of health care services is moderated by the offering of competitive contracts to payers, all employers will benefit to some degree. While competition is clearly increasing, the degree to which this will affect the aggregate expenditure on health care services is not known. A more cynical scenario would have the costs being shifted from larger employers and others who have preferred arrangements to those who do not. Cost shifting in health care services is a refined art that is widely practiced. To some degree it is conceivable that smaller employers, individual policyholders, private pay patients and perhaps even public financing programs could show an increase in the rate of cost increases to compensate for reduced revenue from preferred arrangements. This assumes that the rate of inflation will not significantly decline below what it would have been without preferred arrangements. To the degree it is an accurate prophecy, most employers will not benefit from the flurry of contracts with providers in the long run. If it is true, employers will pay higher taxes with the money they save on health benefit expenditures. The best guess is that what will occur is a melange of both of these scenarios.

   If the employer decides that, on balance, the opportunity for preferred arrangements is worth pursuing, another set of questions arise.

3. *What are the relative advantages and disadvantages of employers contracting directly with providers as opposed to buying preferred arrangements through intermediaries?* Direct contracting has intuitive appeal. Employers can choose those providers with whom they wish to negotiate and can negotiate for a set of provisions which they feel are desirable. They are not limited to whatever packages may be available already. However, considerable effort is required to negotiate any contract with providers. Even simple negotiations like arranging for a discount or having providers accept a fee schedule and not attempt additional collection from the patient can often get tangled up in problems of representations, monitoring procedures and demands that payment be made within a certain time period and that prepayment review

for appropriateness not take place. In most cases, however, what an employer wants is more complicated and to knowledgeably negotiate requires expertise in hospital administration, physician-hospital relationships, health law, utilization review programs, health data systems and financing systems. In most cases the employer will not be paying directly but through carriers and therefore needs also to bring the carrier into the negotiations to understand what is feasible. Not many employers have this expertise on their staff nor will they necessarily be willing to make a substantial investment in expert outside consultants to assist them. Nonetheless, a predominant employer in a city or county may decide that there is enough at stake to undertake direct negotiations.

To stimulate the interest of providers and to obtain the most favorable terms requires significant market clout. In most cases employers wishing to contract directly will therefore decide to combine with other employers in the area. Area coalitions usually have sufficient clout to get a favorable contract, and can share the cost of developing preferred arrangements. However, if the companies that ban together represent a large portion of the area's employed population, legal advice is necessary to be sure that there are no violations of antitrust laws. Trusts that represent a significant number of smaller employers or a single large joint labor-management trust could also wield significant market force to obtain attractive preferred arrangements. Smaller employers may be able to piggyback on the contract provisions already negotiated between providers and large employers or employer groups.

While some employers will contract directly with hospitals, physicians and other providers, it is very likely that the majority of arrangements will be made through intermediaries. Intermediaries are looking at preferred arrangements as a new source of profits. Many are quite aggressive in negotiating with providers, arguing that their access to a large number of employers can increase their business. Some, such as Blue Cross and Blue Shield and some administrators of employee health trust funds and/or mutual employer trusts, have extensive experience developing contracts with hospitals and physicians. Therefore it is likely that the number of prepackaged options available to employers will proliferate. Choosing a prepackaged option avoids many of the problems and potential liabilities involved in negotiating directly with providers.

Employers who consider buying a preferred arrangement through an insurance carrier should also be content with the other services they are receiving from that carrier and other aspects of their contract before signing up. Preferred arrangements tie the employer to the carrier much more closely than do usual relationships with carriers. Employers have incentives under preferred arrangements to push insurees to obtain care from contract hospitals and physicians. Insurees have the same incentives. Some will change their usual arrangements for care to benefit from the opportunity. However, since the contract is through the carrier, its longevity is tied to the basic contractual relationship between the employer and the carrier. Even if the carrier does not perform in satisfactory fashion or raises prices for administrative services at a rate that is considered unacceptable, employers will not be as likely to change. Any change could provide incentives for insurees to go to a different set of providers. The change could therefore both adversely affect employee relations and contribute to discontinuity of care.

4. *If the decision is made to contract through an intermediary, how should an employer assess the stability and financial soundness of the plan and the specific intermediary proposing it?* The first step should be to ask to review the contracts between the intermediary and the providers. A second should be to understand how the money will be handled. Will all payment to the providers go through the administrator of the preferred arrangement plan, or will it go directly to the providers through the employer's usual carrier? Are the sponsoring organization and the individuals involved experienced and well respected? Is there a specific provision which holds harmless the employer and employees if the plan has financial problems?

As employers are faced with a growing list of options for preferred arrangements, they will increasingly pose the following question:

5. *How can the cost impact of competing opportunities for preferred arrangements be best analyzed and compared?* This is usually the toughest question. Just because a discount is offered does not mean that the employer will save money. In many cases an employer may not have a sufficient data base about the historical costs and utilization of different providers by their insuree population to determine whether their company would benefit by a particular preferred arrangement. Lack of adequate data may also make it almost impossible to determine which of several offerings presents the best opportunity for savings. For example, estimates of the benefits and costs of a preferred arrangement for hospital services in an area must include consideration of:

a) The current utilization of all facilities in the same market area;

b) The relative price of each facility (case-mix adjusted, if possible);

c) The cost of providing specific incentives such as reimbursing at 100 versus 80 percent, waiving the deductible under comprehensive policy or providing a bonus of some number of dollars, e.g., $200 or $300 paid directly to the claimant;

d) Estimates of the percentage of inpatient claimants who would come from each of the other area hospitals to the preferred facility, and the cost impact of such changes on overall hospital costs.

Even seemingly straightforward proposals can be difficult to analyze. For example, Hospital A has offered an employer a 10 percent discount on all billings. While the 10 percent discount will also apply to what the patient is billed, the patient will be responsible for fulfilling all other usual co-insurance requirements. A simple way to calculate the savings is to take last year's billings from the hospital, project them ahead for inflation and reduce the projected amount by 10 percent. It is probably also reasonable to assume that the attractiveness of the facility to insurees will not be measurably enhanced by the 10 percent discount on the out-of-pocket portion. However, what is the guarantee that Hospital A, which is offering this discount to a large number of employers, will not increase its prices faster than other hospitals in the area, eliminating or reducing the savings? What is the guarantee that utilization per hospital day or per stay will not increase, since the preferred arrangement provides no incentives to be efficient in the delivery of services? If the company does not have an age-adjusted, case-mix analysis, such increases are likely to go undetected.

Hospital B, vying for the same market as Hospital A, is offering what it claims is a superior deal. The hospital has gotten the agreement of its physician staff to share the risk of a preferred arrangement whereby the hospital will provide services for a flat rate per hospital stay. Hospital B is guaranteeing not to increase rates for at least eighteen months. It claims that in its arrangement, hospitals and physicians have the same incentives to be cost efficient in the ordering and provision of care. In addition, Hospital B is offering to waive any patient deductible and to bill only one-half of what the benefit plan would normally require the patient to pay. At first glance, this offer is much more attractive to the patient than that provided by Hospital A and probably no less attractive to the employer. However, Hospital B is already 23 percent more expensive per stay than Hospital A. With significant insuree discounts, it is likely that Hospital B will increase market share. The new result of contracting with Hospital B, assuming that A and B have similar case mixes, will be an increase in overall hospital costs to the employer.

6. *How can the quality of care of the proposed preferred group be assessed?* Assessment of the quality of care is often difficult. Nonetheless, there are some questions which can help signal some problems:
a) Do the physicians involved have board certification?
b) Have most of the physicians been well established in the community for some time?
c) Have any of the physicians been disciplined by the state board

that is charged with overseeing the quality of physician provided medical care?

d) Have any been sauctioned by the local peer review organization?

e) Are all the hospitals accredited by the Joint Commission on Accreditation of Hospitals?

f) Have any of the hospitals been sanctioned or otherwise criticized by the state licensure and/or certification agency?

Profiles of care (utilization, procedure rates, length of stay, etc.) for each physician and hospital may be maintained by local peer review organizations, at least for publicly financed patients. In general, these profiles are not disclosed to the public. However, hospitals and physicians desiring to obtain contracts with payers may be willing to share these profiles with them. Local planning agencies and state planning and health care regulatory and data agencies may also maintain facility profiles which are generally published or are available upon inquiry.

7. *If a contract is let, how can it best be monitored?* To adequately monitor the effects of a preferred provider arrangement requires a sophisticated data system and considerable analytic expertise. Often there are a number of changes occurring at the same time or overlapping in time with the preferred arrangement. Other changes might benefit additions, changes in co-insurance provisions, introduction of utilization review or greater attention by the employer to encouraging early rehabilitation. At the same time patterns of utilization, intensity of services and prices will be occurring in the locales with the preferred providers. A system that can capture the effect of most of the interacting changes requires both good baseline data and careful programming, including multivariate statistics.

Of particular concern should be the pattern of utilization and costs of the preferred providers. Has the average number of visits per insuree remained the same, increased or decreased when he or she switched to a preferred provider? What has happened to the number of hospital days per 1,000? How do the utilization and cost trends experienced by the payer under the preferred arrangement coincide with what is generally occurring in the community and with the insured population of other area employers? These are but a few of the questions which require sustained attention so that the impact of the contract can be assessed.

RECOMMENDATIONS

For employers that decide to take the leap, a few recommendations may be helpful:

1. Look for opportunities where physicians join together with hospitals

in which they have staff privileges to offer a preferred arrangement, with both groups at financial risk if efficient care is not provided.

2. Look for plans which have devoted considerable attention to the selection of hospitals and/or physicians to join them and have not opened up the plan to any interested physician or any hospital in a market area.

3. Favor preferred arrangements that incorporate a well-defined utilization review program. Make sure that you know who will be performing the review and what it will be looking for. Issues dealt with by the review should include quality of care in many dimensions and both over- and underutilization of physician and hospital services. In general, an independent outside review organization is preferable.

4. Obtain some assurance in the contract that the prices of services for preferred providers will not rise faster than those of other providers in the area, and write in a clear method of ascertaining compliance with this provision.

5. Select plans of providers which already can show a pattern of efficient care and which make efficient use of nonpreferred services that are less expensive substitutes for hospital care, such as home health care and hospice care.

6. Select plans that already have moderately priced services compared to usual and customary prices in the area.

7. Select plans which use providers that are acceptable to your insured population and which are in locations convenient to where they live.

8. Require that the plan administrator or provider group have an information system which can provide you with accurate and timely reports and which can also provide ad hoc reports at your request within a specified period of time. If there is any question about the capacity of the information system, either wait until it can be improved or build in strong financial penalties for nonperformance of the information requirements in the contract.

9. Initiate a carefully conceived educational program for employees, dependents and retirees which defines the nature of the preferred arrangement, gives them specific realistic examples of how the use of preferred providers can reduce their out-of-pocket health care costs and points out that they have freedom of choice to use their existing providers if they so wish.

NOTES

1. "Preferred Providers Proliferate," *Washington Report on Medicine and Health/ Perspectives,* June 20, 1983, p. 3.
2. "A New Cure for Health-Cost Fever," *BusinessWeek,* September 20, 1982, p. 117.

3.   Ibid.
4.   Arthur Young & Company, National Health Care Group. "Preferred Provider Organizations," *Industry Briefing: Health Care*, Vol. 5, No. 11, 1982.
5.   Op. cit. (*BusinessWeek*, 9/20/82).
6.   Lawson, J.C. "Preferred Provider Plans Offer Firms Up To a 10 Percent Discount," *Business Insurance*, March 8, 1982, p. 37.
7.   O'Connor, M.L. "Preferred Provider Organizations: A Market Approach to Health Care Competition," *Hospital Forum*, Vol. 26, No. 6, pp. 16–21.

15 EMPLOYER COALITIONS: COMBINING KNOWLEDGE AND CLOUT

*T*here are over 100 local coalitions whose primary concern is health care costs and other issues of health care delivery. Many of these coalitions are composed exclusively of employers. Others include representatives of the insurance industry, hospitals and/or physicians. Between 25 and 50 percent of local coalitions are sponsored by or affiliated with the Chamber of Commerce.[1] While the goals of each coalition differ slightly, those of the New York Business Group on Health encompass those most usually listed:

1. To provide opportunities for members to learn how to operate their own health-related activities more effectively: health benefit programs, employee health services, health education/health promotion activities, employee counseling programs and the other programs intended to enhance the health, well-being and productivity of their workforces.

2. To work with health care providers—hospitals, nursing homes, health service agencies, physicians and other health professionals—to assure the availability, quality and cost-effectiveness of the health services needed by employees and their dependents and the public at large.

3. To provide an arena in which the business community can develop and articulate its responses to health-related issues and problems, and join forces with labor and with government to resolve them.

4. To monitor the actions of city and state governments with respect to health legislation and regulations, and their provision of health services.

5. To provide a vehicle for the collective exercise of social responsibility by businesses in meeting challenges to the health and well-being of the community.

THE MEMBERSHIP DILEMMA

Membership in the New York Business Group on Health is open to all employers in the Greater New York City area except those providing direct health services, labor unions and government agencies.[2] Strong arguments exist both for limiting coalition membership to purchasers of care and for broadening membership. Limiting membership can create a counterproductive, adversarial relationship. As Robert Carpenter, manager of health care cost containment at Republic Steel Corporation in Cleveland points out, "Not all the cost villains are providers. It's absolutely essential we work with lots of people around the table." The contrary view, expressed by Curt Kelly, president of the Employers' Health Care Costs Coalition in San Diego, is that, "It's too hard to get unanimity with groups that include providers," and that "if insurers are allowed to join, coalitions become a forum for selling insurance."[3]

Effective cost containment cannot take place without affecting the delivery system. Many changes require negotiations between the purchasers of care and the major providers. The advent of preferred-provider arrangements makes negotiations common. The argument is over whether these negotiations should take place at arm's length, which suggests that each side formulate its own strategy and only then come together, or whether greater effectiveness and/or efficiency in the process can derive from open discussions with everyone at the same table. Based both on the good and bad experiences of a number of coalitions, it appears difficult to formulate an overall strategy and action plan that addresses the tougher issues when providers, insurers and employers must all concur. As competition increases, coalitions which include providers will become arenas for marketing competing groups of hospitals, physicians, HMOs and other provider constellations.

Determining which groups should be part of a coalition might best be decided by a core group of employers who take an initial look at the specific characteristics of the area and decide how much difficulty various groups are likely to have achieving concensus on meaningful objectives and specific action plans. For example, in an area that is heavily unionized and where Blue Cross/Blue Shield has a dominant market share, it might be decided that the inclusion of these two groups would be very beneficial. Blue Cross/Blue Shield, concerned about its market share eroding as more companies look to self-administer or switch to other, lower cost third-party administrators, is in turn anxious to show how its data systems and cost containment efforts can be an important part of the solution.

What is clear is that nobody wants to be left out of the growing number of coalitions. In 1982 the AFL-CIO, the American Hospital Association, the American Medical Association, the Blue Cross and Blue Shield Associations, the Business Roundtable and the Health Insurance Association of America agreed to endorse the potentials of voluntary coalitions on a local, state or regional basis and to encourage their members or affiliates to participate as part of the coalitions. Obtaining agreement from these diverse organizations on anything is rare and reflects both the fear of being left out and the important opportunities each feels it has to contribute to cost containment.[4]

Given these inherent problems, coupled with the fact that most coalitions are relatively new and that working relationships require time to build, it is not surprising that by 1984 only a minority of them had established measurable objectives and an even smaller number had a discrete action plan.

While most coalitions are still at a very early stage of development, a few have already had an impressive impact. The Michigan Health Economics Coalition was able to effect legislation supporting reduction of bed capacity in acute hospitals, to forge terms of third-party reimbursement legislation and to have an influence in the development of the Health Alliance Plan, a 75,000-member HMO.[5] The history of the formation of that coalition and the passage of the bed-reduction legislation furnishes a view of how local conditions provide opportunities for accomplishment. In 1976 the big three automakers tried to negotiate increased employee cost sharing. When this failed, a meeting was arranged to initiate discussions on what statewide approaches might be adopted. It was attended by about twenty-five people from the legislature, executive branch, big three auto companies, UAW and Michigan Blue Cross and Blue Shield. Providers were not present. A subcommittee was formed, the activity cleared within the respective organizations to the level of at least vice-president and a goal was set of finding a meaningful change on which all parties could agree and which would address some of the sources of rises in costs and could be implemented through state actions. A report by an independent group focused attention on about 7,400 excess beds in that state, an average of 4.4 beds per 1,000 residents. Since the methodology to identify areas of excess beds was available and a state planning apparatus was in place to help implement a law where it passed, this focus seemed appropriate and was accepted by the group members.

It took fourteen months to final passage in June 1978. During this period extensive negotiations, frequently mediated by state legislators, went on with many affected groups, including the hospital association, bankers concerned about capital indebtedness and those concerned about the possible impact on hospitals with large numbers of black patients and minority employees. Concern for the effect on rural hospitals led to an exemption for service areas with fewer than 25 excess beds and special consideration for hospitals located more than twenty-five miles or thirty minutes from other hospitals. Negotiations with the osteopathic physicians, concerned about the effect on their hospitals, led to concessions but not endorsement by that group. What ultimately made the difference was the strong support of Governor Milliken, the UAW (which represented nearly 20 percent of Michigan voters at that time), the automakers and Blue Cross/Blue Shield (which pays 60 percent of hospital bills, including their administrative role on behalf of Medicare). The following year further negotiations with the hospital association reduced the beds identified as excess from 4,620 using the original formula to 3,800 under a law which amended the formula for such identification.[6]

Special local circumstances, including the domination of one union and a highly concentrated industry, considerable experience on the part of these organizations in lobbying and, despite their opposition to the legislation, a reasonably constructive role played by the Michigan Hospital Association, make it difficult to generalize from this success story. Nonetheless, each area has a variety of local factors which facilitate a particular type of cost-containment effort. Hard work, the willingness to commit significant resources to the cost-containment effort and perseverance appear to be necessary for effective action.

One of the strongest continuing efforts attempted by any coalition to impact the health care system has been in Arizona. The Arizona Coalition for Cost-Effective Quality Health Care, which now has more than 1,000 members, proposed and fought to get legislative passage of a two-year moratorium on hospital building and the creation of a hospital regulatory commission. It failed in 1983, but did secure passage of a law requiring hospitals to use a routine billing format based on DRGs. The coalition's hope is that this will make it easier for patients to compare costs of different hospitals. The Arizona coalition pledged to continue fighting for a hospital regulatory commission through an initiative campaign if there was no agreement reached between them and the hospitals. In response, the Arizona Hospital Association authorized creation of a $1.8 million political fund to combat the initiative, if necessary, and to lobby against any future "hostile" legislative initiative proposed by the coalition.[7]

COMMON COALITION ACTIVITIES

Coalitions exhibit significant variation in their resources. Some have no regular staff assistance whereas others have full time staffing. Among the major activities undertaken by coalitions are:

1.   **Data**   Developing common report formats and merging aggregate data from a number of companies so that each can compare their costs,

utilization, etc. As Table 15.1 shows, companies in the same area may have very diverse patterns of hospital bed days per 1,000 eligibles, surgical rates, outpatient claims, and both average annual per capita costs and costs per hospitalization. This is a very useful activity because it gets companies used to working together in the sharing of data, and points up both the trends in cost increases and the unexplained differences in cost and utilization among companies. Embarking on this exercise also frequently involves benefit managers with their carriers to assure the accuracy of the information received and to obtain additional analyses to identify the reasons for significant differences from the experience of other companies, whether better or worse (Chapter 9).

Some information provided by these analyses may be surprising. One of the first coalition-sponsored studies was the product of the Penjerdel Council of the Greater Philadelphia Chamber of Commerce and the Health Services Council, a coalition of businesses in the tristate area of parts of Pennsylvania, New Jersey and Delaware. It was conducted in cooperation with Blue Cross of Greater Philadelphia, which insures 70 percent of the local working population and their dependents. Some findings included: 10 percent of hospital bed days for workers and dependents is for short-term mental and emotional disorders; substance abusers have hospital admission rates 700 percent higher than others; and the cost of a normal delivery is twice as much in a medical school hospital as in a community hospital.[8]

2. *Health Benefits* Developing model packages for benefit redesign. Several coalitions have developed a model benefit package that includes second opinion surgery and better reimbursement for some outpatient services, especially when they can substitute for inpatient services. The model plan may also contain model co-insurance and benefit features. In some cases the coalitions have worked to develop changes in reimbursement for hospital and physician services. The Utah Health Cost Management Foundation, Inc., of Salt Lake, a broad-based coalition with insurers, providers and government officials, is urging 100 percent coverage incentives to consumers for outpatient surgery, and bonuses for surgery performed during certain hours or slow times. They are also encouraging similar incentives for baby deliveries performed at birthing centers. Information to go into these packages can be derived in part from the experiences of the coalition as captured in the reports comparing members' experiences.

3. *Self-Insurance* Reviewing the pros and cons of self-insurance, based on experience of some member companies and other available data.

TABLE 15.1
VARIATION IN SELECTED UTILIZATION AND COST VARIABLES FOR FOUR COMPANIES LOCATED WITHIN THE SAME AREA

Company	Inpatient admissions per 1,000 covered members per year	Average cost per hospitalization	Average length of stay per OB procedure	Average length of stay per surgical procedure	Outpatient cost per outpatient visit	Annual benefits paid per covered employee
A	124.4	$2849	2.7	5.2	$127.47	$1113
B	169.7	3215	2.9	6.5	175.34	1445
C	132.6	2182	2.0	4.8	103.82	929
D	147.3	2460	2.8	6.1	148.27	1216

4. *Utilization Review*   Developing jointly sponsored programs of utilization review. For example, the Employers Health Cost Coalition of San Diego is sponsoring utilization review which covers preadmission certification, discharge planning and review of ancillary services, including x-rays, diagnostic laboratory tests and items like inhalation and physical therapy, which together can total more than half of a hospital bill. The San Diego Foundation for Medical Care, administrators of this Coalition Action Program, is paid $25 per employee hospital admission (1982).[9]

5. *Legislative Activism*   Becoming active in local and state legislative issues that can affect health care costs and health care delivery. Some coalitions maintain a legislative tracking system for such proposed bills. In a number of cases the coalitions have coordinated a strong lobbying effort for or against specific pieces of legislation. Passage of the Michigan law to reduce hospital beds is one example of effective lobbying. Another is the passage of a California statute that facilitates the development of contracts between insurers and selected preferred providers (1982). In addition, a number of California coalitions have supported a bill to establish mandatory rate review for hospitals within the state.

6. *Local and Regional Activism*   Becoming active as a group in local and regional health planning activities. This may include participation in community planning for hospitals and for other health care services and facilities. When the coalition members feel that the planning process is not working effectively, they can also lobby the state government not to issue certificates of need. Some business coalition advocates feel that it is logical that these organizations supplant the current local health planning apparatuses that contain significant provider membership and have many consumers who do not represent business interests. Although business interests alone may not be able to make planning decisions, coalitions, by appointing members to health planning bodies as representatives of a large group of employers rather than a single company, can have a greater impact on the planning process.

7. *Education*   Providing education to corporate executives on the causes of ill health, the nature and degree of the health care cost problems of employers in the area and reviewing options for what can be done both by individual companies and the coalition to alleviate the problems. Business executives who are hospital trustees can learn how to be more proactive in looking at how the needs of the hospital and those of the community for cost-effective health care are meshing. As one

coalition organizer stated, "Our managers on hospital boards seem to be more loyal to them than to their employers when it comes to trying to control hospital costs."

8. ***HMOs*** Helping to establish cost effective HMOs where no HMOs exist or there is little confidence in the ability of existing ones to save money for employers over time. Several coalitions have already helped to establish HMOs in their area. The willingness of large employers to aggressively market a new HMO to their employees can be a strong boost. Where several HMO options already exist, coalitions can evaluate the history and potential of each, and in some instances coalition members have concurred on which they will aid in marketing. Where coalitions are convinced of the superiority of existing HMO options over available insurance plans, they can launch a cooperative education program, encouraging their employees to carefully consider the advantages of HMO membership.

9. ***Health Promotion*** Developing health promotion programs. Several coalitions surveyed by the National Chamber of Commerce in May, 1981 indicated that they are planning to develop fitness, lifestyle modification and other related programs to offer to their members.[10]

Although local coalitions hold great promise, there are a number of barriers to their effectiveness. Effective coalitions require not only company participation but active support of their activities from top management. Without this level of commitment it is usually difficult to get company agreement on participation in coalition activities that require resource investment, that may anger some provider organizations or that could possibly have an adverse effect on industrial relations. In some cases companies have even had difficulty obtaining approval for their coalition dues. Another barrier is the nature of representation on the coalitions. In some cases the individuals chosen by the members to be their representatives have very limited decision-making authority. They may not be able to commit their company to anything without obtaining clearance from several levels up within the organization. Decisions can be delayed, but more importantly, the discussion of issues at the coalition meetings can be greatly inhibited due to reluctance to advance a company position.

Another problem is the diversity of knowledge of and experience with health care and cost containment among coalition member representatives and others within their respective organizations. Coalition members may range from organizations who have been working diligently to contain health care costs for a number of years to those who have recently recognized the problem and have not yet decided what kinds of internal and external activities they are willing to undertake. It is therefore difficult to achieve consensus on substantive matters that require an active role for coalition members. An essential initial activity is therefore the education of members, followed frequently by a joint data collection effort which provides the opportunity to work collectively without requiring a significant commitment to an action program. These two activities have dominated the early history of many coalitions, and by 1984 only a minority have progressed to specific action projects.

NATIONAL AND REGIONAL COALITIONS

The strongest and most active coalition to date has been at the national level. The Washington Business Group on Health (WBGH), founded in 1974 with only 5 members, has grown to more than 200 members. The WBGH was established by industry to generate, monitor and report on private-sector, health cost-management initiatives. The WBGH has taken a leadership role in all areas of corporate health management: health-related cost management, employee mental health and wellness, health promotion, and disability and rehabilitation. In addition to having worked as staff for the Health Task Force of the Business Roundtable, the WBGH has been instrumental in establishing coalitions in several other cities, including Pittsburgh, Seattle and Columbia, South Carolina. Ongoing activities of the WBGH include:

- Development of position papers representing the views of industry on current issues in health care delivery and financing. Staff frequently deliver testimony at congressional committee hearings on proposed legislation.
- Sponsorship of conferences and educational seminars for the business and health communities on a wide variety of corporate health management issues. Staff frequently make presentations on their particular areas of expertise, including health-related cost containment, private-sector initiatives in health care, disability and rehabilitation and employee wellness programs.
- Provision of technical assistance to emerging coalitions across the country in mobilizing support, obtaining funding and educating others in their communities.
- Monitoring of employers' prudent buyer projects. Through its surveys, conferences, and discussions with employers, WBGH has become a national clearinghouse for these activities.
- Provision of technical assistance and monitoring of employer and insurer utilization review programs. WBGH staff advise employers on the design of utilization data systems, the interpretation and use of data and the relationship of new DRG data analysis for benefit redesign, system reform and the establishment of health enhancement programs.

- Assistance to member companies in the design and implementation of wellness programs, including their integration with the corporate health and medical benefits. WBGH staff coordinates the Evaluation Working Group, composed of companies with strong interest in the evaluation of employee health enhancement activities.

Rapid dissemination of information from the WBGH to the business community has been facilitated by the recent addition of *Healthnet*, an interactive computer service that links WBGH members, local business groups and subscribers. In addition, late in 1983 the WBGH began the newsletter *Business and Health: A Report on Health Policy and Cost-Management Strategies*, which is published ten times a year.

The largest regional coalition is the Midwest Business Group on Health, established in 1980 and with over seventy members by 1982. Organized to help control the cost of health care and slow the growth of employee expenditures, its two initial priorities have been data and local coalition development. The group is helping members work with insurance companies or administrators to get better reports on what their health benefit dollars are buying. This focus is in line with the results of a 1982 survey of members which identified management information as the top goal. A second priority is the development of local chapters, and these have already been formed in Chicago and Rockford, Illinois. Other stated targets, developed on the basis of suggestions from members, include the development of more utilization review, the investigation of pricing alternatives, such as preferred providers and state rate setting, employee education about the high costs of health care and alternative providers and coverages, improved disability management, and education for hospital trustees from member companies.

THE FUTURE

Coalitions have the potential for being an important vehicle to help employers better manage health care benefits. The purchasing power represented by a strong coalition can be particularly important in an era of increasing freedon with respect to the types of arrangements that can be developed among direct purchasers, insurers and providers. Many of these arrangements would be much more difficult, if not impossible, for most companies to develop individually. Coalitions can also develop and refine common data bases that will permit more accurate identification of common problems and tracking of whether the action programs are effective in achieving their objectives.

It is unlikely, however, that coalitions will reduce the amount of activity necessary within a company, since many opportunities to reduce the rate of rise of health care costs must be evaluated with respect to the unique features of every corporate environment, and decisions are made based on what is best for that employer. Changing a benefit package, for example, is always an important decision that each company will wish to consider carefully. It is unlikely coalitions will ever be able to get uniform agreement on specific benefit changes that all agree to put in place simultaneously. Different pre-existing packages, labor relations issues including constraints of labor agreements, other benefit trends, company profitability and cost impacts, as well as other unique historical factors, are all likely to influence both decisions and timing of implementation.

Perhaps the greatest benefit of employer coalitions will be strong, unified positions on local health planning issues and on local, regional and national legislative issues. If business speaks with one voice, it will be a very powerful force. This is especially true if business is willing to use some of its chits with legislators to gain passage of bills that can help contain costs, such as hospital rate setting or contracting with preferred providers. Hospitals and physicians may be more careful about raising prices and increasing utilization if they realize these actions can have important repercussions. What would happen, for example, if in response to rapidly rising physician prices, the coalition recommended to its members that they consider paying at the 75th or 80th percentile under UCR compared with the 90th percentile as is currently done by most? A hospital with already higher than average costs may think twice or more before taking actions that will cause their prices to rise even further in the face of a coalition investigating preferred-provider contracting for its members representing 60 percent of the local population.

To be an effective coalition member, an employer must be represented by an individual with the authority to speak for the company and make some decisions without additional clearance, and with enough knowledge to be an active participant. To be sure the time and money associated with membership are a good investment, it is important to get the coalition to establish clear objectives and, after a suitable period of time, for employer representatives to feel comfortable with joint activities to implement specific, action-oriented programs.

Diversity of membership can have an impact on effectiveness. In general, it is easiest to get agreement and to advance most rapidly when the membership represents the same economic interests. Coalitions that limit their membership to nongovernmental employers seem to have had the greatest success in agreeing on specific activities. However, since government is the largest employer, it is important to include governmental organizations in coalitions. It is also important from the beginning to establish a liaison with the other economic interest groups, labor, insurers, hospitals, and physicians.

NOTES

1. "100 Business Groups May Do Health Planning By 1990," *Modern Healthcare*, Vol. 11, No. 10, 1981, p. 14.
2. The New York Business Group on Health, Newsletter. Vol. 1, No. 9, December, 1981, p. 1.
3. Rundle, R.L. "Businesses Team Up To Contain Health Costs," *Business Insurance*, December 14, 1981, pp. 13, 18.
4. Keenan, C. "Boon or Bane, Business Coalitions Have Entered the Health Care Scene," *Hospitals*, Vol. 56, No. 2, 1982, p. 64.
5. Ibid.
6. Government Research Corporation. *A Report on Coalitions to Contain Medical Care Costs*. Washington, D.C., October, 1979.
7. Keppel, B. "Firms Dun Hospitals for High Costs," *Los Angeles Times*, July 30, 1983, p. 10.
8. Op. cit. (Rundle).
9. Blitzer, C.G. "San Diego Firms Attack Hospital Overcharges," *Business Insurance*, May 17, 1982, p. 3.
10. U.S. Chamber of Commerce. *National Directory of Local Health Action Programs*, 1981.

16 HEALTH CARE REGULATION: A ROLE FOR EMPLOYERS

EMPLOYERS AND GOVERNMENT REGULATION

Hundreds of federal, state and government-affiliated agencies share a sometimes overlapping responsibility for intervening in various aspects of the health care system. Thousands of health care regulations already exist and many new ones are added each year. Regulations, of course, cause distortions in the medical marketplace and affect the cost, quality and availability of health services.

Employers are greatly affected by many of the laws and regulations. A few examples of this impact are:

- The 1973 HMO Act (PL93–222) required that employers with over twenty-five employees offer as a health insurance option enrollment in at least one federally qualified HMO that was operating in the geographic area, and requested that membership be offered to employees.
- Many state insurance departments have mandated coverage to be included in all health insurance policies sold in the state. For example, a number of states require certain catastrophic coverage and some require specific levels of mental health inpatient and outpatient cov-

erage and coverage of treatment for alcoholism and drug abuse. Such provisions directly affect premium costs.

- Rate setting by an increasing number of state governments affects the prices and nature of services in all hospitals and frequently nursing homes.
- Medicaid principles of reimbursement, usually required to be promulgated in regulatory form, effect reimbursement to almost all health care providers and usually shift costs to private payors.
- Approval of new building and renovation requests under state certificate of need statutes have major impacts on present and future hospital costs and on the degree of redundancy built into the hospital system.

Regulations influencing health care service delivery are broadly divisible into those with principally economic objectives and those with primarily social objectives. The private sector has an interest in the social aspects of regulation in its role as a good citizen and influential major taxpayer. From the point of view of good corporate health management, however, there is a strong interest in economic regulation, which attempts to influence the distribution, pricing, quality and delivery of health care.

The seven major types of economic regulation in health care are facility standards, utilization review, capital expenditure controls, health maintenance organizations, state rate setting, federal and state principles of reimbursement for Medicare and Medicaid and regulation of insurers and policies that can be sold. Understanding the origin, nature and effects of each major type of economic regulation can help industry decide where and when to become an active participant in the regulatory process.

Standards of Construction and Operation for Health Care Facilities

The establishment and strength of health care facility standards has been motivated by two forces: government mandate, through licensing laws and conditions of participation in public financing programs, and voluntary activities of health professional groups concerned with quality.

With the passage of Medicare and Medicaid in 1965, the federal government required all participating hospitals to become certified as meeting a uniform set of federally established conditions of participation, which were based primarily on standards established by the Joint Commission on Accreditation of Hospitals (JCAH), a private organization sponsored by organized medicine and the hospital industry. Hospitals which were accredited by JCAH did not have to become certified. The intention of the certification process was to bypass state licensing laws which were viewed as inconsistent and inadequate.[1]

Presently most hospitals are affected by three sets of standards:

1. Licensing laws which control the opening of new health facilities and
   , support and enforce standards of performance among licensed facilities
   in operation;
2. Medicare conditions of participation which require facilities to meet a
   set of minimum standards in order to receive reimbursement; and
3. JCAH accreditation standards which represent a voluntary system of
   professional quality. assurance.

Recognition of duplicative efforts and inconsistencies among the three programs has led to a movement to try to bring the standards into greater conformity.[2] Some states and the federal government generally accept JCAH accreditation as demonstrating that their requirements for licensure and certification have also been met. However, both levels of government also inspect the hospitals on a periodic, random or other basis. Some inspections are specifically to validate JCAH findings. Some hospitals which choose not to apply for JCAH accreditation or do not pass accreditation still may be licensed and/or certified under Medicare.

Review of the Appropriateness of Health Care Facility Utilization

Government-mandated utilization review activities originated in 1966 under Medicare conditions of participation which required all hospitals to establish committees to review the records of Medicare patients to determine appropriateness of admission and length of stay, and to promote efficient use of hospital resources. Review committees had the power to recommend denial of reimbursement for inappropriate care.

By 1970 it was apparent that Medicare hospital utilization review committees were ineffective in controlling hospital use. Data which indicated that physician-controlled review systems had potential for decreasing the costs associated with unnecessary utilization, led to the enactment of P.L. #92–603, the Professional Standards Review Organization (PSRO) program. The PSRO legislation "created a nationwide network of physician-run organizations that had final authority to grant or deny payment for care rendered to federally supported patients under both Medicare and Medicaid."[3] PSROs were mandated to assure that the delivery of all federally financed care is medically necessary, conforms to professional standards of quality and is efficient.[4]

The hospital review system implemented by most PSROs consisted of three interrelated functions:

1. Concurrent review to determine medical necessity for all admissions
   and all extended lengths of stay;

2.   Retrospective medical care evaluations to determine if professionally accepted standards of quality care were met; and

3.   A profile analysis of the patterns of health services utilization and patterns of care in the area based on aggregate patient discharge data.[5]

Were PSROs more cost effective than the utilization review programs they replaced? This question has not been definitively answered by any of the evaluations to date nor is there much available information about the effect of PSROs on quality of care. Lack of conclusive evidence of effectiveness led to a reduction in federal funding for PSROs, and, ironically, mandated utilization review was a dying concept at the same time that the incentives in the new Medicare principles of hospital reimbursement clearly required hospital utilization review.

Capital Expenditure Controls

The availability of health care facilities and services affects their use and the costs of health care. State certificate-of-need (CON) legislation was designed to check the unrestrained growth of the health care system by requiring that new hospital and long term care beds and specialized equipment and services receive approval from a public body comprised of both providers and consumers.

CON legislation was strengthened by passage of the P.L.93–641, the National Health Planning and Resources Development Act of 1974, which mandated that certificate-of-need legislation be enacted by all states by January, 1980. By 1981, forty-nine out of fifty states had implemented CON legislation,[6] although content of CON laws varied widely among states with respect to services covered and threshold amounts for capital expenditures which require review.[7] Congress has raised the thresholds for capital expenditures under the program from $250,000 to $600,000 and with the growth of interest in competition there has been a large jump in CON applications and less resistance to approval in many states. To moderate capital growth, New Jersey mandates a facility-specific cap on the ratio of capital reimbursement to total reimbursement. In 1982 the cap was set at 12.9 percent.[8] Massachusetts, under its all payer rate setting, limited increases in capital-related costs to 1.5 percent of total hospital expenses in 1983.[9] New York State went even further by imposing a moratorium on new health care facility construction.

No study to date has been able to demonstrate that CON is achieving the intended outcomes of decreased aggregate hospital capital expenditures and slower rates of hospital cost increase. There is good evidence to suggest, however, that CON has reduced the dimensions and costs of many hospital expansion projects and prevented others from ever beginning.[10]

Health Maintenance Organizations

The HMO Act of 1973 was passed by Congress to stimulate the growth of prepaid group practices as a lower cost alternative to traditional fee-for-service medicine. The HMO Act authorized grants to public and private nonprofit organizations interested in becoming HMOs for feasibility studies, planning and initial development. In addition, federal loans and loan guarantees were available for initial operating debt for the first three years.[11] Of importance to businesses is that the act also required all employers with more than twenty-five employees who offered health insurance benefits to include HMO coverage as an alternative to traditional insurance if a federally qualified HMO was operating in their locale and requested that membership be offered to employees and their dependents. This provision of the act is known as the "dual choice" mandate.

For an HMO to become federally qualified, however, it had to fulfill some very stringent requirements, including the provision of a much wider array of services than was typically covered under private health insurance, such as preventive services, mental health services, and treatment for drug and alcohol abuse. All federally qualified HMOs were required to use community-rating, were restricted in the amount of co-payment which could be charged to patients and had to offer a 30-day period of open enrollment once a year. The HMO Act also required the disclosure of certain information, maintenance of adequate financial reserves and a quality assurance program.[12]

Soon after the HMO Act was implemented it became obvious that the act's requirements were dissuading organizations from becoming federally qualified HMOs. In 1976, later in 1978 and again in 1981, amendments to the HMO Act were passed to lessen some of the burdensome requirements of the original legislation.[13]

Although the federal HMO program was clearly the impetus for the development of many new HMOs, the impact of the legislation was significantly less than anticipated by its designers. In 1970 there were fewer than 30 HMOs serving 2.9 million people and in 1980 there were 230 HMOs serving 9 million enrollees. However, 85 percent of total HMO enrollees in 1980 were in HMOs that had developed privately.[14] Since federal support for HMOs is no longer available, this figure can be expected to increase in future years.

State Rate Setting, Including All-Payer Plans

Nowhere is the tension between regulatory and competitive approaches more evident than in business attitudes towards moderation of hospital cost increases. While most employers favor pro-competitive approaches to slow hospital inflation, many have supported state hospital rate setting legislation. Long the victims of cost shifting as Medicaid and more recently Medicare have shaved their share of hospital payment, employers have increasingly supported statewide all-payer systems which introduce prospective hospital budgeting with absolute limits for each hospital. Of equal importance, all-payer plans either require that each payer pay the same amount for equivalent hospital services or, more commonly, limit the differential between the various public and private payers.

Maryland, New Jersey, New York and Massachusetts regulate all-payers while Rhode Island, Washington and Connecticut have considerable experience with state rate setting applied to some payers. In 1983 Maine, Wisconsin and West Virginia all passed legislation which, to differing degrees, established or moved toward all-payer systems. In many other states throughout the country employers are pushing for all-payer legislation.

Each of the mandatory programs has adopted its own unique approach to rate setting based either on budget review and negotiation with individual hospitals or the establishment of formulae which determine reimbursement based on hospital performance relative to other hospitals with similar characteristics such as size, location, case mix, teaching versus nonteaching or a combination of these.

All-payer systems provide a mechanism to limit growth in the hospital sector and improve the equity between the public and private sector. However, regulation often tends to reward the least efficient providers, reduce competition and conserve excess capacity. Carolyne Davis, Administrator of the Health Care Financing Administration, claims that "under rate setting, there may be less incentive for private payers to increase cost-sharing, to develop health maintenance organizations and other alternative delivery systems, such as preferred provider organizations and to develop effective utilization review programs."[15]

On the other hand, some proponents feel that state rate setting can provide proper incentives, by allowing hospitals to keep some savings if their costs are lower than their prospectively determined budget. By moving toward a flat-rate reimbursement per DRG or other case mix adjusted index of hospital resources utilized, efficient providers can emerge and less efficient ones drop out of sight. What is ironic is to see employers who spend considerable resources fighting regulation of their businesses be such strong proponents of regulation of the health care industry. Regardless of philosophical leanings, what is clear is that regulatory approaches usually work more quickly than competitive strategies, with considerable intuitive appeal to those frustrated by largely unsuccessful efforts to slow health care inflation and stem the tide of more cost shifting from the public to the private sector.

Several studies which have compared the rates of increase in hospital costs between states with mandatory rate setting programs and those without have found slower rates of increase in the rate setting states by on the order of 3 to 5 percentage points.[16] In one of the most comprehensive assessments of mandatory prospective reimbursement systems completed to date, it was found that "after adjusting for economic and population growth, mandatory prospective reimbursement states maintain a lower rate of growth in hospital revenues of 3 percentage points per year than retrospective reimbursement states."[17]

Medicare and Medicaid

Titles XVIII and XIX of the Social Security Act, Medicare and Medicaid, became operational in 1966 to alleviate economic barriers to health care for the elderly and the poor. The programs are distinct entities. Medicare provides health insurance benefits to 29.5 million persons,[18] including most people over sixty-five, recipients of Social Security Disability Insurance (after twenty-four consecutive months of benefits), and victims of end-stage renal disease. Part A of Medicare, the Hospital Insurance Program, is a social insurance program financed primarily through the Social Security payroll tax. It covers hospitalizations, extended care and home health services and is free to all persons over sixty-five who are entitled to monthly cash benefits under Old-Age, Survivors and Disability Insurance, or railroad retirement programs.*

Part B of Medicare, the Supplementary Medical Insurance (SMI) program, is financed by voluntary premium payments and general revenues, and is available on an optional basis to all persons eligible for coverage under Part A, as well as seniors ineligible for Part A. Part B covers physician services, ambulatory care, home health care, and radiology and pathology services furnished by physicians to hospital inpatients.**

Unlike Medicare, which is strictly a federal program, Medicaid is a joint federal–state program that provides medical assistance to certain categories of low-income people including aged, blind and disabled cash re-

---

*The small percentage of individuals over sixty-five who are not eligible for free coverage may enroll voluntarily for a monthly premium ($113 per month in 1983) if they are also enrolled in Medicare's Supplementary Medical Insurance program.[19] The hospital insurance program provides beneficiaries with 90 days of hospitalization per benefit period, the first 60 of which are free after a deductible roughly equivalent to one day of hospitalization ($304 in 1983). (A benefit period is an episode of hospitalization and/or nursing home care, beginning with the first day not included in the prior benefit period and ending when the person has not been an inpatient for 60 consecutive days.) The next 30 days require a daily co-insurance equal to one-fourth of the deductible ($76 in 1983). If an individual requires more than 90 days of hospitalization in one benefit period, he or she may draw upon the lifetime reserve of 60 days which is furnished to each enrollee at a daily co-insurance of 50 percent of the deductible ($152 in 1983). With respect to nursing home services, the first 20 days of skilled nursing care in a benefit period are free; thereafter, a patient must pay a daily co-insurance equal to one-eighth of the inpatient hospital deductible ($38 in 1983).[20]

**The monthly premium for Part B was $12.20 as of 1983.[21] SMI requires an annual deductible plus a 20 percent co-insurance for all services other than home health care and inpatient radiology and pathology services. These services are not subject to any deductibles.[22]

cipients of SSI and members of families receiving Aid to Families with Dependent Children. States may opt to provide Medicaid to the "medically needy," persons covered under one of the qualifying programs who have enough income to live on (and so are not receiving welfare) but cannot afford their medical care. In 1982, 21.7 million persons received Medicaid.[23]

Although Medicaid is optional for states to participate in, all states have a Medicaid program. Medicaid is administered by the states, who currently receive 50 to 77 percent of program costs in matching federal funds, depending on the state's average per capita income.[24]

The federal government requires that certain basic services be covered in any Medicaid program: hospital care, outpatient hospital services, laboratory services, x-rays, skilled nursing care (for patients over twenty-one), home health care, physician services, family planning, rural health clinic services and early periodic screening and diagnostic testing (EPSDT) for persons under twenty-one. In addition, states may opt to provide a number of extra services, including prescription drugs, dental care, inpatient psychiatric care, etc.[25]

Other than requiring the basic services and certain minimum standards for participating providers, the federal government has allowed states to control their Medicaid programs. States determine the scope of services offered, reimbursement rates for services other than hospital care,* eligibility levels and cost-sharing provisions.**

Medicare and Medicaid are noted for their inability to control program costs. Between its first full year of operation and 1982, Medicare costs rose over 10,000 percent, to $52.2 billion. Medicaid program costs have also grown at an exponential rate, reaching $34 billion in 1982.[27] State Medicaid programs are responding to fiscal and political pressures by eliminating nonmandated eligibility categories, cutting back on optional services and establishing restrictive reimbursement policies.

Starting in late 1983, Medicare began a three-year transition to move to prospective payment for all hospital inpatient services based on diagnostic-related groups. Payments will be increasingly based on national rather than regional rates, although an urban/rural differential will be maintained and adjustments will be made to reflect local labor costs. Every hospital, to protect against provider abuses, is required to contract with a Peer Review Organization. States have the opportunity to ask for waivers to substitute their own cost-control systems.

---

*Medicaid reimbursement for hospital care must follow Medicare's reasonable cost-payment system, unless a waiver to use an alternative system is obtained from the Secretary of Health and Human Services.
**"Nominal" cost-sharing may be imposed on optional services for cash assistance recipients and on any services for the medically needy.[26]

Regulation of Private Health Insurance

Regulation of private health insurance is predominantly a state function which is carried out by an insurance department (or branch of another department or agency) under the direction of an insurance commissioner who is either appointed or elected. Traditionally, the major public policy objectives of regulating the health insurance industry, like all lines of insurance, have been to assure financial solvency of insurers and to protect policyholders against fraud or unreasonable policies through policing of advertising and marketing strategies.

In recent years state legislatures have passed laws affecting health insurers in four main ways:

- Require expansion of the scope of coverage by mandating a specific type of coverage, e.g., treatment of alcoholism;
- Provide policyholders with greater freedom in choosing the services of a nontraditional provider;
- Mandate a variety of health insurance plans for state employees; and
- Strengthen the administrative and fiscal oversight of the insurance industry.[28]

In general, regulation of commercial carriers targets marketing practices (e.g., advertising, clarity of policy coverages and exclusions), while regulation of the nonprofit Blue Cross and Blue Shield organizations deals primarily with the premium rates charged to subscribers. Changes in Blue Cross/Blue Shield rates require prior approval from the insurance department while rates of commercial carriers, reviewed for reasonableness on an ongoing basis in two-thirds of states, do not receive the same degree of scrutiny, nor do they usually require prior approval.[29] Since Blue Cross and Blue Shield have traditionally had a significant market share of the private sector business, delays in approving rate increases and denials of such requests sometimes have been used as attempts to moderate health care cost inflation.[30]

LESSON FOR EMPLOYERS

Political Lessons

A considerable amount has been learned from these regulatory efforts. Employers who are working to control their health care costs and wish to contribute to a national solution to this problem can benefit from understanding these lessons. Armed with knowledge, it is easier to decide when to enter the fray of health care politics. The lessons from past experience should be helpful in clarifying when regulation can contribute to the solution of a problem. The lessons can also help involved employers prepare a more persuasive case for deregulation where the employer believes that type of change is required. Finally, coalitions and other employer groups can benefit from past lessons by learning how to help make existing regulatory programs serve their interests to control costs and assure affordable, high quality care.

Some of the valuable lessons from past regulatory experiences come from the political arena.

1. *Public Support*   It is difficult to muster broad public support for efforts to constrain reimbursement for services provided by health care professionals, particularly physicians. Although perceptions are changing, most Americans, including those in the workforce, have quite limited appreciation of the health care system's complex interrelationships and are unaware of the tenuous relationship between additional inputs into the health care system and improved health status.

   Providers of care are generally persuasive in labeling all possible regulatory economic constraints as destructive to high quality health care. For all these reasons, corporate positions in favor of cost-effective health care can place them without political allies and at risk of being harshly criticized as "uncaring" by providers.

2. *Political and Economic Influence*   The rapid growth of the health care sector has been accompanied by its rapid growth in political and economic influence. For example, of the more than $32 million contributed in 1978 to congressional candidates from political action committees (PACs), more than one-third ($11.5 million) was from health-related PACs.[31] During the 1981 election cycle, AMA–PAC was, as usual, first in PAC campaign contributions with $1.84 million.[32] In recent years both the hospital and nursing home industries have also formed PACs and have been large contributors to political campaigns. Industry has partially mobilized to make its views, concerns and suggestions known through support of the Washington Business Group on Health, which undertakes policy analysis and presents its members' views before appropriate congressional committees and to the administration. The Business Roundtable, a powerful group of top executives from large companies, is also making its members' views known on health care issues that involve lawmaking and/or rulemaking.

3. *Political Will*   Until health care costs had risen to levels which broke federal and state budgets, the political will to address directly many of the fundamental problems in health care delivery was lacking. Instead of passing laws that dealt directly with economic incentives or creating effective mechanisms to slow the rise of health care costs, Congress and state legislators had been content to pass laws which, at

best, dealt with a single portion of the system, a small fraction of the total dollars spent for health care, or only affected costs indirectly. Only recently with the move to prospective DRG-based reimbursement and the changes brought about by TEFRA has Congress acted decisively. Some states have developed reasonably effective hospital cost-control systems, but in most cases only when their Medicaid program expenditures had already led them into large deficits. The private sector can help apply measures so that problems do not become as severe before effective action is taken.

Process Lessons

4.   ***Due Process***   Due process requirements have sometimes impeded effective regulation. The Federal Administrative Procedures Act delineates a specific procedure which must be followed by every administrative agency in promulgating regulations. These requirements are intended to assure a fair open regulatory process with ample opportunity for input by the regulatees, third parties with a strong interest and interested private citizens. However, the exacting due process requirements also assure that the time lag from the beginning of staff work to final rulemaking is protracted and cumbersome, often as long as two years. In addition, very strict process requirements encourage "gaming" the system by representing providers.

5.   ***Criteria for Decision-making***   Many regulatory decisions are made in the absence of clear rules to guide the decision-making process. With health care it is not clear how much is enough or how to distinguish "needs" from "wants" or "demands." Therefore, quantitative rules based on even the best available evidence are apt to be vehemently assailed by providers as inappropriate. When measurable standards do not exist, the onus usually falls on the regulators to justify why, for example, a hospital can't have the rate increase it claims is necessary for high quality care, or why there can't be more nursing homes in an area when there are long waiting lists at existing facilities.

6.   ***Cost-effectiveness***   Increased attention to cost-effectiveness analysis in health care services would greatly improve the quality of discussion on what further resources are "needed" in the health care system. In other relevant health areas, agencies such as the Consumer Product Safety Commission and the Environmental Protection Agency are required by law to balance the array of social and economic consequences in arriving at regulatory proposals. To date, providers generally have argued that no cost increase is too great if better health for a single individual can result. In the final analysis, how much we spend on

health care should not continue to be the sum total of thousands of difficult decisions where those who say no are accused of opting for poor quality care. Better options are a national political decision on how much more spending will be allowed and/or a restructured marketplace with improved provider incentives.

Organizational Lessons

7. **Coordination** Most regulatory responses have not directly addressed the basic problems in health care. In general, they have not changed the incentives nor have they taken a broad view of the problems. As a result, no global, coordinated strategy for dealing with current structural problems has emerged. A fragmented set of regulatory programs and financing schemes exists, each designed to treat a different symptom without addressing its underlying cause. Often these activities work at cross purposes, harming each other rather than acting synergistically. A recent notable exception is the establishment of prospective reimbursement of hospital care under Medicare. This law has changed the incentives for hospitals in the direction of efficiency.

8. **Cost Shifting** Cost-containment regulation leads to cost shifting without clearly constraining costs. For example, state rate regulation, when not applicable to all payers, has led to cost shifting from regulated (public) to nonregulated (private) payers. The acceleration of cost shifting to private payers in the early 1980s contributed to large increases in employer health care costs. Increasingly stringent Medicaid reimbursement policies for hospitals have led to Medicaid paying less than actual costs. Expensive equipment that would require a certificate-of-need to be installed in a hospital may appear instead in a physician's office.

9. **Self-regulation** Governments are generally desultory in regulating their own health care institutions. The Veterans Administration, Department of Defense and Indian Health Service hospitals are exempt from most regulations. State governments are rarely as aggressive in enforcing staffing and facility standards for certification in state-run hospitals as in nonstate facilities, because strict enforcement can mean larger additional expenditures for strained state budgets.

10. **Provider Incentives** Few positive incentives exist for compliance with many regulatory requirements. Rarely is public recognition accorded health care providers who have been exemplary in conforming to economic or even social regulations. More important, there is usually little to be gained from not trying to get away with as much as possible. For

example, under rate setting when next year's budget is based on this year's actual costs, and when you are not permitted to retain budget savings, why work diligently to effect savings?

11. **Regulator Incentives** Strong disincentives exist for regulators to be diligent in carrying out their mandate. Regulators have had few allies, and operate in an environment where regulation is decried as adversely affecting quality of care and doctor–patient relationships. Young, bright and hardworking men and women who enter the corps of regulators with excitement over the potential for public service frequently burn out and usually leave disillusioned by lack of constituency and overwhelmed by the political power of provider groups. Many who remain content themselves with adhering to process requirements and rarely sticking to their initial best judgment in politically charged situations. Often they lack strong leadership and support from above. Most serious, however, is the pervasive climate of increasing mistrust of regulation by the public and denigration of public service as a worthwhile career choice. These factors, frequently coupled with low salaries, poor working conditions and inadequate resources to achieve the regulatory objectives, make it difficult to attract and retain appropriately trained individuals who also have experience within the health care delivery system.

POSITIVE EFFECTS OF REGULATION

Because of the myriad problems associated with health care regulation, it is tempting—for some, irresistibly tempting—to argue against any further regulatory incursions. Yet regulation can work. It can achieve its objectives when they are clearly enunciated. Some examples of success are given below:

1. **Safety and Access** Facility regulations have reduced the risk to patients of fire, decreased the chances that a patient will receive a transfusion with incompatible blood, and increased the likelihood of having access to (although not necessarily receiving appropriate care from) an array of health care professionals with expertise in specialty areas that could improve health outcomes.

2. **Cost Containment** Rate regulation has apparently reduced the rate of rise in hospital costs. In a comparison between six states with hospital rate setting programs and the other states during 1975–1978, Biles et al., found average annual rates of increase in hospital costs of 11.2 and 14.3 percent respectively.[33]

3. ***Standards and Criteria*** Methods and standards of providing care have been greatly influenced by regulation. Record-keeping procedures have been upgraded, facilitating more comprehensive, coordinated care by a number of professionals. As an example, residents of publicly funded nursing homes receive more frequent physician visits due in part to requirements for both periodic visits and the designation of a medical director for some classes of facilities.

4. ***Provider Education*** Regulation has performed an important educational function. For example, participation in PSROs has undoubtedly made physicians more aware of and sensitive to the costs of what they order for patients. The certificate-of-need and Section 1122 processes have greatly increased the ability of hospitals to plan for their future needs in a coordinated, logical and defensible manner. Health professionals and administrators have improved communications regarding their respective needs. Hospitals budget much more carefully today than even ten years ago to assure that they can maximize reimbursement from third-party payers, particularly state and federal governments, and to avoid overruns that can less easily be passed through to the reimbursable cost base. Nursing homes have become more efficient operations and more sensitive to quality of care issues.

5. ***Public Awareness*** Perhaps the single greatest impact of regulatory efforts has been to put health care more in the public eye. If there is a problem in quality of care, if the prices go up at their usual but hard to justify rates or if a hospital or doctor refuses to see and treat a sick patient with an emergency problem, it is on the evening news and the front page of the local paper. Regulatory agencies have requirements for public access to information and for public disclosure that reduce the chances that serious problems in the structure and delivery of health care service will be outside public view. Concern over what the public may think, who may decide to institute a lawsuit and what action the regulatory agencies might take as a result of an exposed problem can all have chastening effects on behavior.

To an increasing extent employers are becoming active participants in regulatory processes and in lobbying for the passage, repeal or amendment of statutes which place constraints within the health care sector. The issues are complex, and as one senior executive in Phoenix said, "It's easy to get swept up in health care legislation and get very involved and committed before you begin to understand how complicated this is and that it requires almost full-time attention." Yet the interest of employers as major purchasers is clearly at stake. For example, further controls on public financing without controls on private financing for health care invariably lead to more cost-shifting to the private sector. An employer who feels that HMOs are more costly because the lower risk employees are enrolling may feel strongly that he or she should not be required to offer an HMO option to his or her employees under federal law. Companies concerned about the overbuilding of hospitals or other health care facilities can help by supporting a tough certificate-of-need process or by supporting a moratorium on hospital construction for some period of time, as has occurred in several states and large cities. Concern about anticompetitive behavior of physicians has led many companies to support the Federal Trade Commission continuing jurisdiction over the health professions.

While few companies have the clout to make all the difference alone, the presence of local coalitions of large employers and a national health-interested membership organization, the Washington Business Group on Health, provides the combined political power for business to influence almost all health legislation at both federal and state levels. However, the presence of coalitions should not imply unanimity of position among employers. Some employers feel that because health care is different than other industries, regulation is necessary and they must fight for state rate setting and other external constraints on price, supply or demand. Others feel that the only solution to the problems of a market with poor incentives for cost-effective care is strong stimulation of a more competitive market.

Regardless of viewpoint, the role of employers in determining regulatory policy is being clearly established. Business is routinely being called upon to testify on important regulatory issues, and as the largest purchaser and an interest group with substantial political muscle, it can make an increasing difference in the types of regulation to which health care providers are subjected. In addition, managers in companies are asking and being asked to serve as members of regulatory bodies, such as rate-setting commissions and mandated bodies that make certificate-of-need and other regulatory decisions. Through these roles and by lobbying for appropriate resources for regulatory agencies to attract qualified personnel and make responsible decisions, employers can also greatly improve the efficiency and effectiveness of the process of regulation. Employers can make sure that regulatory agencies have clear objectives, good management structures and well-defined procedures to help assure that the tax dollars paying for regulation are buying something of value.

NOTES

1.   Somers, A.R. *Regulation: The Dilemma of Public Policy.* Princeton: Princeton University Press, 1969.
2.   Ibid.
3.   Smits, H. "The PSRO in Perspective," *New England Journal of Medicine*, Vol. 305, No. 5, 1981, pp. 253–259.

4. Goran, M. "The Evolution of the PSRO Hospital Review System," Supplement to *Medical Care*, Vol. 17, No. 5, 1979, pp. 1–47.
5. Office of Policy, Planning and Research. *Professional Standards Review Organizations 1978 Program Evaluation*. U.S. Department of Health Education and Welfare, Health Care Financing Administration. Washington D.C.:U.S.G.P.O., 1979.
6. Schwartz, W.B. "The Regulation Strategy for Controlling Hospital Costs: Problems and Prospects," *New England Journal of Medicine*, Vol. 305, No. 21, 1981, pp. 1249–1255.
7. Salkever, D.S. and Bice, T.W. "Certificate-of-Need Legislation and Hospital Costs," *Hospital Cost Containment: Selected Notes for Future Policy*, M. Zubkoff, I.E. Raskin and R.S. Hanft (editors). New York: Prodist, 1978, pp. 429–460.
8. "States Respond to Capital Question," *Washington Report on Medicine and Health*, April 4, 1983.
9. Stowe, J.B. "Strengths and Weaknesses of the Massachusetts Model," *Business and Health*, Vol. 1, No. 5, 1984, pp. 9–13.
10. Howell, J. "Evaluating the Criticisms of Certificate of Need," *Volunteer Voice-Western Center for Health Planning*, Vol. 2, 1981, pp. 1–16.
11. Uyehara, E. and Thomas, M. *Health Maintenance Organization and the HMO Act of 1973*. Santa Monica, California: The Rand Corporation, 1975, pp. 1–10.
12. McNeil, R. and Schlenker, R.E. "HMOs, Competition and Government," *Milbank Memorial Fund Quarterly/Health and Society*, Vol. 53, No. 2, 1975, pp. 195–224.
13. Iglehart, J. "The Federal Government as Venture Capitalist: How Does it Fare?" *Milbank Memorial Fund Quarterly/Health and Society*, Vol. 58, No. 4, 1980, pp. 656–666.
14. Ibid.
15. *Business and Health*, Vol. 1, No. 5, 1984, p. 7.
16. Op. cit. (Schwartz).
17. Hafkenschiel, J.H., Cameron, J.M., Knauf, R.A. and Roth, G.J. *Mandatory Prospective Reimbursement Systems for Hospitals*. California Health Facilities Commission, Report on 82–86, 1982.
18. Gibson, R.M., Waldo, D.R. and Levit, K.R. "National Health Expenditures, 1982," *Health Care Financing Review*, Vol. 5, No. 1, 1983, pp. 1–31.
19. U.S. Department of Health and Human Services. Social Security Administration. *Social Security Bulletin*, Vol. 47, No. 1, 1983, p. 1.
20. *1983 Annual Report of the Board of Trustees of the Federal Hospital Insurance Trust Fund*. U.S. Department of Health and Human Services. Health Care Financing Administration, June 24, 1983.
21. Op. cit. (Social Security Bulletin).
22. U.S. Department of Health and Human Services. Social Security Administration. *Social Security Bulletin Annual Statistical Supplement*, 1980, pp. 13–31.
23. Op. cit. (Gibson et al.).
24. Op. cit. (Social Security Administration, 1980).
25. Op. cit. (Social Security Administration, 1980).
26. Op. cit. (Social Security Administration, 1980).
27. Op. cit. (Gibson et al.).
28. Carlin, P. *Health Insurance Regulation*. Intergovernmental Health Policy Project, George Washington University.

29. Shaughnessy, P. *Regulation of Private Health Insurers*. Denver: Spectrum Research Inc., 1975.

30. Krizay, J. and Wilson, A. *The Patient as Consumer: Health Care Financing in the United States*. Twentieth Century Fund Report, Lexington, Massachusetts: Lexington Books, D.C. Heath and Company, 1974.

31. Winters, H.L. *Consumer Guide to Health Care Costs*. California Health Facilities Commission, 1980, p. 50.

32. Farney, D. "Business PACs Decide to Go on the Defensive." *Wall Street Journal*, June 10, 1982.

33. Biles, B., Schramm, D.J. and Atkinson, J.G. "Hospital Cost Inflation Under State Rate-Setting Programs," *New England Journal of Medicine*, Vol. 303, No. 12, 1980, pp. 664–668.

INCREASING 17
COMPETITION IN
THE HEALTH CARE
SYSTEM

Why not let the health care system behave more like private industry, with limited regulation and encouragement of much more competition? This question is increasingly posed by employers and health care economists. Politicians are more and more inclined to answer the question by loosening the regulatory requirements and allowing increases in competition. Employers are starting to act as prudent purchasers, considering health more and more like many other services they purchase on behalf of their employees.

Moving toward competition can mean many different approaches. Some of the major concepts have been incorporated in legislative pro-competition strategies. Others have been addressed separately, through employer and insurer initiatives on and other changes in state or federal law, or settled through the court system. The following five approaches reflect the major types of activities which are already underway or are being seriously considered as worthwhile efforts to inject more competition into the health care system.

COMPETITIVE APPROACHES

Restructuring Tax Laws

Excessively comprehensive health insurance coverage triggers overuse of health care resources, with accompanying inefficiency and acceleration of

253

health care cost increases. A related premise is that current tax provisions artificially distort demand by eliminating economic incentives to purchase anything less than "Cadillac coverage."

Tax exemption of employer contributions to health benefits create a tax shelter for employees which increases in value as the marginal tax bracket of the employee increases. As a percentage of household income, employee tax benefits have been estimated to be 0.65 percent when annual household income is $10,000 to $15,000, compared to 0.98 percent when annual household income is $50,000 to $100,000.[1] An additional tax benefit accrues to persons who itemize their deductions and are allowed full deductions from federal and state income tax for certain nonreimbursed and itemized medical expenses that exceed 5 percent of adjusted gross income. In federal fiscal year 1983 the Congressional Budget Office estimated that the federal revenue loss from tax exclusions exceeded $25 billion; additional losses of revenues to state governments and the Social Security system push the total losses above $30 billion.

Restructuring these tax benefits could include:

1. Eliminating the tax subsidy for health insurance benefits.
2. Placing a cap on the amount of employer contributions to employee health plans which can be excluded from the employee's taxable income.
3. Requiring the employer to pay a tax on part or all of his or her contribution to employee health insurance benefits.
4. Reducing or eliminating the current tax deductibility of some individual health expenses.

Some movement was seen in the direction of reducing personal deductions when the threshold above which medical expenses can be deducted was increased from 3 to 5 percent. A much greater potential dollar issue, however, is a possible cap on the tax exclusion for employee contributions made by employers to health insurance benefits. Such a cap was proposed by the Reagan administration in 1983 and was actively considered again in 1984. Although the cap was proposed in the name of restraining the rise in health care costs, it was seen by most supporters in Congress primarily as a handy expedient to reduce the $200 billion federal deficit.[2]

The hope of health care system planners and economists is that a tax cap would accelerate the trend toward more cost sharing through higher deductibles, co-payments and coverage restrictions.[3] Consumers would become more price-sensitive and providers would have to respond by moderating prices. The great divergence in prices of different providers would be reduced. Employers would benefit through a lower rate of increase in health insurance premiums. The Congressional Budget Office estimated that a cap set at $150 per month for a family and $60 per month for an individual would reduce calendar year 1987 employment-based health insurance premiums by 13 percent compared to 1982 policies. In turn, this could lower spending on insured medical services by 9 percent as of 1987, relative to spending under then-current policies.[4]

Despite a strong economic rationale for a tax cap proposal, opposition was virtually unanimous. Business did not want to see additional government regulation, and was concerned that a cap could be the first step to taxing fringe benefits. Since the cap would primarily affect unionized companies, who are in general paying the greatest amount per capita for health benefits, it was also argued that these companies would be forced to pay the tax rather than successfully renegotiate union contracts to reduce the benefit levels in order to get under the cap. Health insurance officials argued that hospital costs, one of the major reasons for overall cost increases in the health care sector, would be least affected because employees view protection against hospital expenses as the number one concern. As a result, hospital insurance would be the least changed and many of the important but lesser-cost items, such as dental and vision care, mental health, primary care and preventive measures would be sacrificed. Labor protested that a cap would undermine the hard-won health insurance benefits that they had bargained for over several decades and would cost their members more money with no increases in benefits. Most provider groups were strongly opposed, concerned either that the cap would affect their revenues under insurance or that benefits for their services would be abridged or eliminated. A surprising island of support was the American Medical Association, which supported the concept of a cap, perhaps to counter growing criticism that they were doing little to help contain health care costs.

Despite limited support from health organizations and associations, the issue of a tax cap is likely to be raised again, especially if the large federal deficit persists and the inflation rate increases. If seriously considered, a number of difficult issues will have to be resolved:

- Should the cap be the same in all parts of the country, despite the wide regional variation in health care and health insurance costs?
- Will employees whose current benefits exceed the cap select lower cost alternatives, which could reduce utilization and have a moderating effect on health care costs, or will they simply be willing to pay more and keep the existing coverage?
- What will be the added costs to employers of having to provide and keep track of enrollment in and payments for additional indemnity health insurance plans?
- What, if anything, should be done to reduce the potential tendency to eliminate effective cost saving preventive and primary care services?

Increasing Consumer Choice

Requisite to any competitive market is significant consumer choice. While HMOs have been offered by many employers, they represent a different delivery system and may not provide an acceptable option for the majority of employees who still prefer the indemnity type of insurance plan. To foster competition, many of the "procompetition" proposals offered before the Congress would require employers of a certain size (e.g., at least 200 employees) to offer several different indemnity insurance plans to all employees. Some advocates of this approach feel that each plan offered should be through a different carrier to encourage competition. To encourage employees to select a less comprehensive health benefit package, most of the procompetition bills provide a rebate from employers to employees for the full difference between the employer's contribution and a less costly plan selected. Some of these proposals would go so far as to sever the connection between employment and health insurance by limiting the employer's role to making contributions toward any health plan. Individuals without employer-based insurance would receive tax credits to apply toward the purchase of health insurance.

The intended benefits of multiple choice among plans, equal employer contributions to all plans, and employee rebates for selecting a lower cost option are to:

1. Provide consumers with an economic incentive to select a lower cost plan;
2. Promote competition among insurers to offer the most attractive low cost options; and
3. Encourage providers to organize into competing economic units such as HMOs in order to be able to offer the most comprehensive service package at the lowest price.

According to the Congressional Budget Office, combining a tax-free rebate for employees with a cap on tax-excluded employer contributions to health plans would increase the effectiveness of the competition strategy but would decrease federal revenues.[5] There is considerable merit in pushing employers to provide the same dollar benefit for all employees, at least within a certain class (for example, those exempt employees buying individual policies). Currently some employers pay the same percentage of the total cost of several different plans in the same area, with the result that they are providing a larger subsidy toward the more expensive plans. Already, effective competition from HMOs has spurred providers to develop preferred arrangements (see Chapter 14) which increase competition by offering price discounts and reduced employee cost-sharing.

However, there are many criticisms of the multiple choice mandate.

Many larger employers, in attempts to reduce the rate of increase of health care costs, are already offering several different plans. This trend can be expected to increase as experience is gained and reports of savings circulate within the business community. Forcing employers to offer different plans from different carriers would be administratively cumbersome, increasing costs. Smaller employers would potentially be particularly hard-hit by additional administrative requirements.[6] Another potential problem with mandated multiple choice for employees is that inefficient carriers might be guaranteed a market they presently don't have.[7]

In order for the plan to encourage comparison shopping, employees need to clearly understand the options and trade-offs implicit in different plans. However, employees cannot be expected to compare multiple plans that differ in such complex respects as coverage, deductibles, stop-loss, co-insurance and eligibility. It may be better to have two plans which present fairly clear choices to employees about cost sharing but otherwise are the same or very similar in other features.

Under some of the proposed plans, even self-insured employers would have to offer a number of plans provided by carriers, reducing the cost savings that can be achieved by the employer taking the risk of adverse claims experience. An even greater danger lurks in those procompetition proposals which seek to sever the link between employment and private health insurance. While this arrangement would make it easier and perhaps less expensive for those working-age adults to obtain affordable health insurance coverage, it would eliminate the very strong incentive which employers have to minimize the health problems of their workforce and other insurees. Currently, an employer can develop cost-containment programs which can be effective in reducing its rate of rise of health care costs and can also promote efficiency by reducing inappropriate or unnecessary care. Were insurance independent of employment, the rating system would most likely not be related to the experience of the individual company. Thus the company would have very limited incentives to devote resources to cost-containment activities.

One of the greatest problems associated with a multiple choice system is that consumers tend to select the plan which they perceive to be the most personally advantageous, given their current and short-range anticipated need for medical care. A consensus is that younger, more healthy individuals would tend to opt for less comprehensive, relatively inexpensive coverage while older, less healthy persons would select more extensive coverage, despite its higher cost. Those individuals who anticipate needing an operation or otherwise incurring significant health care expenses have good reason to switch to the more comprehensive plan. The end result of this adverse selection can be that healthier individuals desiring broader coverage are forced to settle for the lower option because the premium for the higher option would have become prohibitive. Rebates exacerbate the situation by creating an economic incentive for healthier persons to settle for minimal coverage.

Experience with multiple choice plans offered by employers is becoming available, but it is yet insufficient to draw firm conclusions about the extent of adverse selection and how the construction of different plan options affects adverse selection. However, experience with multiple-choice health plans offered by the federal government suggests that adverse selection can be a serious problem. In recent years there has been a significant increase in the number of federal employees transferring into the high-option Blue Cross/Blue Shield plan during open season. By the early 1980s, the percentage of employees transferring into the plan had risen from 1 percent to between 5 and 8 percent, and most of these persons were expected to take advantage of special features unique to the Blues, such as liberal mental health benefits.[8]

Some degree of adverse selection may be unavoidable. However, it can be minimized by: 1) limiting switches from the low- to high-cost plans to one enrollment period every two years; 2) initiating waiting periods for coverage under the higher option plans; or 3) charging employees for switching plans.[9] Perhaps the best opportunity to minimize problems of adverse selection is to balance features of the multiple plans through different levels of co-insurance, deductibles, stop-loss and contribution to premiums, so that the plans are likely to be considered reasonably equivalent in value per employee dollar for a wider range of employee circumstances.

Reducing Barriers to Competition

Commercial insurers have traditionally been precluded from entering into contractual relationships with providers by both federal antitrust and state statutes. This has relegated insurers to pay whatever providers charge, limited only by usual and customary area fee profiles for practitioners that were inherently inflationary. In recent years, however, these constraints are falling, and insurers are aggressively developing more favorable contractual relationships which increase their ability to compete for fully insured, partially insured and administrative-services-only business. While it is not yet clear to what degree insurance companies can combine to develop such contractual arrangements without running afoul of antitrust laws, the ability of individual insurers to contract is already making a significant difference in the marketplace.

Relaxation of regulatory requirements for federally qualified HMOs and Blue Cross and Blue Shield plans have also tended to increase the competitiveness of the market for health services and health insurance. HMOs can now rate by class (e.g., by age and sex categories), even if they cannot rate each company separately based upon its unique experience. The Blues have been relieved of their responsibilities to offer first-dollar coverage and insurance to individuals at a loss.

Improving Consumer Knowledge

Only in the last few years have the professional society requirements against advertising of professional services and related prices been struck down in court. While health professionals and institutions are bound by the truth-in-advertising regulations, they are otherwise free to advertise to the public. For the first time consumers are being bombarded by health care advertising. Some examples are:

- A hospital advertises that it will waive co-insurance and deductible charges for patients who come there for a defined set of elective surgical procedures.
- A doctor advertises that as a gay male he is uniquely qualified as a physician to treat sexually transmitted diseases among gay males.
- A dentist prints her fee schedule in a newspaper ad and invites prospective patients to compare prices.
- An instant-care center advertises that its prices are, on the average, 50 percent below the charges for equivalent problems seen at the local hospital emergency room.

Patients are becoming less reluctant to ask in advance about prices of services. While comparison shopping is still the exception, it may be accepted practice within a few years in other than prepaid care settings. Some employers and insurers are facilitating price shopping by offering information to insurees on the relative costs of different providers in their communities. Some insurers will tell an insuree whether the amount requested by a physician for elective services is above the usual and customary level used as the basis for payment decisions. Several employers are distributing lists of area hospital average per diem costs to employees and their families. TELANSWER, a service offered by Los Angeles-based U.S. Corporate Health Management, provides information to insurees via telephone on the relative costs of competing hospitals for most areas of the country. (More information on TELANSWER is provided in Chapter 5.)

Increasing the Capacity of the Health Care System

Increased supply usually means lower prices. The degree to which this operates for physicians, hospitals, dentists and other health resources is still hotly debated. In the hospital sector, increased overall capacity has generally meant lower utilization and higher costs due to the need to cover fixed operational expenses over fewer units of service. Investor-owned hospitals have argued that they are more efficient and should be allowed to build even in areas with an abundance of hospital facilities because, through competition, they can save the community money. Studies to date have not corroborated this last assertion. However, as competition mounts it is likely that there will be opportunities for more efficient hospitals to provide equivalent services at lower costs. The results will probably be closure of many facilities, including some built through the tax subsidization of municipal bonds. If there are already a sufficient number of facilities in an area to promote competition, it would appear wasteful to allow addition of new ones. A more difficult public policy issue arises if an additional facility is proposed for a community with a single institution. If the institution now has a large population of publicly financed patients, it is subsidizing them through higher charges to private payers. A new hospital which would siphon off these private payers but would not accept public-pay patients saves no money for the system. Instead, it requires a greater increase in tax dollars to support these patients than the employers save by having lower prices for their insurees.

The overall impact of the increasing number of physicians per capita on the competitiveness of the health care market is difficult to isolate. Each physician orders many health care services each year, especially hospital and laboratory services. More physicians therefore translates into more services. Some early studies also showed that an increased density of physicians only increased prices, since physicians felt that they needed to charge more per patient to attain their preconceived target income. Moreover, most physicians have the ability to increase demand for their services. For most problems, there are no clear standards for how many visits are required or even what distinguishes necessary from unnecessary laboratory and x-ray services. However, as the concentration of physicians increases there is growing evidence that more physicians are having to reduce their income expectations and are willing to accept salaried positions at lower levels than did physicians entering the job market five or ten years ago. It is difficult to establish a new practice in many choice metropolitan areas. The advent of instant-care centers, physicians who make only home visits, and those with night and weekend hours testify to the increase in competition among physicians. As cost-sharing increases, fee schedules replace usual and customary charges for physician services and a greater percentage of the population opts for prepaid care, physicians will be forced to compete on the basis of price. Studying medicine is also becoming increasingly expensive, with cost a growing barrier to medical school enrollment. Tuition alone in some medical schools exceeds $20,000 per year. Most students emerge as doctors with very large debts.

At some point, price competition and higher entry costs should force down the attractiveness of the medical profession in terms of security and level of income. As federal subsidies to medical students are also reduced, medical school enrollment might be expected to decline. However, the number of current applicants far exceeds the number of available places. A large decline in applications would have to be seen before enrollment declined, unless schools of medicine retained the same academic standards for acceptance. One anticipated consequence of changes in the way that medicine is viewed is a decline in the academic qualifications of applicants to medical school. The potential impact of such a decline on the quality of care should concern us all.

There are already too many dentists to fill the current demand. Yet, a further increase in the number of dentists per capita is anticipated. Among the most popular continuing-education courses for dentists are those on how to use marketing to increase patient volume. Advertising for dentists has become common in many areas. Franchised dental clinics in department stores and in storefront offices testify to increased competition, which is based on both convenience and price. More competition has come despite the large growth in dental insurance and prepaid dental plans that are paid for in whole or part by employers.

Greater competition is also being seen between physicians and other health care professionals. Optometrists account for the majority of the examinations for eyeglasses and contact lenses. Most people go to them because they perform these services at lower cost than ophthalmologists. Chiropractors are consulted instead of orthopedists, at a fraction of the price. Weight problems are increasingly brought to nutritionists rather than M.D.s. Clinical psychologists, psychiatric social workers, psychiatric nurses, and those with master's degrees in counseling are seeing many patients who might once have consulted psychiatrists. In most of these cases the differences in price are substantial. While insurance coverage still tends to favor physicians, there is a clear movement underway to cover nonphysicians.

Whether many of the newer health professionals are saving the health care system money is difficult to answer. While per service cost is decreased, the presence of health professionals and the lower prices both increase demand. The aggregate expenditure for some types of services will probably increase. Therefore, in formulating a procompetitive strategy for nonphysician health professional education and licensure, it is necessary to consider not only quality of care issues and whether there is an unmet need for care, but also what the impact is likely to be on the overall system.

COMBINING COMPETITIVE AND REGULATORY APPROACHES

Many of the competitive strategies that have been advanced may ironically induce expanded regulation in certain areas. Access, patient safety, quality and affordability will all be continuing subjects for regulatory efforts. As incentives are created for efficiency, increased regulatory attention will undoubtedly be focused on assuring that patients receive sufficient care. The DRG-based payment system which is being phased in under Medicare requires a peer review organization to scrutinize the care to its beneficiaries in hospitals. As cost-containment efforts become more pronounced, there is likely to be considerable regulatory activity to assure that insurance coverage provides a certain minimum scope of benefits. As competing care plans are offered to the poor and over-65 age group, some legislated method of spreading the costs for these high utilizers among all those utilizing that plan is likely, due to government financing problems. Public concern for fiscal solvency of the variety of new health plans that are springing up will lead to financial regulation, with reserve requirements, required underwriting principles and reinsurance to pay for care in the event of bankruptcy. As the variety of health care plans proliferate it is also likely that advertising claims may be subjected to regulation regarding what can be said about benefit packages, out-of-pocket costs and other important plan features.

NOTES

1. Congressional Budget Office. *Containing Medical Care Costs Through Market Forces.* 1982.
2. "Tax Exclusion Is Back," *Washington Report on Medicine and Health Perspectives*, January 17, 1983.
3. Op. cit. (Congressional Budget Office).
4. Op. cit. (Congressional Budget Office).
5. Op. cit. (Congressional Budget Office).
6. Pernice, J. and Markus, G. *Health Insurance: The Pro-competition Proposals.* Library of Congress. Congressional Research Service. Issue Brief Number IB81046.
7. Ibid.
8. Simler, S.L. "Government Blunts Competitive Edge of Federal Employees' Health Benefits," *Modern Healthcare*, Vol. 11, No. 12, p. 90.
9. Health Insurance Association of America. *Competition in the Health Care System: An Evaluation of the Pro-Competition Bills.* 1981.

DISEASE *18*
PREVENTION
AND HEALTH
PROMOTION

Marshall Winegar, the senior vice-president at Second National Bank, was one of the bank's greatest assets. Hard-working, personable, with twenty-three years of experience and the wisdom of maturity, he was the engine that kept things running smoothly during a fifteen-year period of sustained growth. At age 49 he had a stroke which paralyzed him and left him unable to communicate. Despite intensive medical care and physical therapy, he will never return to his job, nor even be able to enjoy retirement. The bank estimated the cost of replacement at $250,000, but felt that it would have been willing to invest several times that if the money could have prevented Marshall's stroke.

At a large building products distributor, two of the top six officers suffered heart attacks in one year. One returned to work after three months with a reduced work load at his previous salary level. The other was so impaired that he had to take disability retirement. Overall the cost of these episodes was in the hundreds of thousands of dollars. But in addition to the dollar costs there were major human costs, both to the families of the heart attack victims and to the organization, where for some period of time the other top executives had to assume the ill men's tasks in addition to their own, adding considerable stress, which lead to additional health problems.

The chairman of the board of a large defense contractor, age 58, developed an aggravating persistent cough which wouldn't go away. He was more easily fatigued, making it difficult to follow his usual demanding sched-

TABLE 18.1
TEN LEADING CAUSES OF DEATH—UNITED STATES, 1981

Rank	Cause of death	Rate per 100,000 population*
1	Heart disease	336.0
2	Cancer	183.9
3	Stroke	75.1
4	Accidents	46.7
5	Chronic obstructive pulmonary diseases and allied conditions	24.7
6	Pneumonia and influenza	24.1
7	Diabetes mellitus	15.4
8	Chronic liver disease and cirrhosis	13.5
9	Arteriosclerosis	13.0
10	Suicide	11.9

*Non-age-adjusted.
Source: National Center for Health Statistics, 1981 (provisional) mortality rates.

ule. After a day of tests his doctor gave him the devastating news that he had lung cancer. He had been a two-pack-a-day smoker for forty years.

Each of these three true stories led to the inevitable question, "Could the tragedy have been avoided?" In each case the after-the-fact conclusion was yes, probably. Within one year of these episodes each of the three companies started to develop a program designed to prevent or forestall avoidable illness for all of their employees.

PREVENTABLE HEALTH PROBLEMS

In too many cases tragedy has been the spur to a careful examination of the opportunities to prevent many of the most feared but common health problems that are leading causes of death in the United States.

Table 18.1 depicts the major causes of death in the United States for 1981. Today most of the serious health problems are those that develop slowly over time. During their course, they contribute significantly to disability and to diminution in quality of life by impairing activities of daily living. In many cases, by the time these diseases have made themselves known, their course cannot be altered. For example, lung cancer has a five-year cure rate of about 5 percent, and this has not appreciably improved over the last twenty years. Stroke occurs suddenly, often without clear warning that could lead to treatment to forestall it. Stroke frequently compromises quality of life and in many cases permanently precludes normal functioning. Heart disease often first shows itself as a heart attack, and many

TABLE 18.2
LEADING CAUSES OF DEATH FOR A CONSUMER PRODUCTS MANUFACTURER

Causes of death—1981	Number*	Percentage of total
Diseases of the heart and blood vessels	1447	49
Cancer	390	13
Stroke	120	4
Pneumonia and influenza	103	3
Diseases of digestive tract (includes some cancer)	89	3
Motor vehicle accidents (nonoccupational)	64	2
Other accidents	38	1
Suicide	45	1
Homicide	24	1
Subtotal	2320	79
All other causes	632	21
Total	2952	100

*Includes retirees.
Source: Unpublished company data.

first heart attacks are fatal. One-fourth of all fatalities from heart attacks occur without prior evidence of heart problems.[1] Victims of heart disease are frequently forced to reduce both work and leisure time activities. While these and other major health problems may appear with devastating suddenness, they share a long progressive phase which is without symptoms. They are clearly chronic diseases, even though they appear as if fron the blue.

Employers see a pattern of health problems which is very similar to the national statistics for the adult population. For example, a large consumer goods manufacturer found employee deaths over a period of several years as shown in Table 18.2. Almost four-fifths of the total mortality is accounted for by the nine top-rank order causes of death.

While death is an infrequent occurrence among employees, short- and long-term disability are frequent. Of particular concern are the longer illness-related absences. In one large company, top causes of illness-related absences of at least ten days include a number of different problems such as disorders of bones and joints, disorders of the nervous system, mental disorders, pneumonia, abdominal hernias and gynecological problems. Delivery and pregnancy complications are also high on the list (Table 18.3).

TABLE 18.3
LEADING CAUSES OF ILLNESS RELATED
TO ABSENTEEISM

Rank order by cause (total population)

Rank*	Cause	ICD code
1	Ischemic heart disease	410–414
2	Disorders of bones and joints	720–729
3	Delivery, complications of pregnancy	630–678
4	Accidents	800–899
5	Disorders of the nervous system	320–389
6	Mental disorders	290–315
7	Other musculoskeletal disorders	730–738
8	Pneumonias	480–486
9	Abdominal hernias	550–553
10	Diseases of the uterus, female organs	620–629

*Ranked by days lost from ten-day absences.
Source: Unpublished company data.

Health Risks

From the standpoint of the employer, a salient feature of most of these major diseases and disease groups is that they are *preventable* or postponable to a significant degree. Characteristics which increase our risk for each serious health problem are called health risks or health indicators. Health risks associated with some major disease groups are summarized in Table 18.4.

Most of these health risks are under our control. Many are related to our daily health habits. Only daily attention to health practices can reduce our risks to the desired level. For example, the risk of dying and the severity of injury from motor vehicle accidents can be reduced by about one-half simply by wearing seat belts. Since no one knows when he or she will have an accident, the only way to minimize risk is to get into the habit of wearing a seat belt all the time. Another major cause of disability is dental disease, the most common chronic health problem in both children and adults. Periodontal disease, the medical name for gum disease, is responsible for most tooth loss in adults. It is caused by recession of the gums, facilitated by the plaque which forms on our teeth. Having a dental hygienist remove the plaque two or three times a year will not prevent the progression of gum disease. This can only be accomplished by *daily* stimulation of the gums through proper use of dental floss or some other gum stimulator. Daily practice and the accompanying protection are best assured by developing habits, so that performance does not depend on memory.

TABLE 18.4
PROMINENT CONTROLLABLE RISK FACTORS

Cause of death (1977)	*Risk factors*
Heart disease	Smoking, high blood pressure, elevated serum cholesterol, diabetes, obesity, lack of exercise, Type A behavior
Cancer	Smoking, alcohol, solar radiation, ionizing radiation, worksite hazards, environmental pollution, medications, infectious agents, diet
Stroke	High blood pressure, elevated cholesterol, smoking
Accidents, other than motor vehicle	Alcohol, smoking (fires), product design, home hazards, hand gun availability
Influenza/pneumonia	Vaccination status, smoking, alcohol
Motor vehicle accidents	Alcohol, no safety restraints, speed, automobile design, roadway design
Diabetes	Obesity (for adult/onset), diet
Cirrhosis of liver	Alcohol
Suicide	Hand gun availability, alcohol or drug misuse, stress
Homicide	Hand gun availability, alcohol, stress

Epidemiology

Health risks are not diagnoses. Diagnoses tell you what your health problems are today. A health risk is an individual characteristic which increases that person's chances of acquiring a particular disease, or group of diseases in a defined future time period. Categories of health risks include:

- *health habits,* e.g., smoking, lack of seatbelt use.
- *family history of certain medical problems/conditions,* e.g., diabetes, heart disease, intestinal polyps.
- *physiological or biochemical factors,* e.g., high blood pressure, elevated blood cholesterol.

A health risk can also be created when a person does not get a screening procedure such as a Pap smear or mammography to identify problems before they reach the serious disease stage.

In most cases the relationship between health risks and specific diseases have been elucidated by studying large populations and understanding what distinguishes those who acquire the disease from those who don't. This study of the distribution and causation of disease in large groups constitutes the science of epidemiology. For example, epidemiology has established the strong link between smoking and cancer by finding that smokers are fifteen times more likely to get cancer than nonsmokers.

Other studies have shown that the risk of acquiring lung cancer increases with number of cigarettes smoked per day and the duration of the smoking habit. Epidemiology, therefore, provides information that permits us to estimate how much greater the risk is for a smoker to get lung cancer than a nonsmoker. Individuals who have this information can see how their risk compares to the risk of others. Epidemiology gives only probabilities. Since not all smokers will develop a smoking-related problem, there is truth to the contention that you can't predict with foolproof accuracy which of the smokers will succumb to lung cancer, emphysema, heart disease or any of the myriad of other diseases clearly linked to smoking. Four-pack-a-day smokers may never get any of these. Yet the odds are very strong they will and, in fact, the odds are many times those of a nonsmoker.

Epidemiology also tells us how various risks interrelate. For example, we know that high blood pressure, high blood cholesterol, smoking and diabetes are all independent risks for heart attack. While both high blood triglycerides and high blood cholesterol increase heart disease risk, they have strong interactions such that their combined risk can be incorporated simply by calculating the risk of the high cholesterol alone. By contrast, smoking and drinking excessively give a higher combined risk for acquiring cancer of the esophagus than either adding or multiplying the numerical risk scores for each of these risks.

Health Risk Appraisal

"Your chance of dying of a heart attack during the next ten years is 306 percent greater than for the average man your age." This assessment derived from information on heart attack risks provided by a 63-year old male corporate marketing director (Figure 18.1). Such quantified information about personal health risk is available through the use of health hazard appraisals, perhaps more aptly termed health risk appraisals. Health risk appraisal is usually defined as a method that describes an individual's chances of death or of acquiring specific diseases within a delineated period of time in the future.

Two kinds of information are needed to make a health risk appraisal work. First, information on the risks for particular diseases is required from epidemiological studies. Information necessary includes:

1. *Identification of the specific independent risks for a disease* For example, known independent risks for heart disease are smoking, hypertension, high blood cholesterol, Type A behavior, lack of exercise and strong family history of heart disease.

FIGURE 18.1
SAMPLE SUMMARY RESULTS OF A HEALTH
RISK APPRAISAL

Male
Age: 63

Cholesterol: 297
Blood pressure: 120/60
Smoking: No
Exercise: No vigorous exercise
Weight: 5 percent over top of desirable weight range

Heart attack risk: 306 percent greater than average
Risk can be reduced by more than 350 percent compared to average

---

2. *Quantitative relationships between each risk and the disease over a specified period of time* For example, how much does a high cholesterol value (over 240 mg/dl) increase the risk of heart attack over a ten-year period, all other things being equal? In general, this information is needed separately for males and females and for different ages, since these variables affect the chances of acquiring many diseases. Risks frequently also differ by age. For example, a high cholesterol would indicate a different ten-year risk for heart disease in a 30-year-old versus a 60-year-old. Where studies permit, it is preferable to have some information on how various degrees of a problem (e.g., level of cholesterol) affect the chances of having an adverse health outcome. Figure 18.2 shows how each of these factors affect risk of heart disease due to high total cholesterol.

The other major input into the risk formulae is information on the individual whose risk is to be assessed. In general, information is collected on: 1) age; 2) sex; 3) family history of relevant diseases; 4) *personal health habits* (e.g., type of exercise, intensity, frequency and duration); 5) *risk indicators that can be measured directly* (e.g., systolic and diastolic blood pressure, blood sugar, cholesterol); and 6) *specific preventive health practices* (e.g., frequency of Pap smears for early detection of cervical cancer).

Frequently health risk appraisals combine all of the known risks of death projected forward for a ten-year period and, in combination with other information on death rates by age, sex and race, develop an overall risk of death. From this calculation average expected longevity can be determined. An age is often given for which that individual's overall risk would be the average risk, the so-called "health age" or "risk age." This health age can then be compared to the age for which the individual's risk would be average

FIGURE 18.2
INCIDENCE OF CARDIOVASCULAR DISEASE BY
CHOLESTEROL LEVEL

Incidence per 10,000 population

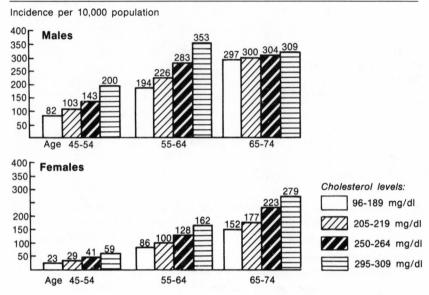

Source: Framingham Study, 18-year follow-up

Source: *Cardiovascular Primer for the Workplace*. National Heart, Lung and Blood Institute, NIH Pub. No. 81–2210. Washington, D.C.: U.S.G.P.O., 1981, p. 9.

if all the identified health risks were reduced to specific levels. The lower age is generally referred to as the "achievable" age.

The instruments available for health risk appraisal number in the hundreds, making selection difficult. One survey of twenty-nine of the most commonly used instruments found variation in price from free to $40 each, in length from twenty-four to several hundred questions, and in scoring, from self-scored to hand-scored by the providers of the instrument to computer-scored.[2] Figure 18.3 provides an example of the information that is provided to someone having taken one of the shorter questionnaires. It differs from most, however, in that it devotes the majority of space to providing information on what the individual can do to reduce some risks.

Health risk appraisals are limited to those diseases where there is a large body of epidemiological data that has identified quantitative relationships between risks and chances of dying or having a specific health problem. In many cases our knowledge of what causes diseases is limited. In some cases risks are known but are not within an individual's control. For example,

FIGURE 18.3
SAMPLE DETAILED RESULTS OF A HEALTH
RISK APPRAISAL

My overall health risk

There are many ways to describe health risk. Two ways which are commonly used are the risk of dying within the next ten years and life expectancy. These are based on the most important known health risks.

Based on my questionnaire responses, my chances of dying in the next ten years are about the same as those of the average man 4 years older than I. Another way to look at my risk is to estimate my life expectancy. My estimated life expectancy is 67 years or 5 years less than the average man my age.

My risk for major diseases

Heart attack: My risk of having a heart attack is 51% greater than average for a man my age. If the risk factors I can control were at ideal levels, my risk of heart attack would be 93% lower.

Average risk:	XXXXXXX
My risk:	XXXXXXXXXXX
Attainable risk:	X

Stroke: My risk of having a stroke is 51% greater than average for a man my age. If the risk factors I can control were at ideal levels, my risk of stroke would be 81% lower.

Average risk:	XXXXXXX
My risk:	XXXXXXXXXXX
Attainable risk:	XX

Motor vehicle accidents: My risk of dying from a motor vehicle accident is 272% greater than average for a man my age. If the risk factors I can control were at ideal levels, my risk of motor vehicle accident would be 82% lower.

Average risk:	XXXXXXX
My risk:	XXXXXXXXXXXXXXX
Attainable risk:	XXXXX

Cancer: My risk of getting cancer is 105% greater than average for a man my age. If the risk factors I can control were at ideal levels, my risk of cancer would be 63% lower.

Average risk:	XXXXXXX
My risk:	XXXXXXXXXXXXXX
Attainable risk:	XXXXX

Type of cancer:	Indicators which increase my risk:	Type of cancer:	Indicators:
Lung	Smoking cigarettes	Pancreas	Smoking cigarettes
Bowel	History of polyps	Mouth/throat	Smoking & alcohol
	Family history of polyps	Esophagus	Smoking & alcohol
Larynx	Smoking cigarettes	Bladder	Smoking cigarettes
	Drinking alcohol		

continued

Risks I can reduce

Smoking: It has been well documented that smoking is a risk factor for many diseases, including heart attack, stroke, emphysema and other lung diseases, and cancers of the lung, mouth and throat, esophagus, larynx, pancreas, and bladder. The size of these risks is a function of many things, including the number of years I have smoked, the type and quantity of cigarettes I smoke, and, for some cancers, the amount of alcohol I drink.

Quitting smoking now could reduce my risk of dying in the next ten years by as much as 40%. My risk of having a heart attack or stroke can be reduced by 54%, of getting lung cancer by 90%, of getting other cancers by 48% and of dying of chronic lung disease by 52%. Since smoking also causes many other health problems such as chronic cough, eye irritation, shortness of breath and fast heart beat, quitting will likely improve how I feel every day.

Type A behavior: I exhibit many behavioral traits characteristic of type A personality, including great sense of time urgency and impatience, great competitiveness, strong desire to excel, aggressiveness and high level of hostility. These traits increase my risk of having a heart attack and make me feel constantly under stress. I can reduce my type A behavior risks by setting aside time to relax, learning to avoid situations that bring out these behavioral traits, scheduling work and leisure time at adequate intervals, making a conscious effort to decide what is important to me and what is not and making decisions on the use of my time accordingly.

Cholesterol: I said I didn't know my cholesterol but thought it was about average. The value 233 was used in calculating my risks. This is above the

range recommended by the UCLA Center for Health Enhancement. I should aim for a reduction of at least 20%. My cholesterol should be as low as I can get it since every additional amount in the blood adds an extra risk of acquiring degenerative diseases of the heart and blood vessels. While hereditary factors influence blood cholesterol, my cholesterol is under my own control to a considerable degree. Saturated fats, primarily animal fats from meats and dairy products, raise cholesterol. Polyunsaturated fats, found primarily in vegetables, lower cholesterol. Limiting cholesterol in my diet by minimizing consumption of meat products and egg yolks may also help in lowering cholesterol. Reducing to my ideal body weight will also help me reduce my cholesterol. If I do all of these things and my cholesterol is still above the recommended range, I should consider consulting my physician to see if additional approaches to lowering my cholesterol are indicated.

Overweight: The desirable weight range for a man my age, height and build is 177 to 195 pounds (80 to 89 kilograms). My weight is 242 pounds or 24% above the top of that range. Increased weight has been established as a risk factor for heart failure and stroke. Excess weight tends to increase risk in large part by its tendency to increase cholesterol and blood pressure as well as putting an added strain on my heart. The best way to reduce my weight is by modifying my regular eating habits, not by going on a crash diet. A program of gradually increasing vigorous exercise which accommodates my overall physical condition will help me manage my weight.

Blood pressure: The lower my blood pressure the better I am. I said I didn't know my systolic blood pressure but thought it was about average. The value

130 was used in calculating my risks. My systolic (higher number) pressure is above the level where many physicians feel concerted efforts to reduce it are indicated. I should see my doctor to have my blood pressure rechecked and the need for treatment evaluated. The things I can do which may lower my blood pressure are: reduce my salt intake; exercise regularly; lose weight; practice relaxation techniques and/or biofeedback to reduce stress. Lowering my systolic blood pressure by 20 points will reduce my risk of having a heart attack or stroke by 25% and will add 1.1 years to my life expectancy.

Other risks under my control: My life expectancy would be increased if I expended 4000 calories per week in vigorous exercise and wore my seat belts all the time.

Source: UCLA Center for Health Enhancement, Education and Research, and General Health, Inc.

a strong family history of cancer is an important quantifiable risk for acquiring that cancer or a related one.

One criticism leveled at these instruments is that they are usually oriented toward assessing risks rather than positive aspects of health. Increasingly, to provide more emphasis on health, these assessments have included more qualitative information that has considerable personal health relevance, even though it may lack statistical validity and is not predictive of future health problems. Scales of well-being which help to assess how an individual feels about himself or herself and compares the responses to a similar group and/or to how he or she wants to feel are popular. Stress indices, depression indices and inventories of social support are frequently included. Also included with increasing frequency are nutritional assessments which help analyze food preferences and/or what an individual recalls having eaten over a one- to three-day period. Analysis then determines the composition of what an individual is eating and what deficiencies exist in his or her diet.

Health risk appraisals and the broader health assessments have a number of important roles in health improvement programs. For the individual they can provide:

1. *Education about his or her chances of getting specific diseases, the known causes of the diseases and what can be done to reduce those that are within individual control*   Since the information derives from a computer program, it appears very objective and is hard to ignore. It is more specific than the risk information generally provided by physicians.

2. **Motivation to make changes** The benefits of reducing risks are shown in terms of decreased chance of dying or of acquiring a number of serious and feared diseases.

3. **A baseline to track future progress** This year's health risk appraisal can be compared with next year's. Decreases in risks due to improved health-related behaviors can be a powerful reinforcement to stick with the changes.

4. **Comparative information** on how he or she is doing versus others of the same age and sex or versus the entire group of participating employees at her/his company.

5. **A permanent record for the person to keep,** eliminating the need to attempt to remember the most important information, as is the case with many encounters with health care professionals.

For the employer the results are equally useful even if only an aggregate company health risk profile is provided to avoid confidentiality problems and improve the chances for honest responses to questions by employees. Among the employer benefits are:

1. **An objective assessment of the "health" of employees as a group** Reports can give the percentage of smokers, the percentage with treated and untreated hypertension, how many are overweight, etc. But in addition, they can suggest how many of the employees are at particularly high risk based on their health habits.

2. **A summary of the major preventable future health problems in quantified terms** This can be helpful in assessing the health impact of doing nothing.

3. **The frequency and severity of a number of reducible problems** to help guide the planning and development of health improvement programs.

4. **A baseline** from which the progress of participating employees as a group can be tracked over time.

As with any new approach there can be problems with the use of health risk appraisal/health assessment.[3] Among the most serious potential pitfalls are:

1. **Inadequacy of the data base used in developing risk information to employees** In some cases individuals developing these instruments

have not exercised prudent judgment in discerning between good and poor data. In others, good data has been inappropriately used, invalidating the printed results.

2. *Provision of logical but unproven information about the benefits of reducing certain risks*   For example, while high blood total cholesterol has been proven to increase risk of heart attack and stroke, no careful study has determined that reducing a high blood total cholesterol reduces risk of dying of these diseases. Prudence is therefore indicated in the interpretation of information fed back to employees.

3. *Failure of the vendor to exercise sufficient quality control in data input and checking results for plausibility.*

4. *Provision of information to individuals without sufficient explanation of how to interpret the information and/or without an opportunity to ask questions when information is unclear.*

5. *Breach of confidentiality by vendors or those responsible for the administration of the instruments.*

Another problem with many of the currently available instruments is that they provide less useful risk information for young and older adults than for the 30-to-60 age group. For example, risk of dying from a specific cause over a ten-year period may be of less interest and relevance to those aged 20 to 30 than information on nutrition, fitness, accidents, drug use, etc. Information on longevity to a person of 69 is much less valuable than suggestions on how to maintain a desired level and range of activities while alive.

In general, health risk appraisals/health assessments appear to be a useful adjunct to other components of a health improvement/risk reduction program sponsored by employers. There is some evidence that even standing alone they can help individuals to make important changes in the direction of health. However, most of those who have reviewed employer programs feel that it is advisable to use these instruments only when the suggestions they contain can be translated into programs offering organized assistance in making those changes. Most employer sponsored health promotion programs use these instruments as an introduction to the program for employees and as a vehicle to suggest which program offerings might be considered the highest priority for each employee.

Being a well-informed purchaser offers the best protection against choosing a poor instrument and/or vendor. Among the questions that should be asked of potential vendors are:

• Where do the data bases and risk estimation equations come from?

- Does the feedback focus primarily on ameliorable risks?
- What quality control mechanisms are in place to protect against errors in entering information, printing out implausible results and delivering a health profile to the wrong recipient?
- What are the safeguards for privacy and confidentiality?
- How clear are the instructions and the feedback information for individuals with a variety of educational backgrounds?
- What is the background of the individual responsible for the data base and with what frequency is a full review of data base, computational programs, feedback content and format performed?
- Does the feedback form contain sufficient caveats to minimize misinterpretation of the information?[4]

If there is nobody within the organization who is competent to judge the quality of resources, an outside expert can be consulted.

Costs of Preventable Diseases

Hundreds of billions of dollars are spent annually in this country on diseases and conditions which are largely preventable. Costs of illness include the "indirect" costs of lost productivity, i.e., absenteeism, turnover, reduced on-the-job performance, as well as the direct cost of medical treatment. Employers bear a major portion of illness costs. They pay the lion's share of the health insurance costs. They also absorb the productivity losses, including those resulting from premature retirement and death. According to estimates by the President's Council on Physical Fitness, premature deaths alone cost American industry more than $25 billion and 132 million workdays of lost production each year.[5]

Estimates of the direct and indirect costs of six of the leading causes of death and disability are presented in Table 18.5. These figures are very conservative in that they are many years old and only represent some of the costs of preventable problems. Nonetheless, they serve to illustrate the magnitude of the dollars spent on preventable health risks and impairment, and the substantial portion of total costs made up by lost productivity—on the order of two to three times the direct medical costs.[6,7,8]

Thus, there are strong financial, not to mention humanitarian, motivations for employers to prevent, or at least slow, the occurrence of these chronic afflictions among their employees. Many workplaces have specific health risks associated with producing certain products and/or providing their services. All employers and employees however, regardless of the nature of their work, have problems associated with a number of prevalent health problems that are amenable to health promotion/disease prevention efforts. The nature of these risk factors, the degree to which they are con-

TABLE 18.5
ESTIMATED COSTS OF HEALTH RISKS/ILLNESSES, 1975*

Diseases/injury/health risk	Billions of dollars		
	Direct	Indirect	Total
Alcohol abuse (a)	12.75	30.00†	42.75
Smoking (b)	8.22	19.14‡	27.36
Cancer (c)	6.41	16.74	23.15
Coronary heart disease (c)	2.49	11.23	13.72
Motor vehicle injuries (c)	4.77	9.66	14.43
Strokes (c)	16.04	41.71	57.75
Totals	50.68	128.48	179.16

*Costs of smoking calculated for 1976.
†Indirect costs for alcohol abuse include costs of motor vehicle accidents, violent crimes, social responses and fire losses in addition to lost production.
‡Indirect costs for smoking include fire losses as well as lost production.
Sources: (a) Berry, R.E., Jr., Boland, J.P., Smart, C.N. and Kanak, J.R. *The Economic Costs of Alcoholism—1975.* Final Report to the National Institute on Alcohol Abuse and Alcoholism, August, 1977. Cited in the National Institute on Alcohol Abuse and Alcoholism, *The Fourth Special Report to the U.S. Congress on Alcohol and Health* from the Secretary of Health and Human Services January, 1981. (b) Luce, B.R. and Schweitzer, S.O. "Smoking and Alcohol Abuse: A Comparison of Their Economic Consequences," reprinted by permission of *The New England Journal of Medicine*, Vol. 298, No. 10, 1978, pp. 569–571. (c) Hartunian, N.S., Smart, C.N. and Thompson, M.S. "The Incidence and Economic Costs of Cancer, Motor Vehicle Injuries, Coronary Heart Disease, and Stroke: A Comparative Analysis," *American Journal of Public Health*, Vol. 70, No. 12, 1980, pp. 1249–1260.

trollable and experience with employer and/or employee-sponsored efforts to reduce this risk and related diseases are summarized below.

BRIEF REVIEW OF MAJOR HEALTH RISKS
High Blood Pressure

Elevated blood pressure is associated with higher rates of death and illness, primarily from stroke, heart disease, and excess pressure in and/or narrowing of major arteries in the body. On the average, individuals with high blood pressure (usually defined as ≥140/90 mm Hg) develop approximately three times as much coronary heart disease, six times as much congestive heart failure, and seven times as many strokes as individuals with controlled or normal blood pressure.[9-12] Those in whom high blood pressure is detected can have it effectively controlled and, depending on their level of elevation, age, sex and race, can reduce their risks of acquiring the major hypertension-related diseases by at least 20 to 50 percent.

Hypertension is a common condition in most worksite populations. Usually high blood pressure (including both controlled and uncontrolled) occurs in 15 to 25 percent of employees. However, employers with a high percentage of older male workers, especially if a significant fraction are black, can have much higher rates. For example, rates of hypertension in white male workers ages 55 to 64 are, on the average, about 42 percent, compared with 59 percent for some ages of black male workers.[13]

A family history increases the chance of a person having hypertension. However, a number of other factors related to blood pressure are under individual control. Obesity is a prime contributor to hypertension. High salt intake in at least a minority of individuals can increase blood pressure. Americans usually consume ten to twenty times their daily requirements for salt, making it difficult to cut down to only what is necessary. However, salt is an acquired taste and reduced salt intake over a several week period will reset the palate to prefer less salt and make foods previously considered only slightly salty very salty tasting. Stress raises blood pressure acutely and in some individuals probably contributes to a sustained blood pressure elevation, although this is difficult to document unequivocally. Some experts feel that exercise can reduce blood pressure. However, the degree to which regular exercise, in addition to its many other benefits, can lead to sustained reduced high blood pressure is not known. Smoking causes an increase in blood pressure. Whether smoking leads to permanent elevations in blood pressure is not established.

In general, worksite-based hypertension detection and control programs have results that are superior to what is achieved in clinical practice, although control rates in both settings are very dependent on the enthusiasm and level of follow-up effort of those running the programs. A voluntary onsite screening, referral and follow-up program conducted by Massachusetts Mutual Life Insurance Company led to an increase in the percentage under control from 36 to 82 percent after one year of operation.[14] Although definitions of hypertension and control vary in different programs which have published their results, in general 80 to 90 percent control can be expected from diligent, well-organized and well-run programs.[15] Costs of programs also vary greatly. The cost of outside medical care per year for hypertension treatment and follow-up under company-sponsored health insurance seems to average between $175 and $250 per hypertensive.[16] However, if only on-site costs are considered, the cost is much less. A carefully done demonstration project conducted at Ford Motor Company showed that the first-year cost of blood pressure screening, referral, follow-up and monitoring was $14.59 (1982 prices) per employee and $37.34 per hypertensive employee. In the same study annual cost per patient receiving on-site treatment for hypertension was $96.19.[17]

A number of models to evaluate cost-benefit have been devised. However, the results are so sensitive to changing assumptions about frequency, level of control achieved, cost of treatment, impacts on disability costs, effects on absenteeism and costs of heart attacks, strokes, etc., that there is no consensus on what conclusions can be drawn.[18] For examples, studies of the effects of hypertension detection and control programs on absenteeism have given conflicting results, and no effect on absenteeism can be projected. What is clear, however, is that control reduces the usual health consequences of hypertension and prevents a large amount of suffering. Many insurance companies, convinced of the reduction in risk if blood pressure is controlled, accept treated blood pressure readings in hypertensives as the basis for calculating insurance premiums.[19]

Smoking

Literally thousands of studies have shown a striking relationship between smoking and excess heart disease, stroke, emphysema, chronic bronchitis, and many cancers, most notably cancer of the lungs, throat, esophagus and bladder.[20-24] As an example, age-standardized coronary heart disease death rates for male smokers of more than two packs per day are 335 percent of that of nonsmokers in the 45-to-54 age group, and 211 percent in the 55-to-64 age group. Cancer mortality ratios for male smokers compared with non-smokers in one study were 8.53 for lung, 3.96 for esophagus, 6.52 for mouth, throat and larynx and 2.55 for bladder.

When State Mutual Life Assurance Company of America compared the death rates of their life insurance policy holders who were smokers with those who did not smoke, they found shocking results in terms of death ratios: 15.0 for respiratory cancer, 14.7 for pneumonia and influenza, 2.7 for the leading causes of death in adulthood—arteriosclerosis, degenerative heart disease and insufficient blood supply to the heart, 8.1 for hypertensive heart disease and hypertension, 8.1 for digestive diseases, and 2.2 for motorcycle accidents.[25]

That smoking is most prevalent among blue-collar workers and those performing service jobs that require limited formal education was verified by an extensive study conducted in the late 1970s. Interviews with 2,528 white male hospital patients aged 41 to 70 revealed that between 43 and 47 percent of the blue-collar workers currently smoked cigarettes, compared to less than 30 percent of the white-collar workers. Over one-third of the professionals interviewed had never smoked, while this was only the case for slightly more than a fifth of the men employed as laborers or craftsmen.[26]

Smoking cessation reduces the risks related to smoking, but some risks decline more rapidly than others. For example, for the average smoker, cessation reduces coronary heart disease mortality risk from 200 to 150 percent of a nonsmoker's rate within one year, and within five to ten years the rate falls to a level similar to that of lifelong nonsmokers.[27] Excess risk of lung cancer due to smoking is reduced by two-thirds after about the first five years and approaches that of a nonsmoker approximately ten to fifteen years after quitting.[28]

A large number of different approaches for smoking cessation have been developed in recent years. Many of these have been applied in occupational settings, but limited information on their effectiveness at the worksite has been reported. In clinical settings initial quit rates are generally 70 to 90 percent in the best programs, be they five consecutive day cessation programs or those conducted over three or four weeks. The best sustained quit rates, measured at six to twelve months, are on the order of 40 percent and occasionally as high as 50 percent. More commonly, however, sustained quit rates are 20 to 30 percent.[29,30] Worksite smoking cessation programs may have a higher dropout rate than clinical programs, since they may conflict with work assignments. Also employees may not have to make a financial commitment to participate in an employer-sponsored program as they would in most stand-alone programs. This tends to increase attrition. On the other hand, a nonsmoking norm in the work setting may increase motivation to quit, ability to quit without an organized program, and lower rates of backsliding.

Smoking is expensive to employers, translating into higher health insurance costs (some studies show a 50 percent greater use of the health care system by smokers), higher absenteeism (an estimated two to three additional days per year), increased accidents at work and a higher rate of disability reimbursable events.[31-33] One estimate of the annual excess cost of smokers is about $400, but given the difficulty in developing accurate estimates that apply to different circumstances, a range estimate of $200 to $500 is probably more defensible.[34,35] Costs of organized cessation programs vary greatly, from services provided free or for a nominal cost by voluntary agencies, to $400 to $500 per person charged by some well-known for-profit groups. On a group basis, good quality programs do not need to cost in excess of $100 to $200 per participant. Whatever estimates are used, there should be a good employer and employee return on investment in smoking cessation programs, both those financed entirely by employers and those cost-shared with employees.

Lack of Exercise

Estimates on the number of American adults getting "regular" exercise vary widely. However, some of the best information on the frequency of exercise comes from a 1980 telephone survey of more than 1,000 randomly selected adults in Massachusetts. Interviewers found that 56 percent claimed to exercise at least twice weekly and over one-quarter claimed daily exercise. The exercise lasted for a median time of 47 minutes, and the most common exercises most recently engaged in were walking (14.0 percent), jogging (12.8 percent) and calisthenics (8.7 percent).[36]

Lack of exercise has been implicated in a number of health problems. Bone loss (osteoporosis), which leads to increased risk of fractures and is particularly common in older women, is increased by lack of exercise. Insufficient exercise to strengthen abdominal muscles and poor flexibility increase the risk of low back problems. Lack of exercise to help maintain flexibility and balance is thought to increase the risk of serious falls in older adults.

Regular participation in exercise programs, including those sponsored by employers, can lead to significant reductions in weight, improved measures of fitness (e.g., faster pulse recovery rates after exercising and better performance on standardized fitness tests), decreases in systolic and diastolic blood pressures, and also reduced skin fold thickness, a good measure of body fat percent.[37-41] How much these indices of health improve is, to the consternation of many part-time, limited exercisers, directly proportional to the frequency and intensity of exercise.

Evidence of the benefits of vigorous sustained exercise has recently been expanded to include reducing risk for heart disease. Growing evidence supports the theory that more physically active individuals have a lower age-specific rate of heart attacks and associated deaths than their sedentary confreres, even when all other heart health risks (blood pressure, smoking status, etc.) are held constant. One recently reported study found that physically fit (measured by physical work capacity assessed by bicycle ergonometry) men between the ages of 35 and 54 were less than half as likely as their nonphysically fit counterparts to suffer from a heart attack in a subsequent five-year period.[42] A study of Harvard alumni found that those expending fewer than 2,000 calories at work and play had a 64 percent higher risk of heart attack than the more active.[43] Another study found that the risk of dying of a heart attack with no prior indication of heart disease was 55 to 65 percent lower in persons who engaged in at least some high-intensity leisure time activities during the prior year.[44] This protective effect may be mediated by an exercise-induced increase in a helpful kind of blood cholesterol called HDL (high density lipoprotein), which appears to assist in removing fatty deposits from artery walls.[45]

Accompanying a regular exercise program, whether employer-sponsored or not, are improvements in energy level, attitude toward job and employer, overall morale and self-perceived work performance. Degree of participation is directly related to level of benefit obtained. For example, men from a variety of occupations randomly assigned to an employer-sponsored exercise program reported improvements in work performance 60 percent of the time compared to only 3 percent for the nonparticipant group.[46] Another benefit of exercise which any regular exerciser will volunteer is a reduction in stress and in feelings of depression, both common concerns problems among employees (see Chapter 19).

From the employer's viewpoint, an additional dividend of exercise

FIGURE 18.4
FACTORS THAT AFFECT PARTICIPATION IN WORKSITE FITNESS PROGRAMS

- Socioeconomic status
- Age
- Sex
- Other health practices
- Proximity to work stations
- Whether job scheduling flexibility permits use during the day
- Type and variety of exercise offerings
- Criteria for entry into the program
- Intensity and type of recruitment into the program
- Costs to employees
- Whether program is open to family
- Nature of supervision
- Personality of supervising professionals
- Availability of tracking system to quantify achievements
- Work load
- Travel requirements of job
- Hours of program operation

programs is a reduction in absenteeism. For example, an unpublished study from the Metropolitan Life Insurance Company found that absenteeism rates in 100 participants in a voluntary exercise program experienced decreases in annual absenteeism from 6.3 to 4.9 days while the control group of 100 nonexercisers showed a net increase from 5.6 to 7.0 days.[47]

In a Toronto insurance company, adherers to an exercise program at the company experienced a 42 percent decline in average monthly absenteeism compared to a 20 percent decline in both the test company overall and a control insurance company in the same city.[48] Some studies suggest that even lower levels of participation can reduce absenteeism, but the results are not conclusive.

As with so many health improvement programs, the toughest question regarding employer-sponsored exercise programs is not whether they work but how to get employees to persist with their new healthful habit. When participation by an individual in an exercise program is defined as having completed a prescreening and at least initiating an exercise program, usual rates of participation are 20 to 40 percent of the workforce. Offsite company-sponsored programs (e.g., YMCA membership) show lower rates, generally in the 10 to 25 percent range. Several well-designed worksite exercise programs have found it difficult to get over one-half of "active" participants to even average two sessions per week.[49,50] A number of the factors that have been shown to affect participation rates are listed in Figure 18.4.

Costs of operating an employer-sponsored program vary considerably, and rules of thumb are difficult to find. An outside jogging track plus some showers and lockers is an investment of a different order of magnitude compared with an indoor exercise facility, a swimming pool or an indoor track. Operating costs also differ greatly, with the major cost being staffing. A full-staffed operation with a facility open ten to fourteen hours a day can incur costs of between $500 and $1,000 per participant per year. Much less intensive staffing is also possible, with commensurate reduction in costs and usually decline in participation. To date there is no good information on the cost effectiveness or cost-benefit ratio of these programs, though many corporations feel they provide a visible benefit that makes a major contribution to productivity and frequently aids in recruitment.

Nutrition

Few health subjects evoke as much controversy as what constitutes good nutritional habits. Most nutritionists, however, are generally supportive of the nutritional goals formulated by the Senate Select Subcommittee on Nutrition in the mid-1970s and then published jointly by the Department of Health, Education and Welfare and the Department of Agriculture in a more general form termed the U.S. Dietary Guidelines. Its seven major guidelines for improving Americans' eating habits are:

- Eat a variety of foods
- Maintain ideal weight
- Avoid too much fat, saturated fat and cholesterol
- Eat foods with adequate starch and fiber
- Avoid too much sugar
- Avoid too much salt
- If you drink alcohol, do so in moderation

Whatever one's perspective, there is little question that nutrition has an important influence on our health. A recent report of the National Academy of Sciences underscored, for example, that a significant but indeterminate portion of cancers were probably related to nutritional habits.

In general, there appears to be agreement that the American diet contains too much fat, currently about 40 percent of total calories, and especially too much saturated fat. High total fat intake appears to increase risk of some cancers, particularly breast and colon.[51] Fats contain more than twice the number of calories per ounce as protein and the much maligned carbohydrates. Saturated fats raise serum cholesterol, which is a well-recognized risk for heart disease, stroke and other blood vessel problems. Many nutritionists feel that whenever possible, total blood cholesterol should be maintained below 200 (milligrams per 100 milliliters). The two best ways to accomplish this are reduction in saturated fat intake and good management of weight, since obesity increases total cholesterol. Reducing the intake of foods high in cholesterol may also help to reduce blood cholesterol, although this effect has been difficult to show in studies of small numbers of individuals fed diets differing only in cholesterol content.

As discussed under hypertension, the amount of salt in the diet of a population correlates reasonably well with the frequency of high blood pressure. For example, the Japanese, whose diet is very high in salt, have more hypertension than Americans.[52] There is some evidence that salt fed to animals at critical times in their early development increases the chances of them being hypertensive as adults. For these reasons a prudent interpretation suggests that we reduce our salt intake.

Americans annually consume almost 100 pounds per person of simple sugars, which not only have many calories unaccompanied by significant nutritional value, but also are the major contributor to both tooth decay and gum disease. Most nutritionists recommend a diet that is high in complex carbohydrates (starches) but low in simple sugars. Diets high both in complex carbohydrates and fiber have been found to be very helpful in the control of adult onset diabetes and have in many cases reduced the need for insulin.[53] Fiber, the nondigestible part of vegetable substances, increases bulk in the stool, reduces constipation and hemorrhoids, some of the most common and distressing medical problems of adulthood, and *may* reduce risks for both diverticulitis (inflammation of small outpocketings of the large intestine) and colon cancer, although neither relationship has been clearly shown to be causal.

Nutrition education programs have been developed at a number of companies but it is difficult to measure their sustained effect on eating habits, much less the effect of changes in eating habits on related health problems. Knowledge of nutrition can be improved by the use of educational materials of various types, including newsletters, nutritional games related to foods served in the cafeteria, tray liners containing nutritional information, etc.[54,55] However, the translation of changes in knowledge and even attitudes into good eating habits is complex and involves what is eaten at home and at restaurants in addition to what is eaten at the workplace.

Overweight and Obesity

Estimates of obesity (excess fat) and overweight (a certain percentage over "ideal" weight) are influenced by definitions which are not standardized since there are different methods of estimating body fat, different decision rules for deciding how much fat is too much and different tables of "ideal" weight against which individual weights are compared. One estimate of the percentage of workers at least 20 percent over ideal weight is 13.6 percent for males and 16.6 percent for females. In women, the frequency of overweight increases with age, from 9.6 percent in those from 17 to 24 to 31.5 percent in the 65+ age group, while in men there is little change with aging.[56] In general, obesity is more frequent in those with limited educational background and lower paying jobs. Unless being overweight is due to unusually large muscle mass, obesity and overweight can be considered as the same problem.

One of the tried and true measurements used to determine individual life insurance premiums is weight (or girth) compared to height. Being overweight increases the risk of death, primarily through increases in blood pressure, cholesterol and blood sugar.[57,58] It also increases the risk of musculoskeletal problems, particularly degeneration of weight-bearing joints, makes exercise more difficult, adversely affects self-image, and may interfere with interpersonal relationships. In general, it is more difficult for obese individuals to get jobs.

While there are strong differences of opinion among scientists about the causes of overweight and obesity, there is general agreement that a number of behavioral approaches have recently increased the ability of individuals to manage their weight. Among the components of an effective weight reduction and management program are usually:

1. maintaining a diary of what foods are eaten, in what quantities, at what times and under what circumstances;
2. altering the usual stimuli that affect food buying and eating habits (e.g., buying from a previously prepared list; not eating while doing other things like watching television; not keeping high calorie foods at home);
3. better managing the eating process (e.g., putting down the utensil between bites, chewing thoroughly, serving smaller portions and not leaving additional food within view); and
4. contracting with oneself or another to promptly reward good eating habits.[59,60]

Results of weight loss programs depend upon many factors, including initial motivation of the participants, their degree of overweight/obesity, the skill of the teacher/facilitator, support from the group, family and friends, financial commitment to weight loss, program quality and the appropriateness of the program to the participant group. For those significantly overweight, a gradual reduction of 15–30 pounds over a six-month period is realistic in clinical programs. While crash diets can lead to rapid weight loss, sometimes as much as one-half to one pound per day if coupled with an ambitious exercise regime, there is no evidence that this group does better over time than those who lose weight primarily through a restructuring of their normal eating habits.[61,62]

While many studies have been reported from clinical settings, the first controlled trial of treatment for obesity at the worksite was published only in 1980. Forty women volunteers averaging 57.1% overweight were assigned to four different treatment groups for a 16-week behavioral program. Those staying with the program for the full period lost an average of eight pounds, but six months later the average loss was down to two and one-half pounds.[63]

Backsliding is the greatest problem in controlling obesity, and many fat people claim to have lost several thousand pounds over the course of many episodic attempts to lose weight, only to regain all they had lost, and sometimes more. Of 26 groups in nine behavioral weight control studies, 14 showed weight gains during the 12 months following treatment. In the 12 which showed continued weight loss, the average additional loss was only 1.6 kg (3.5 lbs.), hardly significant considering initial percentages overweight of 20 to 78 percent.[64]

The largest organized weight reduction program is Weight Watchers, which has served about one-half million people in the United States and 25 foreign countries. The program involves continuing weekly group meetings usually lasting about one hour, which are led by a facilitator who is almost always a graduate of the program. Behavioral concepts have been incorporated into Weight Watchers activities. A number of other commercial and nonprofit volunteer groups have been established to provide skills, training and support in weight management. Most behaviorally oriented programs have one or two sessions per week for a 10- to 20-week period. They use a combination of a recommended diet, behavioral techniques, and suggestions for increasing physical activity. Those under medical supervision sometimes use appetite suppressant drugs, despite the lack of evidence that their use improves the long-term outcome. In judging claimed results from weight-loss programs it is important to understand whether program dropouts were included or excluded from the figures. In one study of a popular commercial weight-loss program, there was a 50 percent dropout rate in the first six weeks and a 70 percent dropout rate within twelve weeks.[65]

Possible health improvements from effective weight management programs include:

1. Reduced incidence of adult-onset diabetes mellitus;
2. Reduced total cholesterol and low-density lipoprotein cholesterol;
3. Reduced blood pressure;
4. Increased ease of exercise;
5. Improved self-image; and
6. Reduced strain on spinal column and weight-bearing joints.

These changes in known health risks can in turn reduce the risk for heart and blood vessel diseases, stroke and some back and hip problems. However, while improvements in health are clear, the relationship between particular amounts of weight loss and reduction in these risks and health problems differs considerably from one person to the next. Therefore, speculation on possible dollar savings based on reduced health risks and clinical problems is not warranted.

GROWTH OF HEALTH PROMOTION/DISEASE PREVENTION PROGRAMS

"Wellness Epidemic Sweeps Companies" headlined a recent issue of *Business Insurance*. That American business is bullish on wellness is increasingly clear.

A 1978 survey of corporate health promotion and risk-reduction activities by the Washington Business Group on Health sent to its 160 member companies, almost without exception among the Fortune 500, yielded 59 responses (36.9 percent). Companies with programs targeted at specific risks ranged from 41 percent offering stress management to 85 percent providing CPR (cardiopulmonary resuscitation) classes.[66]

A 1979 questionnaire about corporate fitness and other health promotion programs was sent by Fitness Systems to major U.S. companies listed in *Fortune* magazine, the 300 top industrials and top 50 of each of the life insurance, commercial banking, utilities, retailing, diversified financials and transportation business sectors.[67] Of the 22 percent of companies from whom a return was obtained, about one-half had diet/nutrition counseling and/or smoking cessation programs, slightly more than one-third had stress management programs, two-thirds had alcohol/drug programs, and one-quarter had physical fitness programs. The extent of employee fitness programs in Canada was surveyed in early 1981 by the Canadian Public Health Association, with the financial support and consultation of Fitness and Amateur Sport Government of Canada. Among the 26 percent of 800 companies responding, fitness programs were reported in 25.4 percent (N = 52), primarily in companies with more than 500 employees.[68]

A 1981 survey of 424 California employers chosen at random from all those with more than 100 employees at one or more sites in California found that 332 (78.3 percent) offered one or more health promotion activities as summarized in Table 18.6. The most frequent activities were accident prevention, cardiopulmonary resuscitation and choke saver. Alcohol and/or drug abuse programs and mental health counseling were in place in about one-quarter of organizations offering any program. Hypertension screening, smoking cessation, fitness, and stress management programs were made available by 10 to 17 percent of these employers. Overall the 424 organizations surveyed offered a total of 938 health promotion activities, and those with some program averaged 2.8 activities.[69]

The largest employers almost invariably had some program activity, with 98.1 percent of the 5,000 + employee group having at least one activity, and averaging 3.9 activities. But one of the most striking results of the survey

TABLE 18.6
EMPLOYER HEALTH PROMOTION ACTIVITIES

Activity	Currently offering program		Planning new programs
	Percentage of programs (N = 938)	Percentage of employers (N = 332)	Percentage of programs (N = 217)
Hypertension screening	4.6	13.0	11.1
Smoking cessation	3.7	10.5	9.7
Weight control	3.4	9.6	7.8
Mental health counseling	8.3	23.5	3.7
Nutrition training	2.3	6.6	6.0
CPR, choke saver	23.9	67.5	11.1
Exercise/fitness	5.2	14.8	12.0
Drug/alcohol abuse	8.4	23.8	8.3
Stress management	5.9	16.6	14.3
Accident prevention	29.2	82.5	4.6
Cancer risk reduction	2.9	8.1	4.1
Other	2.1	6.0	7.4

Source: Fielding, J.E. and Breslow, L. "Health Promotion Programs Sponsored By California Employers," *American Journal of Public Health,* Vol. 73, No. 5, pp. 538–542.

was that many small employers had some program activity. Of those with 100 to 249 employees, 66 percent had at least some program with primarily one activity (31.2 percent) or two (25.0 percent). Most interesting was the rate of acceleration of new programs which the survey brought to light (Table 18.7). Initiation of new activities grew from 11.3 per year during 1962–71 to 111.5 per year for the 1978–1981 period.[70]

Programs vary from a few lectures on nutrition or exercise to extensive multicomponent programs with considerable staffing and the construction of large physical plants to accommodate exercise programs and other health-promoting activities. Programs established in the last five to ten years differ from their predecessors in being more comprehensive, usually involving a number of target health problems, involving full-time staff and being continuous rather than periodic.

Program Examples

It has been estimated that over 500 corporations have set up multicomponent programs to promote wellness. A few of the better-known large comprehensive programs are:

Johnson & Johnson "Live for Life": Established through the interest of the chairman of the board, its objectives are: 1) To provide the means for

TABLE 18.7
CHRONOLOGY OF HEALTH PROMOTION ACTIVITY
INITIATION

Time interval	Number of activities initiated*	Average number of activities added per year
Prior to 1961	68.0**	—
1962–1971	113.0	11.3
1972–1974	45.0	15.0
1975–1977	188.0	62.7
1978–1981	446.0	111.5
	860.0	

*Based on 860 activities for which duration was provided.
**52/68 activities were accident prevention.
Source: Fielding, J.E. and Breslow, L. "Health Promotion Programs Sponsored By California Employers," *American Journal of Public Health*, Vol. 73, No. 5, pp. 538–542.

Johnson & Johnson employees to become the healthiest in the world; and 2) To control and reduce the spiraling illness and accident costs to the corporation.

Guidelines for the program include:

1. Focus on health and the benefits of being well;
2. Gradual introduction of the program to the operating units;
3. Care to construct the program based on sound scientific information;
4. Attention to measurable results; and
5. Emphasis on personal employee involvement.

Major components of the program are a health screen, a lifestyle seminar which introduces employees to the Live for Life concept in depth and a variety of lifestyle improvement programs, such as smoking cessation, exercise, stress management, nutrition, weight control and general health knowledge. These programs are integrated with established medical programs, such as high blood pressure detection and control and employee assistance programs.

Live for Life is primarily a service organization of the corporate staff which provides participating Johnson & Johnson companies with the consulting expertise, training, core program components, professional services and promotional materials necessary for the program's successful implementation at each site. Alteration of the work environment, such as establishment of a company smoking policy, changing the food in the company cafeterias and building exercise facilities, is considered an integral and essential part of the program.[71,72]

TABLE 18.8
"LIVE FOR LIFE" EVALUATION RESULTS

Health screen measure	Percentage of change (baseline—one year)		
	Treatment (N = 737)		Control (N = 680)
Fitness			
Aerobic calories/Kg/week	43	**	6
Weight control			
Percentage above ideal weight	−1	**	6
Smoking cessation			
Percentage of current smokers	−15	*	−4
Stress management			
General well-being	5	**	2
Percentage with elevated blood pressure (≥140/90)	−32		−9
Employee attitudes			
Self-reported sick days	−9	*	14
Satisfaction with working conditions	3	**	−7
Satisfaction with personal relations at work	1	**	−3
Ability to handle job strain	0	**	−2
Job involvement	2		0
Commitment to the organization	0		−2
Job self-esteem	0	*	−2
Satisfaction with growth opportunities	−1		−3

* = Significant at the 5 percent level.
** = Significant at the 1 percent level.
Source: Wilbur, C.S. "Live for Life: An Epidemiological Evaluation of a Comprehensive Health Promotion Program." Johnson & Johnson, New Brunswick, N.J. (unpublished).

Preliminary data comparing about 700 employees who participated in the Live for Life program with about 700 who did not has revealed encouraging results. Table 18.8 summarizes these results and demonstrates that both objective and subjective improvements have been achieved.[73]

___Control Data Corporation "Staywell"___: Control Data Corporation has made a very large investment in improving the health of its employees. Program emphasis is placed both on reducing known health risks and improving health. Considerable effort has been invested in developing state-of-the-art behavioral change programs for smoking cessation, stress management, weight control, improved nutrition and exercise. The program is introduced through a motivational session outlining program goals and opportunities for participation. A health screening is offered as the initial activity, and feedback is provided on health risk and opportunities for risk reduction. A central staff provides program assistance to the operating sites where the program is introduced in phases. Unlike Johnson & Johnson, no gymnasia or special exercise facilities have been constructed. A unique feature is the effort to place as much educational material as possible on computers for interactive learning, taking into account individual learning styles and personal goals.

Kimberly Clark: At its Neenah, Wisconsin plant, Kimberly Clark has constructed a $2.5 million multiphasic testing and physical fitness complex. A full-time program staff of twenty-five under the direction of a specialist in internal medicine provide a medical history and health risk appraisal, multiphasic screening, a physical examination, an exercise test on a treadmill or bicycle and a health review with recommendations. A wide variety of aerobic exercises are encouraged, including jogging, swimming, stationary cycling, circuit training, aerobic dancing, rope jumping, water exercise, cardiac rehabilitation, and cross-country skiing. Health education classes cover smoking control, breast self-examination, high blood pressure control, diet management and cardiopulmonary resuscitation (CPR). An employee assistance program is also available.

Mattel "HEP" (Health Enhancement Program) and "THE" (The Health Enhancement) Program of Tosco Corporation: These two similar programs, both in the Los Angeles area, are operated jointly by each corporation and the UCLA Center for Health Enhancement Education and Research, a part of the UCLA Medical Center. Overall objectives of these health enhancement programs are:

1. To increase employee and spouse's knowledge of factors related to reducible health risks, identify personal risks and to increase their ability to reduce these risks;
2. To achieve and maintain individual lifestyle changes conducive to improved employee and spouse health;
3. To achieve and maintain voluntary program participation, especially among the employees and spouses who are at highest risk for health problems;
4. To provide opportunities and an environment which will support the practice of healthful behaviors;

5. To increase employee productivity potential through decreased illness-related absences, decreased work-related injuries, decreased turnover rates and improved physical and mental well-being;

6. To reduce the number of insurance claims by employees and dependents for hospitalization and physician visits, and to reduce the rate of increase in health care costs;

7. To reduce the number and frequency of disability claims that derive from preventable illnesses and injuries, and

8. To evaluate the effectiveness of intervention strategies aimed toward health promotion in the workplace.

Major components of both programs include:

1. A personal health profile questionnaire;

2. A screening exam covering height and weight, blood pressure, vision screening, resting pulse and blood lipid determinations;

3. A submaximal exercise step test;

4. A feedback session to explain results and suggest individual opportunities for health improvement;

5. High-risk counseling for those most in need of risk-reduction activities;

6. High blood pressure detection, referral and continuing follow-up;

7. Large group educational programs covering a wide variety of health subjects;

8. Cooking demonstrations on how to prepare nutritious meals; and

9. Behavior-change classes on weight management, smoking cessation, stress reduction, establishing a lasting personal exercise program, and nutrition education for all employees as well as those with special problems, such as high cholesterol levels.

In addition, significant changes have been made at Mattel in the choices of food served in the cafeteria and those available in the vending machines. Food content is provided for all common cafeteria selections.

All four of these multicomponent programs share the development of a large number of educational and motivational tools to inform and motivate employees. Attention has been directed toward maximizing participation, which, while very good initially, requires continuous promotion to be maintained. A health newsletter is provided by all these companies to their employees on a regular basis. It generally covers information on the program as well as health information from the companies' own health professionals or a reputable health news service, such as Healthfax.

An increasingly large number of companies have developed programs with many of the same features, but considerable variation exists. For example, IBM's "A Plan for Life" includes courses in exercise, smoking cessation, stress and weight management, alleviation and prevention of back problems, first aid, cardiopulmonary resuscitation, nutrition and risk factor management. These courses are offered free of charge when conducted by IBM, or are provided on a tuition-reimbursement basis if available in co-operation with a group of community resources such as the YMCA, American Cancer Society affiliate, American Heart Association affiliate, etc. This model is readily adaptable to employers who have a widely dispersed workforce where it would be impossible to develop internal programs at every site.

Some companies, such as Safeco, provide an organized informational program without any clinical aspect. Safeco's Health Action Plan for Everyone (SHAPE) is a self-help venture which consists of three basic components:

1. Self-assessment questionnaire and a health-quality profile developed by health care professionals.
2. A "how-to" notebook on lifestyle change.
3. A monthly newsletter that offers tips and encouragement on topics such as diet, smoking cessation, exercise, alcohol abuse and stress.

Use of Incentives to Change Behaviors

A number of companies have experimented with providing incentives for employees to adopt positive health habits. Companies have offered employees from $150 to as much as $5,000 to stop smoking. For example, Iowa-based Pioneer Hi-Bred International will pay $150 to any employee who stops smoking for one year, and an additional $75 if he or she continues not to smoke for another year. The same company has offered employees $5 for every pound they lose down to their desired weight. To be eligible, the individual must be at least 20 percent above recommended weight. Employees who maintain their recommended weight for an additional year get a $75 cash bonus.[74]

Bonne Bell, a skincare and cosmetics marketer in Cleveland for 55 years, offers its employees a number of incentives for healthful behavior: 50 cents per mile for running, up to $200 for quitting smoking and staying off for six months, and $5 per pound up to 50 pounds for weight loss over a six-month period. However, for every pound gained back in the following six months, employees must agree to put $10 into the company's charitable foundation. Ex-smokers who begin to smoke again must agree to put twice their smoking cessation bonus into the same foundation.[75]

Hospital Corporation of America provides incentives for workers to attend its wellness center in Nashville by paying them 6 cents for each mile they bike, 24 cents for each mile they walk or run and 96 cents for each mile they swim.[76]

Fitness Programs

A large number of companies have focused their attention on the benefits of exercise, particularly aerobic exercise. One prototype is Xerox, which has developed at its Leesburg International Center a recreation complex offering an exercise room, squash and racketball courts, basketball and volleyball courts, jogging track, a football/soccer field, a twenty-five-yard swimming pool, badminton courts, tennis courts, and an eighteen-hole putting green. A professional staff is trained in leisure sports, recreational activities, physical fitness and exercise physiology. Participating employees receive a *Fitbook* containing a four-module series of exercises to improve muscular strength and endurance, joint flexibility and cardiovascular fitness, diet guidelines, and information on smoking and substance abuse as well as guidelines for changing individual behavior. A half-hour film covers risk factors and behavioral modification techniques. Employees are encouraged to contract with themselves to reach specific fitness and conditioning goals.

Some indication of the growth in corporate fitness programs is the growth of membership in the American Association of Fitness Directors in Business and Industry from 35 in 1975 to over 3,000 in 1983.[77]

Private Health Insurers and Health Promotion Programs

Both commercial insurers and Blue Cross/Blue Shield have initiated a number of health promotion programs for their employees as well as their subscribers. The commercial life and health insurers have devoted considerable resources to developing innovative programs that can improve health. For example, they have been the originators and principal sponsors of a research-and-demonstration activity to test the feasibility and impact of putting more prevention into primary care. They have convened an outside panel of experts to advise the private insurance industry on choosing priorities for disease prevention and health promotion activities, and, in response to the recommendations of this panel, have established a smoking-reduction program that is available to all their members and their group subscribers.

Metropolitan Life, a leader in health education/promotion for many years, has marketed health promotion services to smaller clients on a fee-for-service basis with considerable success in terms of subscriber interest. The Travellers Insurance Company is providing employers with information for employees, including articles for house organs, audiovisual presentations, a series of booklets and a monthly newsletter on benefits, lifestyle, health care use and self-care. These materials cost one dollar or less per employee per month. Also offered to employees are copies of "Take Care of Yourself" (Vickery and Fries). A number of other insurance companies are marketing, or making available at no extra charge, computerized health risk appraisals (HRAs). Hardly surprisingly, a key target of the HRAs are the individual life insurance subscribers.

Blue Cross and Blue Shield plans have initiated a variety of health promotion/disease prevention activities that range from blood pressure screening and awareness programs to on-site fitness programs (Indiana) to worksite education in stress management, weight control, hypertension, smoking cessation, and fitness (Maine). The Michigan "Blues" have developed a comprehensive "Go to Health" program for their own employees, which includes a computerized HRA and appropriate education and counseling sessions for employees at high risk of cardiovascular disease.[78] Many plans have developed activities initially for their own employees and sometimes their families, but with the hope that they could eventually market them to their subscribing groups.

HEALTH PROMOTION PROGRAMS: KEYS TO SUCCESS

There is no single model for a successful health promotion program. A program can have a single element or have multiple components. It can be targeted at a high risk group, e.g., those who smoke, or it can be aimed at the entire workforce. However, based primarily on oral reports from programs that have worked and those that have not, some keys to success have been identified.

1. *Long-term commitment* A several-year time horizon is necessary to see a sustained impact. Dealing effectively with backsliding requires a sensitive and supportive approach which builds over time and avoids participants' loss of confidence and a feeling of defeatism. Since most employers experience considerable turnover, offering programs on a continuing basis is necessary so that new employees can participate. Making clear that there is a long-term commitment, whether accomplished through making the program a benefit or by other means, encourages those within the organization take the activities more seriously.

2. *Top management support* Programs do not work if not strongly supported by top management. Inevitably there is stiff competition for internal resources. Changes in how things are done within the company, e.g., developing a company smoking policy, requires top support to overcome inertia and minimize roadblocks.

3. *Employee involvement* A program planned and installed by management alone will not enjoy the level of participation and enthusiasm necessary for success. A broad-based group of employees, chosen from diverse job levels and locations as well as on the basis of the respect

with which they are viewed by fellow employees, should participate in the planning. A planning committee is very helpful, not only in the molding of a program, but in suggesting ways of promoting it to employees and their families. Where employees are unionized, involving the union is essential. Without participation, unions frequently feel that the company is trying to usurp the union's prerogatives to provide benefits through collective bargaining and to undermine the allegiance of employees to their union. When information to be collected in the program is considered confidential, union support is particularly important.

4. *Professional leadership*   The individuals in charge of the program should have appropriate training and be enthusiastic. In addition, they should be good role models, both physically and in terms of attitudes toward health and related behaviors. Program leaders should have good interpersonal and group process skills and be well organized. They should have a knack for organizing large groups of people. Finally, these individuals should command the respect of both management and other employees for being very knowledgable about their subject and how to structure programs.

5. *Clearly defined objectives*   Clearly defined program objectives permit an efficient planning and implementation process and are an essential prerequisite to effective evaluation. Programs that have not spent sufficient time ironing these out to everyone's satisfaction often waste considerable resources haggling over which approach is most consistent with fuzzy objectives.◄

6. *Emphasis on careful planning*   Nothing is more essential to a smooth-running program than meticulous planning. In many cases the enthusiasm to have a program operational has led to premature installment, with less than desirable quality, many logistical problems and ultimately a program that is sloppy or amateurish. Clear assignment of overall program responsibility as well as responsibility for each necessary task and subtask often make the difference between success and embarrassment.

7. *Attention to confidentiality*   Most employees consider some of the information that they are asked to provide on their health and health problems confidential. Concerns about who will have access to the information and whether the information can in any way affect their possibilities for promotion or otherwise affect employment can lead to low participation rates and to unreliable information. Decisions on how and by whom this information will be handled must be made early and

the policy made explicit. In some cases, employees' fears, based on prior breaches of confidentiality, may militate for working with a neutral outside professional organization.

8. *Strong and continuing promotional efforts*   Although the programs are designed to help employees, high participation rates require strenuous efforts to promote the program and its benefits to the target groups. A marketing plan is therefore an important element of a successful program, even if the program is entirely financed by the employer and the employee is permitted to participate during working hours. Promotion is particularly necessary after initial enthusiasm has dissipated.

9. *Appropriate assignments of program responsibility*   The person who is given the day-to-day operational authority for the health enhancement program must have a position of trust and responsibility within the organization that permits him or her to call upon other resources of the organization for assistance. Locus of responsibility must be clear to all those involved so that problems are brought to a central location for prompt resolution.

10. *Family involvement*   A major key to employees' success in their endeavors to improve health-related behavior is the involvement of their families. Family support and participation to the greatest extent possible are extremely important to individuals who are attempting to change a lifetime of poor eating habits, quit smoking after several years of two-plus-pack-a-day consumption or develop a regular exercise routine after years of sedentary living. Learning good nutrition, for example, will not lead to change if the family member preparing the food does not have the same knowledge and is not motivated to change what is served at home. In addition, there's a good reason for employers to make participation in the health promotion program available to family members of employees: an average of one-half of the health care costs related to ill health are for dependents.

EVALUATION

Does the program work? What kind of return are we getting on our investment? These are the two basic questions that most employers want answered when they have sponsored health promotion programs for their employees and sometimes for their families.

A number of companies believe they are saving substantial sums of money as a result of their health promotion/disease prevention programs.

For example, New York Telephone estimates that it gains $2.7 million annually from nine health promotion programs made available to 80,000 employees. Their programs include: smoking cessation, cholesterol reduction, hypertension control, fitness training, stress management, screening for breast and colorectal cancers, alcohol abuse control, and training in preventing back injuries.[79]

Campbell Soup estimated a savings of $245,000 over a ten-year period (1969 to 1979) for colorectal cancer screening consisting of sigmoidoscopy every four years after age 44. In addition, 90 percent of Campbell's hypertensive employees are on treatment, and the company believes that as a result, 75 percent of expected strokes per year in the 55-to-65 age group are prevented.[80]

In most cases, company self-reports on the costs and effects of health promotion programs are not based on careful evaluation. Rather, they are gross estimates based on a number of underlying assumptions. For example, Campbell's analysis of the savings attributable to their colorectal cancer screening program hinges on assumptions regarding the number of cases of colorectal cancer that would have occurred in the absence of screening and the direct and indirect costs associated with each case. It also assumes that all cases prevented were due to on-site screening rather than screening that occurred in another setting (e.g., doctor's office or HMO) at the encouragement of an outside health professional. While these estimates of savings due to health promotion measures are useful in showing the value companies themselves have placed on the savings, it is difficult to know if their assertions can be applied to other companies.

Determining Program Success

Whether or not the program works can only be assessed in relationship to the program objectives. Many programs have very broad objectives, such as "To improve the health of employees," and "To reduce health care costs." Translating these objectives into an evaluation strategy is often frustrating and expensive. For example, what does "improve the health of employees" mean? To one manager it may mean decreasing the number of heart attacks every year. To another it may mean reducing the known risks for important diseases in employees. For a third manager it may mean that employees have fewer sick days per year. And a fourth might feel that the best measure is what the employees respond when asked, "Do you feel that the company health promotion program has contributed to improvements in your health?" or "Do you feel more healthy than you did before the program started?" Answering each of these requires a different methodology and attempting to answer some of these questions requires very substantial resources. Some common factors whose changes might be targets for a program evaluation are:

1. **Knowledge of health problems**   Example: Employees better understand the importance of controlling their blood pressure and the relationship of high blood pressure to stroke and heart disease.

2. **Attitudes**   Example: Employees now consider that regular aerobic exercise is important to their health.

3. **Intentions**   Example: More smoking employees state that they intend to give up smoking.

4. **Behaviors**   Example: More employees routinely wear seat belts than before.

5. **Risk levels**   Example: The average total cholesterol is now lower in employees than before.

6. **Illness**   Example: Fewer illnesses per employee per year are reported than before or illness-related absenteeism has declined.

7. **Disability**   Example: Disability overall or for specific categories of preventable problems has been reduced.

8. **Mortality**   Example: Fewer employees are dying per year overall or from specific types of preventable health problems, such as automobile accidents or lung cancer.

9. **Job satisfaction**   Example: Employees report that they are happier in their jobs.

10. **Satisfaction with employer**   Example: Employees report a more positive feeling about their employer.

11. **Productivity**   Example: Employees produce more output per hour, day or week; if no objective measure is possible, employees state that they are more productive.

12. **Health care costs**   Example: Health costs have been reduced or have increased less rapidly.

13. **Turnover**   Example: Turnover and associated costs have been reduced.

While some measures of success can be assessed easily and directly (e.g., blood pressure and weight), most rely on subjective responses (e.g., satisfaction, morale) or have problems with reliability (e.g., self-declared smoking or seat belt use) or are difficult to obtain and/or to analyze (e.g., productivity, health care claims costs). Frequently the outcomes you are looking for are most affected by the business cycle (e.g., turnover), the most recent pay raise (e.g., employer satisfaction) or changes in benefits (e.g., health care costs) than by the program whose impact is being measured. In addition, it is important to decide who is the target group. Is it the entire employee population or those that actively participated in the health enhancement program?

Evaluation of whether a program worked should involve being specific about the criteria for success. For example, three companies reported successful hypertension control programs. Company A reported, "Over 80 percent of our employees participated in screening for high blood pressure and were informed if they had high blood pressure." Company B accurately claimed, "In our program, two-thirds of our employees with high blood pressure have it under adequate control." Company C reported that, "Seventy-five percent of those who were found to have previously unknown hypertension had their blood pressure under control after one year of an intensive follow-up program." The information reported by each company is summarized in Figure 18.5. Based on this set of facts, however, it is very difficult to compare these programs. Can a program such as that of Company B be considered a success if only 20 percent of total employees were screened? Based on the limited information reported, can Company C's program be compared in effectiveness with that of Companies A or B? We don't know, for example, what percentage of those with known hypertension in Company C were under control at the time of initial screening. Can Company A's program be considered successful even though it consisted only of screening and referral, without any follow-up?

What these examples point out is the need for programs to be carefully planned so that each important measure of success can be assessed. The essential point is to decide at the onset what constitutes success and how it can be measured. The types of questions which companies generally will wish to answer are:

1. What was the frequency and severity of the problem at the beginning?
2. To what degree did employees become aware of the program offerings?
3. What were the perceived barriers to participation, initially and over time?
4. Based on clear criteria, what was the participation rate? For example, how many employees exercised at the on-site facility an average of no less than once per week?

FIGURE 18.5
RESULTS OF HYPERTENSION SCREENING PROGRAM REPORTED BY THREE COMPANIES

Company	Percentage of employees screened	Percentage of employees with high blood pressure at initial screening ($>140/90$)				Percentage of employees with hypertension referred and/or treated on-site	Percentage of hypertensive employees adequately controlled at one year
		Previously known		Previously unknown			
		Uncontrolled	Controlled	Uncontrolled	Controlled		
A	82					81	
B	68	3	8	4	6		
C	20					92	75

5. What was the progress initially and over time in the target problems? For example, what percentage of those who enrolled in the weight management class completed it? What was the average weight loss initially, and what was it after four succeeding three-month periods?

Return on Investment

Assessing return on investment or the ratio of benefits to costs is even more difficult than figuring out whether the program worked in nondollar terms. The first problem is frequently deciding what the program cost. In most programs all costs are not carefully tracked. Should the time of each company employee involved in every aspect of program planning be included? Is overhead allocated to the activity? What about depreciation on the space used for the program? If employees participated on company time, is this cost charged to the program? Just keeping track of these costs can be expensive!

Deciding on the benefits in dollar terms is even more difficult:

1. The benefits of the program in terms of reduced health care costs, disability, worker's compensation, etc. may not be available. Few companies have integrated these records so that an assessment of an individual's changes over time can be closely followed.
2. The benefits may not accrue for a substantial period of time, beyond the usual time frame of a program evaluation. For example, based on previous studies, smoking cessation reduces risk for lung cancer, but slowly, over a period of ten to fifteen years.
3. How long an individual or group of individuals will be employees of the company is difficult to predict, so that even if the benefits can be quantified, the proportion that will accrue to the employer who sponsored the program is not known. For example, changes in product mix, business cycles, competition, overall company wage and benefit package, possible incentives for early retirement and other factors are difficult to predict with accuracy.

In general, it appears to be a better strategy to: a) look at the cost of ill health and attempt to estimate what portion of it can be reduced through participation in an employee health promotion program and other related programs; and b) compare this amount with the cost of doing nothing. The difference suggests a range of investment which might be warranted in attempting to improve employee health.

Another good approach is to concentrate on cost-effectiveness analysis, which is directed to answering the question "What is it costing me for a certain improvement in health?" For example, what is the cost per sustained quit in a smoking cessation program? How much is the employer spending per year to achieve adequate control of a previously uncontrolled hypertensive? What is the cost of getting a previously unfit person to a defined level of fitness and maintaining him or her at this level? This type of analysis helps to decide how efficient the program is in achieving its objectives and permits comparisons of different program options. It starts with the assumption that improved health of employees (and their families) is an important corporate goal. And the key evaluation question becomes, "How much improved health is derived from a particular type of program with a particular level of expenditure?"

One of the most difficult aspects of evaluation is attempting to establish cause and effect. Does the evaluation permit stating with some certainty that whatever was found in the way of improvements was attributable to the program? This disarmingly simple question has been the bugaboo of evaluators forever. A few of the major barriers to successful answering of that question are:

1. *Changes in the habits of the entire population from which employees come* For example, from 1965 to 1980 the percentage of male smokers age 20 and over declined 10.5 percent and the average cholesterol went down approximately 10 points (milligrams per milliliter). An employer-sponsored program targeted to these problems showed roughly equivalent changes. Without knowing what changes had occurred in the rest of the population one might consider this program a success. But taking these changes into account, the program may not have had any impact at all. As workers become more aware of what they can do to improve their health and translate motivation into actions, significant changes are occurring. In any worksite evaluation that covers more than a year or two, therefore, some attempt to control for these trends needs to be incorporated.

2. *Program participants are usually different than nonparticipants in many respects* For example, those entering exercise programs sponsored by employers are more likely to be nonsmokers, to be very concerned about health in general, to be younger and to be more knowledgable about the benefits of exercise than nonparticipants. In addition, they may use fewer health care services, may tend to be more loyal employees with lower turnover and may start by having a lower absenteeism rate. Unless a careful baseline assessment of all these factors is obtained, it is impossible to know sometime in the future the degree to which differences in these important measures are due to: a) effects of the program, or b) continuation of pre-existing differences.

3. *Eliminating other reasons for the observed changes* For example, a program to change eating habits of those with particularly high total cholesterol values in the blood was considered a success because in this high risk employee group the average cholesterol decreased by 10 percent over a three-year period. However, no change had been observed from pre- to post-program in the three-day recalls of what was eaten. Further investigation showed that many of this group had enrolled in an exercise program and had not only improved their fitness but also, as a result, lost considerable weight. Losing weight reduces total cholesterol. The correct conclusion was that the observed change was secondary to the exercise and associated weight loss.

THE FUTURE

Employer-sponsored worksite health promotion and disease prevention programs are growing rapidly in large and medium-size businesses. There is theoretical and sometimes experimental evidence that some of these programs can improve some indices of health and reduce health risk. However, the body of evidence to date limits conclusions about what can be expected from specific activities in specific settings. A good dollar return on investment can be obtained from well-organized and well-delivered smoking cessation programs, but the rate of return is unknown for other health promotion interventions. Broad-based intervention that includes attention to both health promotion activities and the health environment at the worksites shows promise of yielding significant changes in health habits. Evaluations of these programs need to pay considerable attention to problems of design and measurement to assure reproducible results.

NOTES

1. "A Layman's Guide to Cardiovascular Disease," *Bostonia*, Winter 1978.
2. *Health Risk Appraisals: An Inventory.* Prepared by National Health Information, Health Promotion, Physical Fitness and Sports Medicine. Washington, D.C. DHHS (PHS) Pub. No. 81–50163, 1981.
3. Wagner, E.H., Berry, W.L., Schoenbach, V.J. and Graham, R.M. "An Assessment of Health Hazard/Health Risk Appraisal," *American Journal of Public Health*, Vol. 72, No. 4., 1982, pp. 347–352.
4. Fielding, J. "Appraising the Health of Health Risk Appraisal," *American Journal of Public Health*, Vol. 72, No. 4, pp. 337–339.
5. English, M.M. "Business Falls In Step With Fitness," *Advertising Age*, February 8, 1982, pp. M–9, M–24.
6. Berry, R.E. and Boland, J.P. *The Economic Cost of Alcohol Abuse.* New York: Free Press, 1977.
7. Luce, B.R. and Schweitzer, S.O. "Smoking and Alcohol Abuse: A Comparison

of Their Economic Consequences," *New England Journal of Medicine*, Vol. 298, No. 10, 1978, pp. 569–571.

8. Hartunian, N.S., Smart, C.N. and Thompson, M.S. "The Incidence and Economic Costs of Cancer, Motor Vehicle Injuries, Coronary Heart Disease, and Stroke: A Comparative Analysis," *American Journal of Public Health*, Vol. 70, No. 12, 1980, pp. 1249–1260.

9. Veterans Administration Cooperative Study Group on Antihypertensive Agents. "Effects on Morbidity in Hypertension, Results in Patients With Diastolic Blood Pressures Averaging 115 through 129 mm Hg," *Journal of the American Medical Association*, Vol. 202, 1967, pp. 1028–1034.

10. Veterans Administration Cooperative Study Group on Antihypertensive Agents. "II. Results in Patients With Diastolic Blood Pressures Averaging 90 Through 114 mm Hg," *Journal of the American Medical Association*, Vol. 213, No. 12, 1970, pp. 1143–1152.

11. Veterans Administration Cooperative Study Group on Antihypertensive Agents. "Effects of Treatment on Morbidity in Hypertension, IV. Influence of Age, Diastolic Pressure, and Prior Cardiovascular Disease: Further Analysis of Side Effects," *Circulation*, Vol. 45, No. 5, 1972, pp. 991–1004.

12. Hypertension Detection and Follow-up Program Cooperative Study. "Five-Year Findings of the Hypertension Detection and Follow-up Program: 1. Reductions in Mortality of Persons With High Blood Pressure, Including Mild Hypertension," *Journal of the American Medical Association*, Vol. 242, No. 23, 1979, pp. 2562–2577.

13. *Cardiovascular Primer for the Workplace*, Health Education Branch, Office of Prevention, Education and Control, National Heart, Lung and Blood Institute, NIH Publication No. 81–2210, January, 1981.

14. National High Blood Pressure Education Program, "At Mass. Mutual Off-Site Care and Good Monitoring Reduce Medical Costs," *Re: High Blood Pressure Control in the Worksetting*, Winter 1980. National Heart, Lung and Blood Institute.

15. Fielding, J.E. "Effectiveness of Employee Health Improvement Programs," *Journal of Occupational Medicine*, Vol. 24, No. 11, 1982, pp. 907–916.

16. Ibid.

17. National Heart and Lung Institute, *The Underwriting Significance of Hypertension for the Life Insurance Industry*, DHEW Publication No. (NIH) 75–426, 1975.

18. Erfurt, J.C. and Foote, A. *Final Report of Hypertension Control in the Worksetting: The University of Michigan-Ford Motor Company Demonstration Program*, June, 1982 (Unpublished).

19. Op. cit. (Fielding, 1982).

20. Malotte, K., Fielding, J.E. and Danaher, B.G. "Description and Evaluation of the Smoking Cessation Component of a Multiple Risk Factor Intervention Program," *American Journal of Public Health*, Vol. 71, No. 8, pp. 844–847.

21. *Smoking and Health*. A Report of the Surgeon General Office on Smoking and Health, U.S. Department of Health, Education and Welfare, DHEW Publication No. (PHS) 79–50066, Washington, D.C.: U.S.G.P.O., 1979.

22. Cowell, M.J. and Hirst, B.L. *Mortality Differences Between Smokers and Nonsmokers*, State Mutual Life Assurance Company of America, October 1979.

23. "Changes in Cigarette Smoking and Current Smoking Practices Among Adults: United States, 1978," *Advance Data*, Number 53, September 19, 1979, National Center for Health Statistics, Department of Health, Education and Welfare, 1979.

24. Hammond, E.C. and Seidman, H. "Smoking and Cancer in the United States," *Preventive Medicine,* Vol. 9, No. 2, 1980, pp. 169–173.
25. Op. cit. (Cowell and Hirst).
26. Covey, L.S. and Wynder, E.L. "Smoking Habits and Occupational Status," *Journal of Occupational Medicine,* Vol. 23, No. 8, 1981, pp. 537–542.
27. Op. cit. (*Smoking and Health*).
28. Salonen, J.T. "Stopping Smoking and Long-Term Mortality After Acute Myocardial Infarction," *British Heart Journal,* Vol. 43, No. 4, 1980, pp. 463–469.
29. Op. cit. (Malotte, K. et al.).
30. Leventhal, H. and Cleary, P.D. "The Smoking Problem: A Review of the Research and Theory in Behavioral Risk Modification," *Psychological Bulletin,* Vol. 88, 1980, pp. 370–405.
31. Op. cit. (Luce and Schweitzer).
32. Kristein, M.M. "Economic Issues in Prevention," *Preventive Medicine,* Vol. 6, No. 2, 1977, pp. 252–264.
33. Op. cit. (*Smoking and Health*).
34. Kristein, M.M. "How Much Can Business Expect to Earn From Smoking Cessation?" Presentation at the National Interagency Council on Smoking and Health's National Conference: *Smoking and the Workplace,* January 9, 1980, Chicago, Illinois.
35. Op. cit. (Fielding 1982).
36. Lambert, C.A., Netherton, D.R., Finison, L.J., Hyde, J.N. and Spaight, S.J. "Risk Factors and Lifestyle: A Statewide Health Interview Survey," *New England Journal of Medicine,* Vol. 306, No. 17, 1982, pp. 1048–1051.
37. Yarvote, P.M., McDonagh, T.J., Goldman, M.E. and Zuckerman, J. "Organization and Evaluation of a Physical Fitness Program in Industry," *Journal of Occupational Medicine,* Vol. 16, No. 9, 1974, pp. 589–598.
38. Kannel, W.B. and Sorlie, P. "Some Health Benefits of Physical Activity, The Framingham Study," *Archives of Internal Medicine,* Vol. 139, No. 8, 1979, pp. 857–862.
39. Durbeck, D.C., Heinzelmann, F., Schaeter, J., Haskell, W.L., Payne, G.H., Moxley, R.T., Nemiroff, M., Limoncelli, D.O., Arnoldi, L.B. and Fox, S.M. "The National Aeronautics and Space Administration—U.S. Public Health Service Health Evaluation and Enhancement Program," *The American Journal of Cardiology,* Vol. 30, No. 7, 1972, pp. 784–790.
40. Barnard, R.J. and Anthony, D.F. "Effect of Health Maintenance Programs in Los Angeles City Firefighters," *Journal of Occupational Medicine,* Vol. 22, No. 10, 1980, pp. 667–669.
41. Bjurstrom, L.A. and Alexious, N.J. "A Program of Heart Disease Intervention for Public Employees," *Journal of Occupational Medicine,* Vol. 20, No. 8, 1978, pp. 521–31.
42. Peters, R.K., Cody, L.D., Bischoff, D.P., Bernstein, L. and Pike, M.C. "Physical Fitness and Subsequent Myocardial Infarction in Healthy Workers," *Journal of the American Medical Association,* Vol. 249, No. 22, 1983, pp. 3052–3056.
43. Paffenbarger, R.S., Wing, A.L. and Hyde, R.T. "Physical Activity as an Index of Heart Attack Risk in College Alumni," *American Journal of Epidemiology,* Vol. 108, No. 3, 1978, pp. 161–175.
44. Siscovick, D.S., Weiss, N.S., Hallstrom, A.P., Inui, T.S. and Peterson, D.R. "Physical Activity and Primary Cardiac Arrest," *Journal of the American Medical Association,* Vol. 248, No. 23, 1982, pp. 3113–3117.

45. Wood, P.D. and Haskell, W.L. "The Effect of Exercise on Plasma High Density Lipoproteins," *Lipids*, Vol. 14, No. 4, 1979, pp. 417–427.

46. Heinzelmann, F. and Bagley, R.W. "Response to Physical Activity Programs and Their Effects on Health and Behavior," *Public Health Reports*, Vol. 85, No. 10, 1970, pp. 905–911.

47. Garson, R.D. Unpublished data. 1977.

48. Cox, M., Shephard, R.J. and Corey, P. "Influence of an Employee Fitness Programme Upon Fitness Productivity and Absenteeism," *Ergonomics*, Vol. 24, No. 10, 1981, pp. 795–806.

49. Op. cit. (Yarvote).

50. Op. cit. (Durbeck).

51. *Diet, Nutrition and Cancer*. Assembly of Life Sciences, National Research Council, Washington, D.C.: National Academy Press, 1982.

52. "Value of Low-Sodium Diets Questioned," *Science*, Vol. 216, No. 4541, 1982, p. 38–39.

53. Anderson, J.W. "The Role of Dietary Carbohydrate and Fiber in the Control of Diabetes," *Advances in Internal Medicine*, Vol. 26, 1981, pp. 67–96.

54. Zifferblatt, S.M., Wilbur, C.S. and Pinsky, J.L. "Changing Cafeteria Eating Habits," *Journal of the American Dietetic Association*, Vol. 76, No. 1, 1980, pp. 15–20.

55. Schmitz, M. "Cafeteria Nutritional Options: Changes and Choices," presented to the American Public Health Association. Montreal, Canada, October 17, 1982.

56. Op. cit. (*Cardiovascular Primer for the Workplace*).

57. Sorlie, M.S., Gordon, T. and Kannel, W.B. "Body Build and Mortality," *Journal of the American Medical Association*, Vol. 243, No. 18, 1980, pp. 1828–1831.

58. *Build and Blood Pressure Study, 1959*. Society of Actuaries, Vol. I, pp. 1–268, Chicago, 1959.

59. Bellack, A.S. "Behavior Therapy for Weight Reduction," *Addictive Behaviors*, Vol. 1, No. 1, 1975, pp. 73–82.

60. Stunkard, A.J. "Behavioral Medicine and Beyond: The Example of Obesity," in *Behavioral Medicine: Theory and Practice*, edited by O.F. Pomerleau and J.P. Brady. Baltimore: Williams and Wilkins, 1979.

61. Jeffery, R.W., Wing, R.R. and Stunkard, A.J. "Behavioral Treatment of Obesity: The State of the Art," *Behavioral Therapy*, Vol. 9, 1976, pp. 189–199.

62. Brownell, K.D., Heckerman, C.L., Westlake, R.J., Hayes, S.C. and Monti, P.M. "The Effect of Couples Training and Partner Cooperativeness in the Behavioral Treatment of Obesity," *Behavioral Research and Therapy*, Vol. 16, No. 5, 1978, pp. 323–333.

63. Stunkard, A.J. and Brownell, K.D. "Work-site Treatment for Obesity," *American Journal of Psychiatry*, Vol. 137, No. 2, 1980, pp. 252–253.

64. Stunkard, A.J. and Penick, S.B. "Behavior Modification in the Treatment of Obesity," *Archives of General Psychiatry*, Vol. 36, No. 7, 1979, pp. 801–806.

65. Volkmar, F.W., Stunkard, A.J. and Bailey, R.A. "High Attrition Rates in Commercial Weight Reduction Programs," *Archives of Internal Medicine*, Vol. 141, No. 4, 1981, pp. 426–428.

66. Washington Business Group on Health, A Survey of Industry Sponsored Health Promotion, Prevention and Education Programs (compiled by A. Kiefhaber, A. Weinberg and W. Goldbeck), Washington, D.C., December 1978 (unpublished).

67. Fitness Systems, Corporate Fitness Programs: Trends and Results, Los Angeles, 1980 (unpublished).
68. Results of the Employee Fitness Program Survey, Canadian Public Health Association, Canada, September 1981.
69. Fielding, J.E. and Breslow, L. "Health Promotion Programs Sponsored by California Employers," *American Journal of Public Health*, Vol. 73, No. 5, 1983, pp. 538–542.
70. Ibid.
71. R.S. Parkinson and Associates. *Managing Health Promotion in the Workplace*, Palo Alto, California: Mayfield Publishing Company, 1982, pp. 89–92.
72. Hollen, G. in *A Summary of an Insurance Industry Conference on Health Education and Promotion*, H.J. Kotz and J.E. Fielding (editors). Health Insurance Association of America, American Council of Life Insurance, 1980, pp. 27–29.
73. Wilbur, C.S. "Live for Life: An Epidemiological Evaluation of a Comprehensive Health Promotion Program," Johnson and Johnson. New Brunswick, N.J. (unpublished).
74. *Employee Health and Fitness*, Vol. 4, No. 9, 1982, p. 105.
75. *Advertising Age*, February 8, 1982, pp. M14–15.
76. *Money*, December 1981, p. 90.
77. Personal communication with Mark Michaelson.
78. Cunningham, R.M. *Wellness at Work*, Blue Cross and Blue Shield Associations, 1982.
79. Berry, C.A. *An Approach to Good Health For Employees and Reduced Health Care Costs For Industry*. Health Insurance Association of America, Washington, D.C., 1981.
80. Ibid.

MENTAL HEALTH *19*
AND PRODUCTIVITY
AT WORK

STRESSORS, JOB PERFORMANCE AND HEALTH

Employers exert a major effect on the mental health and related productivity of their workers. While the effects are difficult to quantify, the recent fascination with the Japanese model of employer-employee relationships and modus operandi testify to the perceived importance of the work environment on performance.

The climate that is developed within the organization greatly affects an employee's mental health, physical health and attitude toward his or her job and employer. Molding the most productive workforce requires attention to minimizing deleterious mental health effects of factors intrinsic to the job. Of equal importance is working to create a psychological environment which maximizes health and productivity. Both of these are continuing challenges. Unlike some other aspects of corporate health management, success in creating a positive environment is the responsibility of every manager. Important roles may also be played by the medical department, the training department, and other mental health professionals with sensitivity to these issues and the ability to provide advice and feedback to every supervisor on creating and maintaining a positive work environment.

TABLE 19.1
CHARACTERISTICS OF FIGHT-OR-FLIGHT RESPONSE

- Increased heart rate (pulse)
- Elevated blood pressure
- Faster breathing
- Tensing of muscles in arms and legs
- Increased perspiration
- Dilated pupils
- Increased awareness through all senses
- Restlessness
- Jaw clenches
- Digestion ceases and blood diverted to muscles
- Shaking
- Feeling keyed up

Characteristics of Stress

Stress is a common denominator for psychological problems associated with the work setting. For a feeling that everyone experiences and is a universal subject of discussion, stress is difficult to define. Stress as commonly used connotes something undesirable and unwanted that requires personal attention to manage effectively. A stress response is characterized mentally by a sense of free-floating anxiety and physically by the "fight or flight response" summarized in Table 19.1. Whether stress has increased, as many claim, is difficult to know because the questionnaires by which we measure stress are of recent vintage. Yet there is little doubt that the rate of change in the external environment has accelerated, and changes are important sources of stress, which are collectively called stressors. As pointed out by Ken Dychtwald, a psychologist interested in aging and health,

In the past fifty years we have entered the atomic age, the space age and the computer age; life expectancy has increased by more than fifteen years, more than eighty new nations have appeared worldwide; the global population has more than doubled; and the gross national product has quadrupled . . . Fifty percent of all married couples get divorced and . . . 80 percent of all people who get divorced remarry. The average American changes jobs every three years and moves every five; 36 million Americans move every year.[1]

Despite its negative connotation in common usage, stress should not be considered an unmitigated negative factor. Rather it is an important creative force that promotes growth and innovation in our lives and in our companies. Dr. Han Selye, one of the foremost authorities on stress, considers it "the

FIGURE 19.1
THE STRESS CURVE

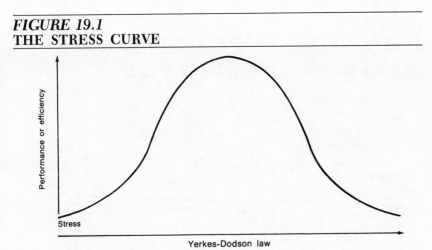

Yerkes-Dodson law

Source: Benson, H. and Allen, R.L. "How Much Stress is Too Much?" *Harvard Business Review,* Sept.–Oct., 1980 (Original citation: Yerkes, R.M. and Dodson, J.D. "The Relation of Strength Stimulus to Rapidity of Habit-Formation," *Journal of Comparative Neurology and Psychology,* 1980, p. 459.)

spice of life." It is stress that pushes everyone to think of ways to solve problems, that spurs re-examination of old assumptions, habits and interpersonal relationships. Many workers feel that they do their best work when under significant amounts of stress. And exciting creative periods also tend to block out negative influences. As one chief executive stated, "If I have a particularly easy week, I can feel an ache or pain, but if I get very busy, I feel really much better."

Despite the aspects of stress, the stress curve (Figure 19.1) first described by Drs. Yerkes and Dodson, shows that too much stress overwhelms our adaptive capacities and adversely affects performance. And, as displayed in Figure 19.2, stress is a common denominator for psychological problems associated with the work setting. The sources of stress include both work and other environments in which the worker lives. Considerable stress is a condition of employment for almost every job. It is worn as a badge of courage by most managers. To dismiss it as a problem is tantamount in some companies to admitting that the job doesn't challenge you. When surveyed about health improvement programs that they would like instituted by the employer, workers frequently rank stress management at the top of the list.

It is impossible to estimate the overall dollar impact of stress in an organization. However, understanding the known effects of stress on both

FIGURE 19.2
STRESS, JOB PERFORMANCE AND HEALTH

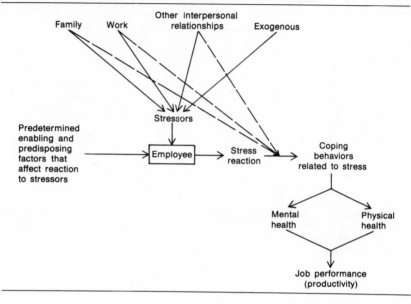

health and productivity leaves the impression that the cost is extremely high, and certainly high enough to warrant concerted efforts to reduce its effects on employees and their families. Stress has been related to a large number of health conditions. Abundant evidence points to the effects of major life events on the risk of falling seriously ill with a number of health problems.[2] However, the daily work and outside stressors also exact a major toll. Some of the health conditions that are more common in individuals who have a high level of stress and/or do not manage stress well are listed in Figure 19.3.

Employers and Stress

The challenge for employers who want a creative, challenged and productive workforce is to create conditions that allow most employees to walk up the left side of the Yerkes-Dodson curve but decrease the chances that they will slide down the right side.

While suggesting a pat prescription for a global problem is hazardous, it seems clear that major opportunities exist to make sure that the organi-

FIGURE 19.3
PARTIAL LISTING OF DISORDERS WHOSE PRESENCE OR SEVERITY IS STRESS-RELATED

Duodenal ulcer	Allergies
Sexual dysfunction in both men and women	Asthma
	Acne
Ulcerative colitis	Backache
Sleep disturbances	Alcoholism
Eczema	Heartburn
Migraine	Chronic fatigue
Spastic colon	Accidents
Depression	Hypertension(?)
Obesity	Coronary Artery Disease*

*Related to Type A personality.

zational stress thermostat is not set too high. Among the strategies likely to support this objective are the following three:

1. Develop a physical and psychological organizational climate that does not produce excess stress (distress) in a significant proportion of employees. Critical to the achievement of this objective is assuring that organizational structure, career development opportunities, job content and reward systems do not engender unnecessary stress.
2. Provide opportunities for employees to learn effective personal stress management techniques to better cope with stressful situations.
3. Train managers as well as other employees to recognize the early warning signs of excess stress both in themselves and co-workers and provide referral channels to confidential resources for professional assistance.

Developing an appropriate organizational climate starts with the style of management. If chief executive officers are inappropriately demanding, or have unrealistic expectations, they are likely to generate excess stress which ramifies throughout their organizations. If they live the job, have no outside interests and can't abide those for whom a personal life is important, the social pressure engendered by taking even a well-planned vacation or not staying late every night may lead to serious problems, with reduced productivity a predictable outcome.

A sense of job security, an important buffer to other job stresses, is an important attribute of a healthful organizational climate. While few U.S. companies essentially guarantee lifetime employment (IBM, a notable exception, has been rewarded for its commitment to job security with an intensely loyal workforce), some develop a reputation for purging those who lose out in internal political battles. While corporate politics is unavoidable, the degree of displeasure with factionalism and the effort undertaken to get all employees committed to a clear corporate philosophy can greatly affect each employee's feeling of security.

Part of a good corporate climate is a fair, open and respected system for disgruntled employees to air their grievances. If an employee feels mistreated, victimized by an unfair performance appraisal, or subjected to excessive demands by a supervisor can he or she receive a sympathetic hearing from a more senior manager?

Type A, Shortcut to Stress Creation

Distinction between personality Types A and B is of more than academic importance. Type A has been well-documented as an independent risk factor for coronary heart disease and is often described as "Type A coronary-prone behavior." In a study of 1674 initially coronary-free middle-aged men and women followed over eight years, Type A's of both sexes were two to three times more likely to develop coronary heart disease than Type B's after controlling for other well-established risk factors such as age, sex, cholesterol level, blood pressure and smoking.[3]

Type A's tend to be men and are more likely to be employed in higher occupational status jobs (professional, technical and managerial). They are most stressed by work situations that lead to perceived decreases in control, tend to be more job-involved than Type B's and may put themselves into positions where their supervisors have high expectations for achievement. Type A's may also be less perceptive of low level environmental problems, tend to be dissatisfied with the work of subordinates, in whom they engender stress by expecting high levels of productivity, and tend to suppress symptoms of fatigue from overwork until they are severe.

Recently, efforts have been undertaken to help Type A's moderate their distinctive pattern of existence using behavior modification approaches. As in most such efforts the first steps are appraisal of individual strengths and weaknesses as well as the implicit goals in the way life is being lived currently. As the originators of the Type A concept, Drs. Friedman and Rosenman underscore the question each person must ask himself or herself repeatedly: "What apart from the eternal clutter of my everyday living should be the essence of my life?"[4] Logical next steps are establishing goals and related priorities to distinguish the real important aspects of life from the others that often consume more time.

Employee Burnout

A burned-out employee is at best a marginal producer. Yet, most often prior to becoming burned-out, this same employee was one of the most productive. Frequently, burnout could have been avoided. Burnout may be viewed as the end product of continuing excess stress resulting from family pressures, job problems, environmental demands or any combination thereof. Job burnout has been described as "a debilitating psychological condition brought about by unrelieved work stress which results in: 1) depleted energy reserves; 2) lowered resistance to illness; 3) increased dissatisfaction and pessimism; and 4) increased absenteeism and inefficiency."[5] Even rough estimates of the cost to industry of employee burnout have not yet been derived, perhaps because it is nearly impossible to distinguish burnout-related problems from the broad realm of mental/emotional/behavioral problems. However, burnout has been labeled by some as "business's most costly expense."[6] Burnout can sharply diminish productivity (through increased absenteeism as well as lost enthusiasm, creativity and desire to achieve), impart a negative influence on co-workers, increase illness rates and contribute to deteriorating interpersonal relationships.

Burnout does not emerge spontaneously; it progresses in stages, most of which are characterized by readily recognizable symptoms. Veninga and Springer have classified five progressive stages of burnout as the "Honeymoon," "Fuel Shortage," "Chronic Symptoms," "Crisis" and "Hitting the Wall."[7] During the "Honeymoon" (usually the beginning of a new job) the employee is enthusiastic about his or her assignment and eager to perform at a high level. Nonetheless, the earliest stage of burnout begins at this point because the employee must expend adaptation energy in adjusting to new surroundings, co-workers, norms, etc. During this initial stage of employment, the employee will develop habits of dealing with stress which, if effective, may deter further burnout. However, if this opportunity is not used to develop positive coping strategies, higher levels of burnout may well ensue.

After the "Honeymoon" is over, the worker may experience a feeling of job disappointment described as a "Fuel Shortage." Work pressures mount, and frequently the employee oscillates between confusion and fatigue in one mode, and attempting to maintain enthusiasm and control in the other. The employee bounces between the two states without remaining at either extreme for prolonged periods of time. Five early symptoms (warning signals) of burnout may become manifest during the latter part of this stage: job dissatisfaction, inefficiency at work, fatigue, sleep disturbances, and escape activities such as excessive eating, drinking and/or smoking and/or using various prescription drugs such as Librium or Valium.

"Chronic Symptoms," the next stage of burnout, is characterized by the presence of more obvious physical and psychological symptoms, such as chronic exhaustion, physical illness, anger and depression. During the "Crisis" stage these symptoms become critical. The burning-out employee may become obsessively frustrated, pessimistic, self-doubting, and develop an "escapist mentality" (i.e., searching for actions or behaviors, such as getting pregnant or committing suicide, which allow an escape from the unhappy work situation). "Hitting the Wall," the most extreme stage of burnout,

FIGURE 19.4
UNDERSTANDING AND MANAGING STRESS: STRESSFUL WORK CONDITIONS

There frequently are day-to-day conditions at work which we find stressful. On the items below, indicate how often each source of stress is true for you by circling the appropriate number.

1.	Others I work with seem unclear about what my job is	1 2 3 4 5
2.	I have differences of opinion with my superiors.....................	1 2 3 4 5
3.	Others' demands for my time at work are in conflict with each other ..	1 2 3 4 5
4.	I lack confidence in "management"	1 2 3 4 5
5.	"Management" expects me to interrupt my work for new priorities....	1 2 3 4 5
6.	There is conflict between my unit and others it must work with.......	1 2 3 4 5
7.	I only get feedback when my performance is unsatisfactory...........	1 2 3 4 5
8.	Decisions or changes which affect me are made "above" without my knowledge or involvement.......................................	1 2 3 4 5
9.	I have too much to do and too little time to do it	1 2 3 4 5
10.	I feel overqualified for the work I actually do......................	1 2 3 4 5
11.	I feel underqualified for the work I actually do	1 2 3 4 5
12.	The people I work closely with are trained in a different field than mine...	1 2 3 4 5
13.	I must go to other departments to get my job done	1 2 3 4 5
14.	I have unsettled conflicts with people in my department.............	1 2 3 4 5
15.	I have unsettled conflicts with other departments...................	1 2 3 4 5
16.	I get little personal support from the people I work with............	1 2 3 4 5
17.	I spend my time "fighting fires" rather than working to a plan........	1 2 3 4 5
18.	Management misunderstands the real needs of my department in the organization	1 2 3 4 5
19.	I feel family pressure about long hours, weekend work, etc...........	1 2 3 4 5
20.	Self-imposed demand to meet scheduled deadlines...................	1 2 3 4 5
21.	I have difficulty giving negative feedback to peers..................	1 2 3 4 5
22.	I have difficulty giving negative feedback to subordinates	1 2 3 4 5
23.	I have difficulty in dealing with aggressive people..................	1 2 3 4 5
24.	I have difficulty dealing with passive people.......................	1 2 3 4 5
25.	Overlapping responsibilities cause me problems	1 2 3 4 5
26.	I am uncomfortable arbitrating a conflict among my peers............	1 2 3 4 5
27.	I am uncomfortable arbitrating a conflict among my subordinates......	1 2 3 4 5
28.	Academic and administrative roles are in conflict...................	1 2 3 4 5
29.	I avoid conflicts with peers	1 2 3 4 5
30.	I avoid conflicts with subordinates	1 2 3 4 5
31.	I avoid conflicts with superiors	1 2 3 4 5
32.	Allocation of resources generates conflict in my organization..........	1 2 3 4 5
33.	I experience frustration with conflicting procedures	1 2 3 4 5
34.	My personal needs are in conflict with the organization..............	1 2 3 4 5
35.	My professional expertise contradicts organizational practice..........	1 2 3 4 5
36.	Administrative policies inhibit getting the job done	1 2 3 4 5
37.	Other..	1 2 3 4 5

Key: 1 = Never; 2 = Rarely; 3 = Sometimes; 4 = Often; 5 = Always.

Source: Jenny Steinmetz, "The Stress Reduction Program at University Teaching Hospital, U.C. Medical Center, San Diego," as cited in "Occupational Stress," November 3, 1977. Reprinted by permission of the Institute of Industrial Relations, UCLA.

occurs when burnout has become so entwined in other problems, such as drug abuse, alcoholism or other mental illnesses, that it is indistinguishable as a separate problem.

Employee burnout is a problem which occurs in all types of businesses. And, contrary to pervasive belief, employees at every level of the organizational hierarchy may experience burnout, not just executives and managers. There may, however, be certain requirements inherent in managerial positions—more time constraints, greater job demands, responsibility for many other employees, for example, which are more conducive to job burnout than other categories of employment and which tend to attract a certain class of individuals more prone to burnout.

Certain work traits, including self-drive, overachievement, workaholism, high motivation, and excessively high self-expectations are considered to be predisposing factors to burnout. Ironically, these types of people (highly self-motivated, creative, enthusiastic overachievers) are, for obvious reasons, exactly the type of individuals businesses are desirous of hiring when they recruit new employees. Such persons start off as efficient, highly productive and successful members of the organization. Unfortunately, failure from within an organization to recognize the connection between constant overachievement and burnout (i.e., how easily the former may be converted to the latter) and to exercise proper precautionary measures against burnout can lead to an exhausted workforce whose achievement, desire and creative talents are defunct.

In addition to all of the "work distressors" listed in Figure 19.4, other features of the work environment (including attributes of the job itself) which are conducive to job burnout include:

- static, monotonous job conditions or, conversely, unrealistically demanding job assignments
- inadequate recognition, appreciation or compensation for employee's contribution
- failure of top management or supervisors to support the efforts of their staff
- excessively long working hours for continuous stretches of time
- frequent dealings with emotionally draining situations, such as mediating between employees or between managers and employees, etc.

MINIMIZING STRESS-RELATED PROBLEMS

Social Support: People as Prevention

Probably the single most important characteristic of the work environment is the social support provided by relationships at the job. Social support may include emotional, appraisal, informational and instrumental support (aid in kind, money, labor or time-modifying environment).[8] Social support mitigates the effects of stress. Social support also protects against a wide range of serious health problems. A study of 4,725 men and women ages 30 to 69 living in Alameda County, California found that social ties, including marriage, contacts with friends, church membership and formal and informal group associations significantly affected the risk of dying over a nine-year period. Combining these factors into a social network index, researchers found that people with the least connections were 180 to 460 percent more likely to die than persons with the most connections. These findings persisted even when corrected for self-reported health status at the beginning of the study (i.e., deleterious health practices, such as lack of exercise, smoking, and significant alcohol consumption), income and use of preventive health services.[9]

Evidence is accumulating that social support at the job can play an important role in reducing the adverse health effects of stress. One study found that strong social support from peers relieved job strain and also seemed to reduce the effects of job stress on blood pressure, number of cigarettes smoked and rate of smoking cessation activities.[10] Somatic complaints such as ulcers, rashes and aches and pains do not appear to increase with increasing job stress in the presence of strong social support from co-workers, but the frequency increases drastically in the absence of such support.[11]

It is also possible that the success of job enrichment activities that develop task-related groups to perform cooperatively a broader range of activities may derive in part from the social network of closer interaction that reduces the adverse impacts of job stress on productivity. Participants perceive a reduced level of job stress and are more satisfied with their jobs. Much of the support which can cushion job stressors comes from supervisors. However, in a work setting where jobs are performed by semipermanent teams, or in professional and managerial ranks, peer support may be of equal or greater benefit.

Strong social support appears to reduce work stress. In a study of over 2,000 men in twenty-three occupational categories, "men with high support from either supervisors or co-workers generally reported low role conflict, low role ambiguity and low future ambiguity, high participation and good utilization of their skills."[12]

Social supports at work can also help to buffer effects of stressors which arise from outside the work environment but which nonetheless adversely affect work performance. A worker beset by marital problems may not find a sympathetic and supportive supervisor of much help in dealing with the problem at home, but he may feel more comfortable at work and avoid impaired job performance. Normal job stresses, which might combine with those from home to carry an employee through to the wrong slope of the

Yerkes-Dodson curve, may not be perceived as additional stressors in the presence of co-workers who are considered as helping resources.

One priority in developing more supportive networks on the job is to target those who have significant interaction with a large number of employees. Managers and union stewards would thus be high on the list and are also good candidates because they more frequently than others possess innate social support skills and tend to provide support to co-workers. Through training they can become more aware of the importance of their social supports to their co-workers. More importantly, most can be effectively taught the skills to provide support in a nonthreatening manner. Training should include communication skills such as active listening and the best ways to provide the emotional support so instrumental in reducing stress. Equally important is organizing the work setting so that conditions conducive to social and physical isolation are minimized. Company support of activities which tend to cement relationships among co-workers (e.g., sports leagues, company outings, hospitable lounges that permit one-on-one conversation) can help to build social networks.

However, training, cooperative work groups and communication skills will only be effective when the organization rewards the behaviors that help build social networks. Is a supervisor rewarded for his or her accessibility to employees? For being a supportive figure, taking time to fully appreciate the employees' concerns and provide both emotional support and counsel? For building a strong team rather than developing a cadre of independent stars? For sacrificing some potential short-term improvements in productivity at the price of very high job stress to a longer term goal of building a stable, productive and mutually supportive team? As James House, an authority on occupational stress, explains, "The United States tends to be a very pragmatic, instrumental society, and especially among men who dominate most work environments, the kinds of skills (empathy, consideration, sensitivity) involved in giving or receiving social support, are often devalued as "soft," "nonproductive," or "nonmasculine."[13] This attitude, however, may be so expensive that were its costs known it would be immediately deemed unaffordable.

Job Design, Responsibility and Stress

Factors intrinsic to a job may also act as stressors. Lack of adequate ergonomic design can create both physical and mental stresses. A chair without a lumbar support, a CRT with poor or excessive contrast or a high noise level can all provoke a stress response over time, as can inadequate lighting or ventilation. Exposure to physical danger is a natural stressor. Shift work can be a strong stressor, especially for those in rotating shifts who have inadequate time to adjust their biological clocks to new work schedules, so that they are tired when working yet suffer from insomnia during their intended sleep period. Those on a permanent night shift can align their circadian rhythms with their work schedule, but do tend to feel isolated from the mainstream of society. Work quantity is a potent stressor at both ends of the spectrum. Excess quantity of work or a job that is too difficult can lead to stress symptoms which may be manifested as high absenteeism, low motivation, escapist activities (such as drinking or overeating) and/or physical symptoms. Lack of sufficient work or monotonous, understimulating work can also lead to amotivational states or ill health.

In a Swedish study of stress among salaried employees, more than one-third frequently experienced stress related to one or more of the following characteristics of work:

- inability to leave the work process even for short periods during the working day
- work processes that were highly preprogrammed or controlled by others
- great responsibility at work (qualitatiye overload)
- few or no options for own initiatives or decision making
- great demands on concentration and attention
- too great a work load (quantitative overload)
- insufficient time to carry out the work tasks
- confused or conflicting work roles; inadequate or insufficient instructions and support from superiors
- monotonous work
- unqualified assignment (qualitative underload)[14]

A number of the most influential stress factors have to do with a worker's perceived role in the organization. Lack of clarity about one's job is a common and strong stressor, as are conflicting demands which imply conflicting priorities. Many researchers have related these characteristics of the job to higher levels of health risk and higher rates of illness. In a study of 1,540 managerial employees, four job factors were positively correlated with heavy smoking, high blood triglycerides and total cholesterol, high blood pressure, obesity and peptic ulcer. The factors were high degrees of ambiguity or extreme rigidity with respect to job tasks, very high or very low degrees of conflict associated with the job and either extreme amounts of responsibility (especially in managing others) or little responsibility.[15]

A major six-year study of work factors and heart and blood vessel diseases, including stroke, found that a hectic and psychologically demanding job (based on employee self-report) increases the risk of developing symptoms of coronary heart disease and of dying of heart and blood vessel disorders. Low decision latitude, particularly limited decision-making required for the job and low personal freedom (to make a private phone call during working hours, to receive a private visitor for ten minutes at work, and to leave the job for one-half hour without telling the supervisor) were also associated with increased risk of cardiovascular disease. Workers in jobs with minimum education requirement coupled with low personal freedom schedules and high job demands were found to have risks approximately fourteen times those of individuals without these personal and job characteristics. This careful study concludes that, "low decision latitude or environmental constraints on the worker's ability to decide how to respond to environmental demands appears to be an independent CHD (coronary heart disease) risk factor. . . . It is constraints on decision-making—not decision-making itself—which appears to pose a new risk factor for psychological strain and CHD." Considerable decision latitude, it was postulated, may help to cushion elevated psychological job demands.[16]

Career development stressors include not only the most obvious job issues, such as job security, thwarted ambition and underpromotion, but also overpromotion and lack of congruence between the work, remuneration and status within the organization. In short, chronic work frustrations and a disparity between where employees are in their jobs and where they feel they should be in the organization are common persistent stressors that can adversely affect productivity and sometimes are associated with depression. Each person has a different work-stress profile. Figure 19.4 is designed to help workers assess personal stressors and their relative magnitude.

To optimize stress and to minimize related mental and somatic problems, which exact a high dollar and psychological toll on an organization, requires, in summary, a number of important organizational characteristics. Creation of a work environment that does not support unusual degrees of internal politicking, provides appropriate opportunities for participation in decision-making, stresses the importance of good industrial relations and good communication skills and does not provide excessive limitations on behavior at work can be of inestimable value.

Equally important is a career development plan that creates a sense of job security to the degree practicable, assures that employees know what is expected of them and leaves them neither underloaded nor overloaded, evaluates performance objectively based on these expectations, and rewards individuals so that a congruence of pay, job rank and status within the organization is achieved. Promotions need to be well thought out in terms of the individual's capacities. Opportunities to develop social supports for each employee are likewise of great importance.

In sum, good human resources management is one of the best guarantees of a mentally healthy workforce. While virtually all employers agree that people are their most important asset, many do not invest in their organizational environment to the degree warranted by potential savings. Very few have made any systematic attempts to assess the cost of major stressors and the savings achievable by trying to alter them.

TABLE 19.2
QUESTIONS AND SCORES* USED IN THE FRAMINGHAM TYPE A BEHAVIOR SCALE

Type A questions	Responses and scores			

For all men and women: Traits and qualities that describe you:

		Very well	Fairly well	Somewhat	Not at all
1.	Having a strong need to excel (be the best) in most things	1	.67	.33	0
2.	Usually pressed for time	1	.67	.33	0
3.	Being hard-driving and competitive	1	.67	.33	0
4.	Being bossy or dominating	1	.67	.33	0
5.	Eating too quickly	1	.67	.33	0

For all working men and women: Feeling during an average day in your regular line of work:

		Yes	No
6.	Have you often felt very pressed for time?	1	0
7.	Has your work often stayed with you so that you were thinking about it after working hours?	1	0
8.	Has your work often stretched you to the very limits of your energy and capacity?	1	0
9.	Have you often felt uncertain, uncomfortable, or dissatisfied with how well you were doing in your work?	1	0

For all men and women:

		Yes	No
10.	Do you get quite upset when you have to wait for anything?	1	0

*To score the questionnaire: Add up scores for questions 1–10 and divide by 10 to get each person's score. Cutpoints for Type A and Type B behavior are .399 for men 45–54 and 55–64, .333 and .299 for women 45–54 and 55–64, respectively.

Source: Haynes, S.G., Feinleib, M. and Kannel, W.B. "The Relationship of Psychosocial Factors to Coronary Heart Disease in the Framingham Study III: Eight-Year Incidence of Coronary Heart Disease," *American Journal of Epidemiology*, Vol. 111, No. 1, 1980, pp. 37–58.

Personality factors, such as whether a person is an introvert or extrovert, also may influence the degree to which individuals are affected by stressors. For example, one researcher found that introverts tend to perform better under stable conditions where environmental stressors are at a low level, while extroverts tend to seek external stimulation and remain more stable in the presence of a more stressful environment.[17]

Type A personality and behavior has been the most intriguing and most frequently studied personality factor. Behaviors considered Type A include ". . . excessive competitive drive, aggressiveness, impatience and a harrying sense of time urgency."[18] Table 19.2 provides a shorthand method for identifying Type A behaviors. Type B is defined as the absence of Type A behaviors, characterizing those who feel less driven to accomplish, who can relax without guilt, don't feel compelled to challenge others, and so on. While the distinctions between extreme Type A and Type B individuals are reasonably clear, most of us are somewhere in between and scales have been developed to assess where an individual fits along the spectrum. Even skilled psychologists disagree 20 to 35 percent of the time whether a person is best described as Type A or Type B.

Teaching Stress Management

While only a small percentage of employees are classic Type A's, many—frequently the majority—can benefit from dealing better with the problems of the workplace that cause distress. Many employers feel that offering stress management classes for employees on or off site is a good investment. Content of these activities should include:

1. Increasing awareness of those particular stressors at or associated with work which cause distress (Figure 19.5);
2. Teaching how to avoid unnecessary stressors that may contribute to excess stress;
3. Teaching the skills to react more positively to natural or unavoidable stressors;
4. Demonstrating and providing practice in the techniques to reduce the effect of stress on soma and psyche; and
5. Pointing out how to minimize dysfunctional coping responses and maximize those that tend to help achieve personal equilibrium.

The same stressors affect different individuals in disparate ways and to varying degrees. Excessive noise may engender constant anxiety in one worker and have little or no effect on another. Frequent promotions may cause decompensation in one manager but be a source of positive stimulation to another with equivalent talent but different background and personality characteristics. Each individual is the best judge of what causes his or her

FIGURE 19.5
WORK DISTRESSORS

- Unhealthy and/or unsafe work environment
- Lack of congruence between the work, renumeration and status within the organization
- Job insecurity
- Ambiguous responsibilities or rigid responsibilities
- Excess responsibility or insufficient responsibility
- Poor interpersonal relationships with boss, colleagues and/or subordinates
- Social isolation
- Constant job variation or complete job stagnation
- Lack of personal freedom
- Inadequate support staff, consultation, assistance, etc.
- Insufficient or excessive upward mobility within the organization
- Inadequate compensation (salary, benefits, etc.) and/or recognition

excessive stress and can increase cognizance by maintaining a diary of stressors, degree of stress, and physical and psychological reactions. A personal stress profile can be easily derived and can serve as a baseline to be reviewed after stress management techniques have been practiced.

Frequently distressors (stressors which evoke feelings of distress) can be avoided, often with only a little forethought. Common work distressors are agreeing to an unwanted extra assignment; scheduling meetings so closely that running from one to another and often arriving late and/or leaving early are the rule; failing to document poor performance of a subordinate who is likely to file a grievance when terminated; and failing to set aside time for important phone calls and therefore having to interrupt meetings to take them.

An American Management Association questionnaire survey of 2,700 top and middle managers found that both levels agreed on the leading causes of stress at work:

- heavy workload/time pressures/unrealistic deadlines
- disparity between what has to be done on job and what the manager would like to accomplish
- the general political climate of the organization
- lack of feedback on job performance

The same managers responded that the three major stressors outside of work were:

- financial worries
- problems with children
- physical injury, illness or discomforts[19]

Encouraging positive thinking can be a useful preventive approach to dealing with work problems. It is certain that many things will not go as planned. The question is, how do employees react to the inevitable bad news? Is every such event the cause for depression, pointing the finger at others and ranting sufficiently to disrupt many other activities? Or do they say, "Nothing I do can prevent the problem at this point but I can come up with the best solution given the circumstances"? The ability to find the opportunity in every crisis can reduce the severity of the reaction to stressors, and can mobilize creativity which will lead to enhanced self-esteem, self-confidence and greater psychological stability.

Stress management techniques are useful in effectively dealing with the psychological and physiological responses to stressful situations. They also can help prevent the illnesses that appear to result from chronic stress. Among the most common mental and psychophysiological techniques employed to reduce stress are relaxation, meditation, biofeedback and guided imagery. All of these can reduce stress at least temporarily and break the constant anxiety and other manifestations that are the antecedents of illness. Common to all of these techniques is thorough concentration, closing down the usual mechanism by which the brain processes information. The result is a "relaxation response," which is the antithesis of the "fight or flight" response to stress.[20] Heart rate and blood pressure decline, as does oxygen consumption, respiratory rate and muscle tension, while the alpha waves signifying relaxation increase. Everyone can easily learn this relaxation response which includes sitting quietly in a comfortable position; closing the eyes; consciously relaxing muscles; breathing slowly and easily, focusing attention on the breathing and maintaining a passive attitude toward the environment and one's own distracting thoughts.

After some period of practice, a relaxed state is achieved in a few seconds, with considerable stress reduction occurring within seconds to a few minutes. Practicing it at the time of greatest stress, even when others are around, can be extremely effective and can help prevent dysfunctional reactions to the stressful circumstances. Some managers have claimed it is the only thing that gets them through some meetings with their boss!

A variation or enhancement of this technique is biofeedback, which uses electronic sensors (of skin temperature, alpha waves, skin conductivity, etc.) to give conscious feedback of body processes which are usually unconscious. As a consequence, the capacity of a person to control these processes is enhanced. While biofeedback does not necessarily increase the effectiveness of various relaxation techniques it can help convince doubting individuals that they can voluntarily control physiological responses and that psychosocial stress elicits physiological responses. In addition, it can serve as a monitor for the effectiveness of relaxation techniques.[21]

Effects of Stress Management on Productivity

Employer-sponsored stress management training is being provided by many employers. The 1981 health promotion survey of California employers found that 13 percent of employers were providing stress management classes. Among employers of 750 or more employees at the surveyed site, 24 percent sponsored such activities.[22] Only a few systematic studies have been reported that directly bear on the impact of occupational stress management on level of stress, other risk factors for illness, or utilization of health benefits or other employer-sponsored health services. Nonetheless, stress reduction programs appear to help minimize the impacts of stress on illness, turnover and disability. In one study, employees who scored either high or low on a stress-level questionnaire received a twenty-minute interview with a psychologist covering workplace stressors, life events, coping skills and maintaining inappropriate balance with regard to time for work, family and self. Referrals were made as necessary including to short-term psychological counseling within the company, to outside psychologists and psychiatrists and to behavioral programs such as weight control programs, relaxation classes and exercise programs. Average absenteeism after the intervention group decreased from over 5 days in the six months prior to program participation to fewer than 3 days in the following six months, while in the control group it increased from approximately $2\frac{1}{2}$ to over $3\frac{1}{2}$ days. Eighty percent of those responding to a follow-up questionnaire six weeks after the interview checked that their coping abilities had improved in handling job and/or personal stress since the interview and 85 percent agreed that the interview was helpful in learning how to prevent or reduce symptoms related to stress.[23]

At the Converse Rubber Company 126 volunteers were randomly divided into three groups. The first group was taught the relaxation response, the second was instructed to sit quietly, and the third group received no instructions. The first two groups were asked to take two daily fifteen-minute relaxation breaks. After eight weeks Group 1 had significantly fewer symptoms, less reported illness, greater self-reported job performance (physical energy, strength of concentration, problem handling, overall efficiency) and better level of satisfaction and sociability than Group 3, with Group 2 reporting intermediate values.[24] Group 1 also exhibited a significantly greater decline in both mean systolic and mean diastolic blood pressure than groups 2 or 3 (7.9 mmHg versus 3.1 mmHg and 0.3 mmHg for systolic and 6.7 mmHg versus 2.6 mmHg and 0.5 mmHg for diastolic). This pattern of changes prevailed for both sexes, all ages and all initial levels of blood pressure, with the greatest blood pressure declines in those with the highest initial values.[25]

A study of 154 New York Telephone employees found that those using any of these stress management procedures showed clinical improvement in self-reported stress symptoms compared to waiting list controls, and these effects were seen regardless of whether subjects practiced their techniques frequently or occasionally.[26] An interesting study at Equitable Life Assurance Society provided biofeedback and other stress management to fifteen subjects with headaches and fifteen with general anxiety. After an average of thirteen sessions of biofeedback (of electrical stimulations from the forehead muscle) there were not only statistically significant decreases in symptoms and increased work satisfaction but a decline in visits to the on-site health center from an average 5.75 in the three pre-treatment months to 1.76 during the three post-treatment months. Pre-treatment, the average annual corporate cost was estimated at $3,350 from visits to the Employee Health Center ($473), time away from job for Health Center ($57), work influence due to symptoms ($2,207) and interference with the work of bosses ($73), co-workers ($543) and subordinates ($42). By contrast, the average annual post-treatment corporate costs per subject were estimated at $533, an 84 percent reduction, which is further estimated to provide a cost/benefit ratio of 1:5.52.[27,28] Although this analysis relies on a number of assailable assumptions, it suggests a possible methodology for corporations to begin assessing the value of some investments in stress reduction for employees with chronic stress-related symptoms.

Coping Skills, Good and Bad

The final approaches to reducing the burden of excess stress involve improving coping styles. Unfortunately, coping may frequently be poorly suited to reducing stress levels and resolving the situation that caused the reaction. All too common examples include becoming belligerent toward family, friends or co-workers, increasing habits deleterious to health such as smoking, drinking and overeating, brooding over the problem in isolation, taking tranquilizers or trying to ignore the problem by absorbing oneself in other activities. The failure to address a problem more constructively, however, increases the risk of developing stress-related illnesses and operating in all spheres at greatly diminished capacity. Employees can be helped to employ more useful coping mechanisms through a discussion of functional and dysfunctional coping styles followed by practice in using the better set of options. These include dealing with the problem head-on but with a constructive attitude, talking it over with family, close friends, clergy or a professional counselor, rethinking the problem carefully to see new possible solutions, and, if there was a mistake made, accepting responsibility and learning from it. These activities are likely to facilitate resolution of the stress reaction and help a person develop the confidence that comes with having mastered a difficult personal problem.

In addition to the mental ways of dealing with stress, there are other effective steps that can be taken. Aerobic exercise has a well-established ability to reduce both physiological and psychological effects of stress. Ingestion of caffeine may, by contrast, increase the physiological signs of stress. The average cup of coffee has 75 to 100 mg of caffeine and even a few cups can cause tremulousness, insomnia, headaches and even ulcers. Therefore caffeine can contribute to both stress reactions and some of the physiological problems of chronic stress.

Creating a Positive Work Environment

Creating a work environment which fosters the prevention of employee stress-related problems, including burnout, should be a high priority for personnel administrators, management and human relations/employee relations departments. Recognition and understanding by top management of these effects on employees and their connection to the work environment (causes, impacts, preventability) is an important initial step toward prevention. A demonstrated commitment from top management is usually necessary if the job environment is expected to minimize the occurrence of burnout and many other stress-related physical and mental health problems.

Positive steps in this direction include:

1. *Encouraging and facilitating the ability of employees to acquire new skills and fresh ideas by attending classes, training sessions, conferences, etc. and insuring an opportunity for workers to apply their learning on the job* Provide the opportunity for job change, e.g., new assignments, staff realignment, advancement whenever possible. One of the major causes of job burnout is the feeling by employees that they have mastered their present position and are stuck in a job with little or no possibility of advancement (intellectual or skillwise, as well as financial). Failure to feel challenged depletes motivation and contributes to many employees terminating their jobs in order to seek more stimulating opportunities. Those employees lacking the skills or motivation requisite to finding more challenging work outside the organization may remain in their present capacity to avoid the economic insecurity and hardship of unemployment. However, it is probable that such individuals will be tardy or absent more often, be less concerned about the quality of their work and may even substitute new behaviors such as excessive drinking, smoking, eating or use of drugs to occupy their time and provide them with an escape from an unpleasant but unalterable situation.

2. *Avoiding the exploitation of the company's most valuable employee resources* Naturally, managers tend to rely most heavily on those individuals who have proven themselves to be talented, industrious and able to accept large amounts of responsibility. However, employees who tend to be overachieving workaholics with unrealistically high self-expectations are susceptible to accepting significantly greater amounts of work than they can possibly complete within given time constraints, even working twelve to sixteen hours a day, seven days a week, which many workaholics are apt to do. According to industrial psychologist Harry Levinson, "the best people are more vulnerable to becoming burned-out people." Therefore organizations should avoid relying consistently on the same employees to bail them out of crises and should definitely prohibit employees from working extreme numbers of hours per day to complete even urgent projects on a continuing basis.[29] Setting people up for failure with unrealistic demands is highly counterproductive, as it instills feelings of guilt, self-reproach, pessimism, and poor morale in those employees who feel they have failed because they are unable to meet their employer's (and frequently their own) expectations.

3. *Providing a reward system which recognizes the positive contributions of all employees* Rewards should be commensurate with the degree of responsibility required and the benefits bestowed on the company. Positive strokes, whether in the form of cash bonuses, time off with pay, lunch with the boss, etc. are essential to maintaining high employee morale and feelings of self-satisfaction. Appropriate recognition and appreciation from the company provide positive incentives for employees to continue a high standard of job performance, and assurance of being rewarded can motivate other employees to improve their productivity.

4. *Being candid with potential employees about the requirements of jobs they are applying for* Highlight any exceptionally psychologically demanding or stressful aspects of the work. While knowledge of possible stress areas won't eliminate future pressure, providing this information can indicate to the potential employee that the company is aware of the problem of burnout and is concerned with its prevention.[30]

5. *Fostering an organizational climate where employees feel comfortable discussing any job frustrations with personnel managers (or other appropriate staff members)* Encourage managers to think how these frustrations can be mitigated. If this cannot be achieved, encourage mutually beneficial alternatives, such as transfer, promotion or creative job redesign. Company policy should emphasize that employees do not have to fear job termination for expressing their feelings honestly.

6. **Providing visible support for employees so they are not made to feel as if they were shouldering an unreasonable amount of responsibility alone** Workers should be able to feel confident that their reasonable criticisms of company policies or procedures will be supported by their supervisors in discussions with higher echelons of management. Employees who are constantly required to take unpopular public positions on controversial issues on behalf of the corporation can easily become burned-out if they feel unsupported and alone in their actions.

7. **Planning recreational activities or retreats where employees have an opportunity both to socialize outside of the work environment and share their thoughts, feelings and perceptions about work** Meeting in a more relaxed atmosphere away from all typical interruptions at work can facilitate the development of long-range plans, goals and objectives and can provide a nonthreatening environment for expressing personal concerns or frustrations regarding work.[31] Encouraging this type of communication can assist in limiting work stress problems by helping some workers to alleviate pent-up frustrations about work and by demonstrating that they're not alone in their feelings. Often constructive ideas for dealing with these problems are brought forth at such sessions. In addition, a more relaxed setting can be conducive to developing closer interpersonal relationships (social networks) among employees.

NOTES

1. Dychtwald, K. *Forum*, June, 1981.
2. Holmes, T.H. and Rahe, R.H. "The Social Readjustment Rating Scale," *Journal of Psychosomatic Research*, Vol. 11, No. 2, 1967, pp. 213–218.
3. Haynes, S.G., Feinleib, M. and Kannel, W.B. "The Relationship of Psychosocial Factors to Coronary Heart Disease in the Framingham Study," *American Journal of Epidemiology*, Vol. 111, No. 1, 1980, pp. 37–58.
4. Friedman, M. and Rosenman, R.H. *Type A Behavior and Your Heart.* Fawcett, 1974.
5. Veninga, R.L. and Springer, J.P. *The Work/Stress Connection: How to Cope With Job Burnout.* Boston: Little, Brown and Company, 1981.
6. Nelson, J.G. "Burnout—Business's Most Costly Expense," *Personnel Administrator*, Vol. 25, No. 8, 1980, pp. 81–87.
7. Op. cit. (Veninga).
8. House, J.S., *Work, Stress and Social Support.* Reading, Massachusetts: Addison-Wesley, 1981.
9. Berkman, L.F. and Syme, S.L. "Social Networks, Lost Resistance, and Mortality: A Nine Year Follow-up Study of Alameda County Residents," *American Journal of Epidemiology*, Vol. 109, No. 2, 1978, pp. 186–204.
10. Caplan, R.D., Cobb, S., French, J.R.P., et al. "Job Demands and Workers Health," U.S. Department of Health and Human Services, Publication No. (NIOSH) 75–160. Washington, D.C.: U.S.G.P.O., 1975.

11. House, J.S. and Wells, J.A. "Occupational Stress, Social Support, and Health," in A. McLean, G. Black and M. Colligan (editors) *Reducing Occupational Stress: Proceedings of a Conference*, DHEW (NIOSH) Publication 78–140, 1978, pp. 8–29.

12. Pinneau, S.R., Jr. "Effects of Social Supports on Occupational Stresses and Strains." Unpublished Paper, 1965, quoted in House, J.S. *Work Stress and Social Support*. Reading, Massachusetts: Addison-Wesley, 1981.

13. Op. cit. (House, 1981).

14. Wahlund, I. and Nerell, G., "Work Environment of White Collar Workers— Work, Health, Well Being." Stockholm: The Central Organization of Salaried Employees in Sweden, as cited in Levi and Lennart, *Preventing Work Stress*. Reading, Massachusetts: Addison-Wesley, 1981.

15. Weiman, C. "A Study of Occupational Stressors and the Incidence of Disease/ Risk." *Journal of Occupational Medicine*, Vol. 19, No. 2, 1977, pp. 119–122.

16. Karasek, R., Baker, D., Marxer, F., Ahlbom, A. and Theorell, T. "Job Decision Latitude, Job Demands, and Cardiovascular Disease: A Prospective Study of Swedish Men," *American Journal of Public Health*, Vol. 71, No. 7, 1981, pp. 694–705.

17. Welford, A.T. (editor) *Man Under Stress*. London: Taylor and Francis Ltd., 1974.

18. Op. cit. (Friedman, M. et al.).

19. Kiev, A. and Kohn, V. *Executive Stress*, An American Management Association Survey Report, Chicago: AMACOM, 1979.

20. Benson, H. *The Relaxation Response*. New York: William Morrow, 1975.

21. Schwartz, G. "Stress Management in Occupational Settings," *Public Health Reports*, Vol. 95, No. 2, 1980, pp. 99–108.

22. Fielding, J.E. and Breslow, L. "Health Promotion Programs Sponsored by California Employers," *American Journal of Public Health*. Vol. 73, No. 5, 1983, pp. 538–542.

23. Seamonds, B.C. "Stress Factors and Their Effects on Absenteeism in a Corporate Employee Group," *Journal of Occupational Medicine*, Vol. 24, No. 5, 1982, pp. 393–397.

24. Peters, K., Benson, H. and Porter, D. "Daily Relaxation Breaks in a Working Population: I. Effects on Self-reported Measures of Health, Performance, and Well-being," *American Journal of Public Health*, Vol. 67, No. 10, 1977, pp. 946–953.

25. Peters, R.K., Benson, H. and Peters, J.M. "Daily Relaxation Response Breaks in a Working Population: II. Effects on Blood Pressure," *American Journal of Public Health*, Vol. 67, No. 10, 1977, pp. 954–959.

26. Carrington, P., Collings, G.H., Benson, H., Robinson, H., Wood, L.W., Lehrer, P.M., Woolfolk, R.L. and Cole, J.W. "The Use of Meditation-Relaxation Techniques for the Management of Stress in a Working Population," *Journal of Occupational Medicine*, Vol. 22, No. 4, 1980, pp. 221–231.

27. Op. cit. (Schwartz, G.).

28. Manuso, J. "Stress Management and Behavioral Medicine: A Corporate Model," unpublished paper presented to the American Psychological Association, Los Angeles, 1981.

29. Levinson, H. "When Executives Burn Out," *Harvard Business Review*, Vol. 59, No. 3, May-June, 1981, pp. 72–81.

30. Ibid.

31. Ibid.

20 EMPLOYEE ASSISTANCE PROGRAMS

*F*or many years occupational alcoholism programs have been established by employers to assist their employees in overcoming problems with alcohol. A relatively recent phenomena are broader employer-sponsored mental wellness programs which deal with a wide range of employee personal problems in a systematic fashion. "Employee assistance programs" is the generic term for the variety of counseling services made available to troubled employees and frequently their families through a company-sponsored program. Employee assistance programs (EAPs) have their roots in the early occupational alcoholism programs, and many still consider EAPs to be synonymous with alcoholism counseling and rehabilitation programs. Most EAPs, however, provide a broad range of services, including counseling for troubled employees with marital and family problems, job-related problems, and other emotional disturbances, as well as for alcohol and drug abuse problems. Many EAPs offer financial and legal assistance as well.[1-3] That most EAPs have been established relatively recently was documented by a survey published by the American Society of Personnel Administrators (ASPA) in 1981. One-third of responding ASPA members with an EAP reported that the program had been in existence for one year or less, and more than half indicated their programs were less than four years old.[4]

Rapid growth of employer-sponsored programs to prevent and alleviate a range of employee problems which can affect job performance has been spurred by several discoveries. One is the very high cost of these problems if they are not dealt with at an early stage. On average, employees with a range of serious personal problems are less productive than employees without these problems. Alcoholic employees, for example, have absenteeism rates two to three times greater than other employees.[5] Emotional problems, regardless of the source, lead to higher rates of accidents, illness, absenteeism, and long-term disability. Other consequences include more health care utilization and accompanying insurance costs, and higher expenditures under worker's compensation and private disability insurance.

Of all personal problems affecting job performance, alcoholism has been the most studied to date. One authority on alcoholism and its impact on business has depicted the cost of alcoholism to industry as having the following components:

- absenteeism, lost time from the job, lateness, reduced efficiency, and increases in overtime pay
- faulty decision-making and increased on- and off-the-job accidents
- reduced morale of fellow workers, friction among employees, and impaired consumer and public relations
- early retirement, premature disability and death, increased personnel turnover, and loss of skilled and valuable employees
- the costs of alcohol control programs and added costs of insurance programs, including life, health and disability insurances, worker's compensation and public liability and property damage insurances[6]

Many of the above costs to industry can be subsumed under the broader heading of lost productivity. In 1975 the annual cost of lost productivity attributable to alcohol problems on-the-job was estimated at nearly $20 billion.[7] In 1984 dollars this cost is probably around $40 billion. The cost to industry of lost productivity due to other on-the-job abuse of drugs such as amphetamines and cocaine, an increasingly recognized problem among all echelons of the workforce, was estimated to be $16.6 billion annually in 1983 dollars.[8]

Employee health benefits and lost productivity costs are not the only factors that have stimulated the growth of employee assistance programs in industry. Other powerful influences have included:

- changing social attitudes toward mental illness that have lessened the stigma of having mental health problems and receiving treatment for them
- demands by some unions for mental wellness programs
- expansion of the range of mental health problems which have been adjudicated or legislated as compensable under worker's compensation
- the requirement by some states that health insurance benefits cover mental/behavioral problems[9]

EAP RESULTS

The steady growth of EAPs has been fueled by positive reports by many companies that have experience with these programs. Among the reported success stories are the following:

1. AT&T has reported that 70 to 80 percent of the employees receiving help through the company's EAP are markedly improved or completely rehabilitated. For 150 employees referred to AT&T's EAP whose work records were reviewed for the two years before and two years after they entered the program, a "dramatic" improvement was discovered in number of accidents on- and off-the-job, tardiness, absenteeism, use of the company medical facility and performance judgments by their supervisor. For example, comparing two-year periods before and after program participation, the total number of on-the-job accidents declined from thirty-six to five. Estimating a cost of $4,000 per accident, the company believes its savings related to these 150 employees to have been over $100,000. [10]

2. Potomac Electric Power Company's (PEPCO) Employee Advisory Service (EAS) was evaluated by comparing the work performance of 112 program participants one year before and one year after participation and also comparing the performance of this group with that of a random sample of 49 PEPCO employees who did not use the EAS. During the year prior to participation, the EAS-related employees had significantly more visits to the Medical Department, extended sick days and lost-time accident days than the control group. A comparison of the subsequent year, however, showed that the EAS participants had fewer extended sick days and lost-time accident days than the nonparticipants. PEPCO found average reductions of 1.67 visits per year to the Medical Department, 6.73 extended sick days per year, and 0.92 lost-time accident days per year among EAS participants. They calculated their one-year savings per participant to be $664, or $74,000 for the total group. At an annual cost of $37,500 to run the EAS, the benefit-to-cost ratio was almost two-to-one. [11]

3. An evaluation of General Motors' EAP found that, during the first year after program entry, grievances and disciplinary actions against participants were reduced by about one-half, lost time was down 40 percent, sickness and accident benefits 60 percent, and on-the-job accidents 50 percent. For every dollar spent on treating employees in the EAP, General Motors reported that two dollars were being returned to them within a three-year period. In those programs where a trained, full-time management staff member was involved with the program, there was a $3 return on every dollar invested during the same three-year period. [12]

4. The most comprehensive and carefully conducted cost-effectiveness analysis of employee assistance programs to date studied the impact of EAPs on several work performance variables in four companies. Program participation was associated with declining absenteeism in those groups with excessive absenteeism prior to participation in the EAP, reaching near normal level by the second half-year after intervention. In all four companies, grievances and disciplinary actions declined substantially among clients after program participation. On-the-job accidents, visits to the company medical facility, and worker's compensation benefit payments were also reduced after program intervention. The only variable which did not, in general, show a reduction was the use of sickness and accident benefits. (It was speculated by the authors of the study that significant sickness and accident benefits were used in the rehabilitation process, and that a longer time period might be necessary to establish whether costs and use would be reduced.)[13]

EAP COMPONENTS

Employee assistance programs generally consist of:

- A written company policy and defined procedure toward employees with personal problems affecting job performance;
- Supervisory identification, confrontation and referral of troubled employees or, alternatively, self-referral or referral by a friend or loved one;
- Confidential counseling and treatment services available either inhouse or through carefully chosen external sources of care; and
- Referral to alternative community services as necessary.

Employee assistance programs are predicated upon the following basic premises:

- The occupational setting provides an excellent opportunity to identify employees with emotional/behavioral problems because reduced effectiveness on the job is often an early manifestation.
- When employees' problems are adversely affecting their job performance, regardless of whether or not the problems originated at work, employers have a legitimate right to require employees to address these problems to the degree necessary to attain adequate job performance.
- Early intervention (i.e., while the worker is still employed) increases the likelihood of successful rehabilitation.[14]

- The desire to protect employment is a powerful motivator for employees with problems to participate in counseling programs and/or other necessary therapy.

EAP MANAGEMENT

Aside from adherence to the same general concepts and basic elements described above, employee assistance programs vary widely with respect to specific modes of operation. Companies initiating an EAP may choose among a broad array of options in determining optimal staffing and referral patterns. A format should be selected which is consistent with program goals and objectives and which matches the corporate environment in terms of both organizational requirements (e.g., Will company time be provided for counseling and treatment? Will services be offered on-site or off-site?), and employee needs (e.g., special problems unique to employees of that company, access, acceptability, languages other than English, etc.).

Availability of resources, both in-house and in the community, will largely determine the type of program which is instituted. Planners of an EAP must carefully assess organizational capabilities to determine whether the program should be managed by in-house personnel or would be better provided by an external resource such as a consulting firm, private provider, or community agency. Outside resources available to a company will depend in part on location. Large metropolitan or suburban areas will generally offer a wider assortment of mental health resources than usually available to smaller and more rural areas. Types of external resources to be considered include: individual providers (psychiatrists, psychologists, occupational social workers); public and private hospitals with psychiatric services; private mental health clinics; community mental health centers; private and public alcoholism/drug treatment facilities; and many other social services.[15] Internal resources available to an EAP will depend on the size of the company, number and size of operating units, geographic distribution and the presence or absence of a medical department or other appropriate counseling resource. Among the model staffing and referral patterns that should be considered in planning an EAP are the following four:[16]

1. In-house psychologists or social workers assess employee problems and refer troubled employees to appropriate community resources for assistance. This staffing pattern works well in a large, centralized organization where full-time positions can be justified.
2. An outside consulting service is contracted with to see troubled employees, assess their problems and refer them to local resources. Sometimes a hotline is installed which the employee can call to receive some

counseling over the phone and, if necessary, be referred to a local agency or facility. While telephone assistance may in some cases be less effective in resolving a problem without need for referral, this style of EAP can be extremely useful for companies with many small operating units, branches, or field forces at scattered locations. Before contracting for a hotline service, there are a number of important questions that should be answered to the employer's satisfaction:

- What are the qualifications and training of the hotline counselors? What backup do they have if they encounter problems that require immediate consultation with a psychiatrist?
- How familiar is the contractor with available resources in each of the locales under consideration? How does the contractor keep abreast of changes in these resources, e.g., types of service, hours of service, fee requirements, location, phone number, etc.?
- Is the phone line staffed twenty-four hours a day, including weekends and holidays?
- What is the experience of other employers who have used such services?
- How do the use rates compare with the rates for programs with on-site counselors?

3. A company contracts with a consulting service, facility or agency to provide diagnostic and some treatment services on- or off-site. Employees requiring additional assistance are referred to other sources of care. Programs run by an outside agency often are perceived as more private since EAP records are not kept by any company employee. Employees may be more willing to participate in a program run by a neutral third party. One advantage of this model is that employers are not required to hire full-time dedicated staff; rather, they are left with the flexibility to contract for the amount of staff needed. However, a contract which pays for every service creates an economic incentive to provide additional treatment. A contract with a capitation fee, by contrast, provides incentives to refer to other resources. Employers should exercise caution in contracting for initial evaluation of troubled employees by organizations that run inpatient treatment programs, since strong economic incentives exist for referral to their programs.

4. A company develops a network of community health and social service resources at each site. A network might include a mental health center, alcoholism and drug abuse treatment and rehabilitation facilities, family counseling services, etc. As part of employee education regarding the company's policy toward mental health problems, all employees are furnished with a resource guide detailing the services available, and

self-referral is strongly encouraged. If occupational nurses are present at these locations, they may play an important role in referring employees directly to appropriate community sources.

These four examples are merely illustrative of different program staffing arrangements. Any outside resources to be used for either basic counseling or referral must be carefully appraised by a knowledgable individual to determine the range of services offered, experience of staff, how clients are received and treated, appropriateness of costs for various services, and reimbursement procedures.[17] In addition, there is a strong need for employee feedback on the quality of services provided by outside resources. The list of referrals will be honed through experience with referral and follow-up. Employers within the same geographic region (e.g., city or county) may wish to jointly develop a resource network, since each effort consumes considerable time and energy.

FEATURES OF SUCCESSFUL PROGRAMS

Despite the variety of program types and staffing arrangements, experience has demonstrated that certain components are common to the successful programs:

1. Top management, line managers, and selected employees all participate in the establishment of a written corporate policy toward employee emotional/behavioral problems which can hinder job performance. Company policy clearly and specifically delineates the organizational philosophy and is disseminated to all employees. All managers receive a copy of the procedure to be followed when identifying, confronting, and referring employees whose personal problems interfere with job performance.

    When the company is union-organized, union participation in the establishment of this policy is essential to an efficacious program. In the absence of a formal employee organization, involving interested employees at different levels within the company will facilitate acceptance of the policies and willingness of employees to use the services.

2. The employer explicitly recognizes alcoholism and drug abuse and mental health disorders as treatable illnesses, and incorporates the following principles in the corporate policy statement:
    a) Employees with emotional/behavioral problems will be treated no differently than employees with any other type of illness.
    b) Employees utilizing employee counseling services will not be dis-

criminated against in any way in promotional opportunities, job security, etc.

c) All employee assistance records will be completely confidential.

d) Employees will be allowed sick leave to participate in any rehabilitation or treatment effort.

e) Employees must assume responsibility for cooperating with rehabilitation efforts; disciplines up to and including job termination may result if a personal problem continues to adversely affect work performance.[18]

The corporate policy should leave no ambiguity regarding the company's philosophy by destigmatizing mental illness within the corporate culture and strongly encouraging employee self-referral into the program. Letting employees know that any contact they (or their family members) initiate with the EAP will be kept in strictest confidence by EAP personnel and not communicated to supervisors is essential to gaining employees' confidence in using the program.

3. Strong, visible support and commitment to the EAP is continuously provided by the chief executive officer (CEO) and other top level managers. CEOs can play a major role in assuring a successful program by creating an organizational environment which is supportive of employees seeking help with personal emotional problems. Ensuring that senior and middle management fully understand the program's goals, objectives and mode of operation, providing adequate funding and proper personnel, and requiring good planning and on-going monitoring and evaluation are all appropriate CEO concerns and responsibilities.[19] In most cases these responsibilities are best discharged through the appointment of an experienced and well-respected employee assistance professional.

4. Supervisors, who have a pivotal role in an EAP, receive education about the importance of the program to maintaining a healthy workforce and training in the appropriate procedures for identifying, confronting and referring "problem" employees. Supervisors carefully document evidence of declining work performance and draw upon this documentation during confrontation with the employee. Supervisors restrict their comments to poor job performance; they do not attempt to diagnose the employee's problem, but rather suggest that the employee seek whatever counseling and/or therapy is necessary from available resources. The supervisor suggests the EAP as one good source of such assistance. If the employee rejects assistance and/or does not show signs of improvement within a reasonable period of time, the supervisor takes appropriate disciplinary action.[20]

5. EAP staff is comprised of experienced individuals with strong counseling skills and, preferably, a minimum of a master's degree in psy-

chology or a related discipline. In order to avoid malpractice and other liability issues, it is desirable that the counselor not only have a relevant professional degree, but also be licensed and/or hold the proper credentials.

Interactions between the client and the counselor are the crux of the EAP. Counseling helps the employee understand the main problem, which may or may not be the presenting problem. For example, an employee referred to his company's EAP for financial problems may have an underlying cocaine habit or a drinking problem which contributes to financial irresponsibilities. A good EAP counselor fulfills a number of roles:

- Helps the employee understand the dimensions and severity of his or her problem;
- Decides jointly with the employee what level of intervention is necessary: a) no further counseling or referral, b) one or two additional sessions that are likely to result in the employee being able to cope with the problem without referral, or c) immediate referral due to the nature of the problem (e.g., serious suicidal thoughts or alcoholism) or due to the need for longer term therapy than is available through the EAP;
- Provides appropriate referrals that meet employees' needs for more specialized or more sustained assistance in dealing with their underlying problem;
- Maintains a current list of high quality resources and updates it on a continuing basis. Referrals should be made based on the nature of the problems, employee preferences and convenience to the employee. In addition, the counselor must take into account the employee's financial resources and the level of coverage provided under the company-sponsored benefit plan. Emphasis should be on providing outpatient referrals whenever possible rather than on more expensive inpatient care which has generally not been found to be more effective. Therefore, the benefit plan should provide economic incentives for the use of outpatient services. Otherwise inpatient treatment remains the most attractive alternative from the employee's financial viewpoint.

Most EAPs cover any member of the employee's family because when a family member is experiencing a serious personal problem, it usually affects the employee as well. Counselors should be prepared to provide referrals for couples and for groups of family members who need to work together to solve a shared problem (e.g., drug use by a teenage child of an employee);

- Helps the employee identify general supportive resources such as co-workers, family members, church groups, close friends, social groups, etc.; and
- Follows up periodically on employees seen (with employee's permission) to ascertain: a) the degree to which the presenting and the un-

derlying problems have been ameliorated or resolved; b) if the referral was made, whether the employee completed the referral and the employee's reaction to the referring resource and degree of satisfaction; and c) the need for other referrals or additional sessions with the EAP counselor. Information from these follow-ups should facilitate continuing refinement of the referral list.

Backup consultation, with either a doctorate-level psychologist or a psychiatrist, must be available to EAP counselors for those instances when they feel that an employee's problem exceeds their level of training or when they simply desire a second opinion on a case. A good counselor recognizes his or her limitations.

EAP counselors should not use the EAP as a vehicle to increase their own practice through self-referral. Counselors should sign as part of their contract a statement that they will neither self-refer nor refer to any individual or organization from whom they will receive any direct or indirect remuneration based upon the referral.

6. Employees periodically receive easy-to-understand written materials and audio-visual presentations about the company's policy toward employee alcohol and drug abuse problems and other emotional disturbances which affect job performance, and about the company EAP. Information about the EAP should be provided to every employee at the time of hiring and also should be disseminated throughout the company on a regular basis. Information should also be sent to the home address of every employer, addressed to the family. The program description should include the services available, counseling hours, location of counseling services, and the name and phone number of the program director (i.e., the counselor, or a senior counselor). Strong written assurances to the effect that visits to the EAP will be completely confidential and that no sanctions will be imposed for participation in counseling must be prominent in all materials and presentations. In unionized companies joint endorsement by management and labor will foster a higher level of confidence and program use.

PROGRAM EVALUATION

Careful evaluation of employee assistance programs will provide employers not only with a measure of the effectiveness of their program, but with an assessment of its strengths and weaknesses. Evaluation results are one of the best tools for improving a program and assuring its continued effectiveness.

Surveys of employers with EAPs indicate that the vast majority of employers consider their programs to be successful,[21] and claims such as "60 percent successfully rehabilitated, absenteeism reduced from fourteen to four days per annum,"[22] and ". . . medical visits to the company clinic were reduced more than 200 percent"[23] are commonly bandied around in personnel administration literature. However, with the exception of a few careful studies, there is a paucity of substantive evaluation data on the effectiveness of EAPs. In addition, much of the data which is available pertains strictly to occupational alcoholism programs, in large part due to a greater experience with these programs.

Apart from the fact that most EAPs have only been in existence for a limited time, lack of good evidence on the effectiveness of EAPs stems from the failure of most companies to conduct carefully designed evaluations of their programs. In addition, methodological defects in many of the evaluations completed to date make their findings somewhat tentative. Barriers to companies undertaking and completing comprehensive evaluations of their EAPs include:

- absence of well-defined program objectives which are amenable to evaluation
- concern with employees' right to privacy and confidentiality of medical records
- unavailability of necessary data
- lack of willingness to commit resources for a meaningful assessment of their program

In addition, many companies have been convinced of their program's effectiveness through anecdotal evidence, and have decided that there is no need for a formal evaluation.[24]

Despite the general tendency not to conduct extensive and methodologically precise evaluations, a recent survey of the Fortune 200 companies found that most companies with EAPs do keep some statistics and/or other measurements of the success of the program for evaluation and possible program improvement. Among these companies, the leading sources of data used for evaluating program success are: 1) employee performance evaluations; 2) absenteeism records; 3) tardiness records; 4) disciplinary actions taken; 5) productivity reports; 6) accident records; and 7) records of paid insurance benefits. Other sources of information include attrition records, spoilage and breakage records, utilization, and disability claims.[25]

An earlier survey by the Washington Business Group on Health of their member organizations' EAPs found that twenty-six of fifty respondents had a system for evaluation. Twenty-two of the twenty-six organizations recorded absenteeism, sixteen recorded disciplinary grievances, seventeen recorded sick leave benefits, twenty-three recorded monthly number of requests for services, nineteen recorded utilization of medical benefits and twenty-two used additional measures of program success. Improved employee productivity was ranked by respondents as the number one program benefit, followed by reduced absenteeism, improved employee morale, lowered hospital/surgical/medical utilization and lower insurance premiums.

NOTES

1. Kiefhaber, A. and Goldbeck, W. "Industry's Response: A Survey of Employee Assistance Programs" in *Mental Wellness Programs for Employees*. R.H. Egdahl and D.C. Walsh (editors), New York: Springer-Verlag, 1980, pp. 19–26.

2. Ford, R.D. and McLaughlin, F.S. "Employee Assistance Programs: A Descriptive Survey of ASPA Members," *Personnel Administrator*, Vol. 26, No. 9, 1981, pp. 29–35.

3. 1982 Survey of Fortune 200 Companies Re: Employee Assistance Programs (unpublished).

4. Ibid.

5. Pell, S. and D'Alonzo, C.A. "Sickness Absenteeism of Alcoholics," *Journal of Occupational Medicine*, Vol. 12, No. 6, 1970, pp. 198–210.

6. Follman, J.F. *Alcoholics and Business: Problems, Costs, Solutions*. New York: AMACOM, 1976.

7. Berry, R.E., Boland, J.P., Smart, C.N. and Kanak, J.R. *The Economic Costs of Alcohol Abuse and Alcoholism*. Final Report to National Institute of Alcohol and Alcohol Abuse. Contract No. ADM28–76–0016. Boston. Policy Analysis. 1977.

8. "Taking Drugs on the Job," *Newsweek*, August 22, 1983.

9. Barrie, K., Smirnow, B., Webber, A., Kiefhaber, A. and Goldbeck, W.B. "Mental Distress as a Problem for Industry" in *Mental Wellness Programs for Employees*. R.H. Egdahl and D.C. Walsh (editors), New York: Springer-Verlag, 1980, pp. 3–18.

10. "How Two Companies Curb Drug Abuse," *Occupational Hazards*, Vol. 45, 1983, pp. 93–96.

11. "Electric Company EAP Works," *EAP Digest*, Vol. 1, No. 5, 1981, p. 46.

12. Remarks by Thomas A. Murphy, Chairman, General Motors Corporation, October 5, 1979 at the Association of Labor-Management Administrators and Consultants on Alcoholism, Inc.

13. Foote, A., Erfurt, J.C., Strauch, P.A. and Guzzardo, T.L. *Cost-Effectiveness of Occupational Employee Assistance Programs, Test of an Evaluation Method*. Worker Health Program, Institute of Labor and Industrial Relations, The University Of Michigan-Wayne State University, 1978.

14. Trice, H.M. and Roman, P.M. *Spirits and Demons at Work: Alcohol and Other Drugs on the Job*. New York State School of Industrial and Labor Relations, Cornell University, 1972 and 1978.

15. Op. cit. (Barrie).

16. Sonnenstuhl, W.J. and O'Donnell, J.E. "EAPs: The Whys and Hows of Planning Them," *Personnel Administrator*, Vol. 25, No. 11, pp. 35–38.

17. Ibid.

18. Rowntree, G.R. and Brand, J. "The Employee with Alcohol, Drug, and Emotional Problems," *Journal of Occupational Medicine*, Vol. 17, 1975, pp. 329–331.

19. Carr, J.L. and Hellan, R.T. "Improving Corporate Performance Through Employee Assistance Programs," *Business Horizons*, Vol. 23, 1980, pp. 57–60.

20. Op. cit. (Rowntree).
21. Op. cit. (1982 Fortune 200 Survey).
22. Sager, L.B. "The Corporation and the Alcoholic," *Across the Board*, Vol. 16, No. 6, 1979, pp. 79–82.
23. Busch, E.J. "Developing an Employee Assistance Program," *Personnel Journal*, Vol. 60, No. 9, 1981, pp. 708–711.
24. Op. cit. (Barrie).
25. Op. cit. (1982 Fortune 200 Survey).

MANAGING 21
DISABILITY
COMPENSATION

DISABILITY: A GROWING PROBLEM

Striking increases have occurred over the past two decades in the amount and degree of disability reported by the American public. Between 1966 and 1976 the number of individuals permanently limited in their activities increased by 37 percent to 30.2 million, despite only a 10 percent growth in population. During the same time period the prevalence of individuals suffering from a long-term disability which made it impossible for them to carry out their main activity increased by 67 percent, from 213 to 355 per 10,000 population. Between 1966 and 1974 there was a 106 percent rise among men aged 45 to 64 in the prevalence of severe disability restricting ability to carry on main activities, and a corresponding 78 percent rise among middle-aged women.[1]

In the decade from 1966 to 1976 there was a 221 percent increase in persons claiming to be severely disabled by diabetes (i.e., main activity impossible), a 172 percent increase in persons claiming to be severely disabled by hypertension (without heart involvement) and a 105 percent increase in persons reporting to be severely disabled by arthritis and rheumatism (Table 21.1).[2] Not only have there been enormous increases in self-reported disability over a relatively short time span, there have been major shifts in the ranking of disability problems (Table 21.1). Between 1966 and 1976 arthritis and rheumatism replaced heart conditions as the number one

TABLE 21.1
EVOLUTION BETWEEN 1966 AND 1976 IN THE UNITED STATES OF THE NUMBER OF PEOPLE CLAIMING TO BE LIMITED IN THEIR ACTIVITY (LONG-TERM DISABILITY) BECAUSE OF SPECIFIC CAUSES

Selected causes	1966 N	Rank	1976 N	Rank	Variation 1966–1976 N	Percentage
Total U.S. population	191,537,000	—	210,643,000	—	+19,106,000	+10
All categories of restriction of activity						
All causes	21,984,000	—	30,175,000	—	+8,191,000	+37
Heart conditions	3,600,000	1	4,737,000	2	+1,137,000	+32
Arthritis and rheumatism	3,248,000	2	5,069,000	1	+1,821,000	+56
Mental and nervous conditions	1,711,000	3	1,479,000	6	−232,000	−14
Visual impairments	1,222,000	4	1,629,000	3	+407,000	+33
Hypertension (without heart involvement)	1,187,000	5	2,082,000	4	+907,000	+75
Asthma—hay fever	1,065,000	6	1,448,000	7	+383,000	+36
Diabetes	562,000	7	1,539,000	5	+976,000	+174
Hearing impairments	403,000	8	754,000	8	+351,000	+87
Main activity impossible						
All causes	4,078,000	—	7,469,000	—	+3,391,000	+83
Heart conditions	964,000	1	1,748,000	1	+784,000	+81
Arthritis and rheumatism	639,000	2	1,307,000	2	+668,000	+105
Visual impairments	466,000	3	538,000	3	+72,000	+15
Mental and nervous conditions	400,000	4	523,000	4	+123,000	+30
Hypertension (without heart involvement)	187,000	5	508,000	6	+321,000	+172
Diabetes	163,000	6	523,000	4	+360,000	+221
Asthma—hay fever	158,000	7	224,000	7	+66,000	+42
Hearing impairments	126,000	8	142,000	8	+16,000	+13

Source: Colvez, A. and Blanchet, M. "Disability Trends in the United States Population 1966–1976: Analysis of Reported Causes," *American Journal of Public Health,* Vol. 71, No. 5, 1981, pp. 464–471.

cause of disability. Together musculoskeletal and heart conditions accounted for one out of every three cases of limited activity. Diabetes moved from seventh to fifth place. Visual and hearing problems and hypertension also ranked higher in 1976 than in 1966.

Extrapolation from a 1978 Social Security Administration survey of 12,000 civilian, noninstitutionalized, randomly selected adults aged 18 to 64 indicated that 165 out of every 1,000 adults (21 million Americans) were limited in their ability to work due to a chronic condition or impairment.[3]* Slightly more than one-half of the 21 million disabled adults (10.7 million) were severely disabled (unable to work at all or unable to work regularly), and the remainder were partially disabled (unable to do the work they did before becoming disabled, or unable to work full-time).

The 1972 Survey of Disabled and Nondisabled Adults by the Social Security Administration found that 143 out of every 1,000 adults (age 20 to 64) were limited in their ability to work.[4] While the two figures are not entirely comparable due to inclusion of a younger age group in the more recent survey, it is reasonable to conclude that there was approximately a 15 percent increase in the frequency of (self-reported) cases of disability over the six-year period. Other national data show that between 1969 and 1978 there was a 40 percent increase in self-reported disability prohibiting work among men aged 45 to 64.[5]

Chronic disabling conditions are not the only disorders affecting the health of the workforce. Acute illnesses and injuries resulted in about 339 million lost workdays in 1980, an average of 3.5 days off the job for each worker in the country.[6] Respiratory conditions and injuries were the leading causes of lost workdays due to acute conditions, followed by infective and parasitic diseases, digestive system conditions and other acute disorders (Table 21.2).

The impact of all these statistics is unmistakable. Industry is facing an increasingly disabled population from which to constitute its workforce. In addition, it faces a higher tax burden to help support those who never enter the working world. Workers who become disabled create an enormous economic burden for employers, not only in terms of lost productivity (higher absenteeism, diminished morale, often earlier retirement from work, etc.), but also in their aggregate impact on disability compensation costs and health insurance premiums. Employers' contributions to Social Security Disability Insurance (SSDI), workers' compensation and privately purchased disability insurance have risen steadily as a percentage of payroll over the past decade, to the point where together they presently account for approximately 3 to 5 percent of payroll depending on state, type of industry, private insurance arrangements, etc.[7-9]

---

*For purposes of the survey, disability was defined as any self-reported limitation in the kind or amount of work (or housework) resulting from a chronic health condition or impairment lasting three or more months.

TABLE 21.2
WORKDAYS LOST DUE TO ACUTE CONDITIONS IN THE UNITED STATES, 1980

	Work-loss days per employed person		
Acute conditions	All ages 17 and over	Ages 17–44	Age 45 and over
Both sexes			
All acute conditions	3.5	3.7	3.1
Infective and parasitic diseases	0.3	0.3	0.2
Respiratory conditions	1.5	1.6	1.4
Digestive system conditions	0.2	0.2	0.1
Injuries	1.1	1.1	0.9
All other acute conditions	0.5	0.5	0.4
Male			
All acute conditions	3.3	3.6	2.5
Infective and parasitic diseases	0.2	0.2	0.2
Respiratory conditions	1.3	1.4	1.1
Digestive system conditions	0.2	0.2	*
Injuries	1.2	1.5	0.6
All other acute conditions	0.3	0.3	0.5
Female			
All acute conditions	3.7	3.7	3.8
Infective and parasitic diseases	0.3	0.3	*
Respiratory conditions	1.8	1.8	1.8
Digestive system conditions	0.2	0.2	*
Injuries	0.9	0.7	1.2
All other acute conditions	0.6	0.7	0.4

*Figures do not meet standards of reliability or precision.
Note: The data refer to the civilian, noninstitutional population. An acute condition is one which lasted less than three months and which involved either medical attention or restricted activity. A "work-loss day" is a day on which a currently employed person seventeen years of age and over, did not work at least half of his normal workday because of a specific illness or injury. In some cases the sum of the items does not equal the total shown, because of rounding.
Source: U.S. Department of Health and Human Services, National Center for Health Statistics, *Current Estimates from the Health Interview Survey: United States—1980*, as cited in Health Insurance Association of America. *Source Book of Health Insurance Data, 1981–1982*, Washington, D.C., p. 75.

These increases reflect greater numbers of individuals claiming disability as well as more generous disability benefits from some programs such as SSDI. Since disabled workers use more health care, they also contribute to higher health care premiums. In 1977 nearly one-fourth of the severely disabled and one-seventh of the partially disabled were hospitalized compared with only 2 percent of the nondisabled. Hospitalizations are much longer among the disabled. Over 85 percent of the severely disabled who were hospitalized in 1977 had lengths of stay over four days compared with less than 65 percent of the nondisabled.[10]

Who Are the Disabled?

Data collected by the Social Security Administration are particularly useful in identifying common characteristics of disabled adults. Major findings from the most recent Social Security Administration survey of the disabled reveal a number of features of interest to employers:

1. Adults 55 to 64 years of age were ten times more likely to be severely disabled than adults age 18 to 34, with the rate of severe disability nearly doubling with each successive ten-year age interval after 34.
2. Men and women reported 17 and 18 percent rates of disability respectively. Ten percent of women (6 million) considered themselves to be severely disabled compared with 7 percent (4.5 million) of men.
3. Severe disability was reported by 13 percent of both the black population and persons of Spanish descent, 8 percent of the white population and 6 percent of members of other races.
4. Education correlated inversely with the degree of disability. The severely disabled had an average of ten years of education compared with an average of twelve years for the partially disabled and the nondisabled.
5. The 1977 median family income for the severely disabled was less than half that of nondisabled persons; 28 percent of the severely disabled had annual incomes less than $5,000 compared with 11 percent of the nondisabled.
6. Highest disability rates were found in the West (19 percent) and South (18 percent), followed by North Central (16 percent) and Northeast (15 percent). The highest rate of *severe* disability was reported in the South (11 percent).

Armed with this information, employers can predict trends in their own disability cases as the result of demographic changes in the workforce. Based on these statistics, the anticipated aging of the workforce will lead to a large increase in disability.

More recent national data substantiate the above findings of the Social Security Administration, particularly those regarding family income and level of education (Table 21.3). In 1980 persons with an annual family income of between $3,000 to $5,000 lost almost 50 percent more workdays per person than workers whose family income was $25,000 or greater. Individuals with less than nine years of education had more than twice as many days off the job due to illness or injury than persons with sixteen or more years of schooling.[11]

TABLE 21.3
DISABILITY DUE TO CHRONIC AND ACUTE CONDITIONS, BY SOCIAL AND ECONOMIC CHARACTERISTICS IN THE UNITED STATES, 1980 (DAYS PER PERSON)

Characteristics	Restricted activity	Bed disability	Work-loss
Sex			
Both sexes	19.1	7.0	5.0
Male	17.1	5.9	4.9
Female	21.0	8.0	5.1
Family Income			
Under $3,000	35.6	12.7	4.7
$3,000–4,999	38.9	14.5	6.0
5,000–6,999	28.4	10.3	6.1
7,000–9,999	23.8	8.6	5.7
10,000–14,999	19.4	7.1	5.7
15,000–24,999	15.1	5.5	5.5
25,000 and over	12.7	4.5	4.2
Educational Attainment			
Under 9 years	41.7	14.5	6.7
9–11 Years	22.4	8.1	5.9
12 Years	17.7	6.1	5.0
13–15 Years	17.8	6.3	4.4
16 Years and over	13.6	4.5	3.3
Occupation			
White Collar	11.8	4.0	4.1
Professional, technical and kindred	11.6	3.9	3.8
Managers and administrators (except farm)	11.7	3.8	3.8
Clerical and kindred	12.4	4.5	4.8
Sales	10.7	3.3	3.4
Blue Collar	12.9	4.2	6.3
Craftsmen and kindred	10.7	3.7	4.8
Operatives and kindred	14.3	4.4	7.5
Laborers (except farm)	15.1	5.7	7.5
Service	14.8	5.4	5.6
Farm	10.6	3.0	3.6

Note: The data refer to the civilian, noninstitutional population. A "restricted activity day" is one on which a person cuts down on his usual activities for the whole of that day because of an illness or injury. A "bed disability day" is one on which a person stays in bed for all or most of the day because of a specific illness or injury. A "work-loss day" is a day on which a currently employed person, seventeen years of age and over, did not work at least half of his normal workday because of a specific illness or injury. "Family Income" includes the total income of each member of the family. "Educational Attainment" refers to the number of years of schooling the individual has received.

Source: Data from the National Center for Health Statistics, U.S. Department of Health and Human Services, published in Health Insurance Association of America's *Source Book of Health Insurance Data, 1981–1982.*

As illustrated in Table 21.3, type of employment is another factor related to disability in the working population. In 1980, blue-collar workers had 50 percent more days off the job per person than white-collar workers.[12]

Disability and Sex Differences

A recent study by Metropolitan Life Insurance Company, utilizing 1977–79 statistics, found that women under age 45 incur more disability days than men, while the opposite is true after age 45.* When all ages were combined, men and women were found to acquire the same number of disability days per year. Although women had a much higher disability rate (about one in five employees per year) than men (about one in eight employees per year), men were found to miss an average of $9\frac{1}{2}$ weeks of work, while women only missed 6 weeks. These differences reflect the greater severity of disabling conditions among men and the persistence of misclassifying pregnancy as a disability. Pregnancy was, in fact, the leading cause of disability of women under age 45. Among younger men accidents headed the list of causes. The greatest causes of disability for women over 45 were accidents, genitourinary diseases and bone diseases. Heart disease was the leading cause of disability among males over 45.[13]

CONTROLLING THE COST OF DISABILITY COMPENSATION PROGRAMS

Disability income benefits supported by employer contributions provide employees with at least partial replacement of income lost when they are unable to work as a result of injury, illness or pregnancy. A multiplicity of programs provide benefits to disabled workers depending on the nature and severity of the impairment, the term of disability, present and past employment, and numerous other variables. In 1979 the Health Insurance Association of America (HIAA) identified forty-two different compensation systems for disabled workers and their dependents.

Attempting to piece together all of the systems that may apply to the employees of any one organization is complicated by several factors. First, a disabled worker is often eligible for compensation from more than one source. An employee permanently disabled by a work-related injury may receive workers' compensation, employer-sponsored disability benefits and a range of individually purchased disability insurance-compensation programs, such as individual disability income, no-fault automobile disability benefits and mortgage disability insurance. Because each individual compensation system operates under its own basis for benefit, qualifications for

---

*Disability was defined as an illness lasting longer than one week.

payment, amount and duration of benefit and offset provisions, achieving coordination among the various components is difficult.

Second, disability may be defined in many different ways. While a worker may be considered disabled (and therefore eligible for benefits) under the requirements of one program, he or she may not be so considered under the requirements of another. Many cases end up in litigation to determine whether or not a worker is actually "disabled" under different programs. Because the adjudications are based on subjective judgments, it is frequently impossible to predict which cases will be awarded benefits. Finally, laws and definitions regarding disability vary widely between states, so employers with personnel in several states have an especially difficult task in working with disability problems.

To become conversant with all the existing disability programs and new program options or rearrangements is a time consuming and therefore expensive undertaking for an employer. However, as discussed in chapter 1, disability is already a major health cost to employers. With an aging workforce a common situation, disability can be expected to increase greatly even if the size of the workforce remains constant. No expenditure in the personnel area is more sensitive to the relative economic incentives for employees to return to work or to remain off the work roll than benefits provided under disability plans. In other words, a major portion of these costs is controllable.

The only way to minimize undesirable incentives is to painstakingly plan a disability program. Knowing about eligibility, size and duration of benefits and rehabilitation opportunities under existing governmental and company-sponsored programs and how they interface is an essential prerequisite to establishing where gaps exist, where combined benefit levels are too high or low and whether proper incentives have been incorporated.

Social Security Disability Insurance

Social Security Disability Insurance (SSDI), which is financed through a federally administered trust fund, is the largest single program of disability insurance. Revenues for SSDI are generated by a tax on covered employment.* In 1981 nearly $17 billion was paid to 2.8 million primary beneficiaries and their dependents.[14] SSDI benefits are available to workers and their dependents after five full months of inability to work. The single criterion for eligibility is permanent and total disability, defined as, "incapability, expected to last for at least one year, of substantial gainful work at any job that exists to a substantial extent in the U.S. economy."[15] In addition to regular Social Security eligibility, disability insurance coverage requires recent employment and contributions to Social Security, usually twenty out

---

*Nearly 90 percent of the workforce is employed in a job covered by Social Security.

of the most recent forty calendar quarters. At age 65 disability benefits are converted to retirement benefits. In 1981 the average monthly benefits for disabled workers, spouses and children were $413, $122 and $123, respectively.[16] In 1982 the maximum monthly benefit was $716.70 for an individual and $1,075 for a family.[17] After twenty-four consecutive months of benefits, disability recipients become eligible for Medicare.[18]

According to a 1982 Congressional Budget Office Study on disability compensation, approximately one-fifth of the labor force—22.3 million workers—are either without coverage or uninsured by major programs for non-occupational-related total disability. Nearly 20 percent of the 91 million workers contributing to Social Security are ineligible for Social Security disability benefits due to insufficient recent attachment to the workforce. Of the 9.4 million workers not covered by SSDI, almost 45 percent are uninsured for total nonoccupational-related disability.[19]

Workers' Compensation Programs

State workers' compensation programs, which are designed to provide disability income to employees who are disabled by a work-related injury or accident, cover about 90 percent of the workforce.[20] Workers' compensation programs paid almost $11.9 billion in benefits in 1979.[21] All employees of covered employers are eligible for benefits, and there are no qualifications for payment. Because workers' compensation programs are state mandated and benefits are determined by statute, benefit levels vary widely. Generally, however, benefits are set at two-thirds of the average weekly wage prior to disability, and are subject to a maximum weekly payment. As is the case with SSDI, workers' compensation benefits are nontaxable income. Full, unrestricted hospital and medical coverage is provided for the duration of need, and frequently a rehabilitation benefit is provided which may include a maintenance allowance and special funding. Normally, benefits received from workers' compensation reduce Social Security, group disability and no-fault automobile disability benefits.[22] Together SSDI and workers' compensation may not exceed 80 percent of "average current earnings" prior to disability. And under the Omnibus Budget Reconciliation Act of 1981, workers' compensation benefits must offset SSDI unless a reverse offset provision was passed by state law on or prior to February 18, 1981.[23] Chapter 22 provides an extensive discussion of trends and issues in workers' compensation.

Short-Term Benefits for Nonwork-related Disability

Employer-sponsored coverage for short-term nonoccupational-related disability usually consists of paid sick leave for a specified number of days, beginning promptly or within three days of disability and/or short-term group disability insurance, which generally requires a waiting period of three days to a week. According to estimates by HIAA, at the end of 1980 approximately 65.4 million persons were covered by short-term disability benefits: 39.7 million through commercial coverage,* 23 million by paid sick-leave plans and 2.7 million by other formal arrangements (union-administered plans and the Federal Mutual Benefit Association).[24]

Virtually all employers participating in the 1983 Hay-Huggins survey offered some type of short-term illness and disability plan, not unexpectedly, because participating employers tend to be large and to provide a broad scope of benefits. A majority of employers (56 percent) covered all employees under their short-term disability plans (including uninsured sick leave) although separate group plans for salaried employees were also common (27 percent). Employees of most companies (54 percent) were immediately eligible for benefits upon employment. Nearly all of the remainder of companies required one to twelve months of employment prior to eligibility.[25]

With respect to the total labor force, coverage for short-term disability is spotty. In 1978 approximately 90 percent of public employees were protected against illness or injury lasting under six months (or the first six months of long-term disability), compared with 60 percent of employees in private industry. Excluding areas with mandatory temporary disability insurance (TDI) laws (California, New York, New Jersey, Rhode Island, Hawaii, Puerto Rico and the railroad industry), only 47 percent of private sector employees were protected by sickness benefits. In addition, the majority of government workers are covered for short-term disability exclusively by sick leave plans which provide significantly larger income replacement rates than insurance benefits, the predominant mode of protection in private industry. As a result, the percentage of income loss provided to government workers against short-term disability (84.3 percent in 1978) is much higher than the comparable private industry figure (23.3 percent in 1978).[26]

Sick Leave Plans Sick leave plans in private industry are far more common among white-collar workers than production workers. A 1980 survey of employee benefits by the Department of Labor found that nine out of ten professional, administrative, technical and clerical participants were protected by sick leave, compared with only four out of ten production workers. Sick leave provisions were also more generous for white collar employees, who frequently had from ten to thirty days of paid sick leave while the production workers typically had from five to ten days. As shown in Table 21.4, more than two-thirds of surveyed sick leave plans limit their coverage to a specified number of sick days per year while most of the remaining plans provide coverage on a per disability basis or through some combination

---

*This figure excludes persons covered under administrative-services-only arrangements, minimum premium plans and other forms of self-insured disability plans.

TABLE 21.4
PAID SICK LEAVE: PERCENT OF FULL-TIME EMPLOYEES BY TYPE OF PROVISION, PRIVATE INDUSTRY, 1980

Provision	All participants	Professional and administrative participants	Technical and clerical participants	Production participants
Total	100	100	100	100
Provided sick leave	62	91	89	39
Sick leave provided on:				
An annual basis only*	47	62	69	32
A per disability basis only†	9	18	13	3
Both an annual and per disability basis	5	7	6	3
No specified maximum number of days‡	2	5	2	1
Not provided sick leave	38	9	11	61

*Employees earn a specified number of paid sick leave days per year. This number may vary by length of service.
†Number of days of paid sick leave is renewed for each illness or disability. This number may vary by length of service.
‡Unlimited sick leave, provided to employees as needed.
Note: Because of rounding, sums of individual items may not equal totals.
Source: U.S. Department of Labor. Bureau of Labor Statistics, 1981. Employee Benefits in Industry, 1980, p. 11.

of the two. Less than 1 percent of plans did *not* specify a maximum number of paid sick leave days.[27]

Short-Term Disability Insurance Two-thirds of *all* disability income policies (short- and long-term) marketed by commercial insurance companies are sold on a group basis through employers, and one-third are sold on an individual basis. Policies vary widely in terms of what perils are insured, amount and duration of benefit, exclusions, waiting periods and eligibility requirements. Policies may be designed to cover loss of income resulting from accidental injury alone or from accidental injury and illness. Since it is easier to feign illness, few policies cover solely illness due to moral hazard.[28]

Short-term disability policies generally stipulate "inability to perform 'own' occupation" as the basis for benefit. Group policies usually exclude work-related injury or illness from coverage; otherwise benefits are reduced by worker's compensation payments.[29] Short-term disability insurance typically provides benefits for 13 or 26 weeks, although some plans provide benefits for 52 or 104 weeks. Since most disabilities are of short duration, employers' risk declines with longer contracts, and their premiums decrease accordingly. For example, the premium for a 26-week contract is less than twice that of a 13-week contract.[30] A survey by HIAA of new group disability insurance policies found that for policies written in 1981, in over 80 percent of cases (i.e., employer groups) covering almost 70 percent of workers, short-term benefits were provided for 26 weeks.[31]

Although short-term disability benefits may be set at a fixed weekly amount, generally they are set at a percentage of salary which is restricted to a maximum weekly benefit. Table 21.5 indicates the percentage of income replaced by short-term disability insurance as reported by the HIAA survey. In the majority of plans, between 60 and 67 percent of gross earnings is replaced by short-term disability benefits. However, 70 percent of employees have two-thirds or more of their gross earnings replaced and 37 percent of employees will receive 70 percent or more of gross earnings should they become disabled. Almost 70 percent of employees were limited to a maximum weekly benefit under $200. More than half of employees were covered by policies providing an average weekly benefit less than $100 and 78 percent could expect to receive no more than $159. Only 16 percent of employees could expect to receive a weekly benefit of $200 or more if disabled.[32]

Evidence from the Hay–Huggins survey suggests that about half of companies offering sick leave/salary continuation plans base the benefit on length of service, with the remainder of companies either paying uniform benefits to all employees or basing the benefit on accumulation of days.[33] The short-term disability program at Pitney-Bowes, Inc. of Stamford, Connecticut is an excellent example of an innovative approach to establishing the level of short-term disability benefits provided. The company's disability benefits for nonexempt employees are tied to length of service and atten-

TABLE 21.5
SHORT-TERM DISABILITY INSURANCE

Percentage of income replaced	Percentage of cases	Percentage of employees
Less than 35	7.7	3.7
35–45	1.1	0.3
50	5.9	8.4
55	0.3	0.2
60–63	25.8	17.3
66–67	41.0	32.9
70 or more	18.2	37.2
Maximum amount of weekly benefit		
Less than $70	14.0	33.4
$70–$109	22.9	11.8
$110–$149	7.1	5.2
$150–$199	10.6	19.3
$200–$299	23.0	15.8
$300 or more	22.4	14.5

Source: Health Insurance Association of America. *New Group Disability Insurance*, Washington, D.C., 1981.

dance record, and for sales representatives are based on length of service and sales performance. For example, a nonexempt employee with three to five years of service and an excellent attendance record (absences of 4 days or less per year in two out of the last three calendar years) will receive 75 percent of pay for remaining days of disability (up to 104 days) above the first 6 days of disability that are fully covered. In contrast, employees with only a satisfactory attendance history (absences of 6 days or fewer during the calendar year) will receive only 65 percent of pay after the first 6 days. Only nonexempt employees with at least three calendar years of seniority are eligible for the excellent attendance premium. If the attendance record is unsatisfactory, benefits may be denied.[34]

For policies written by insurers in 1981, 71 percent of employees share some portion of the premium cost for short-term disability insurance. However, employers pay the majority of premium cost for 87 percent of employees. Less than 2 percent of employees pay the entire premium and 27 percent of employees contribute nothing toward the premium.[35]

Coverage Under Long-Term Disability Insurance

Short-term disability often leads to long-term disability. As discussed in more detail below, attention to rehabilitation in the early period of disability is the key to minimizing the period that the disability will endure. Long-term disability insurance presupposes that the disabled employee will not be returning to work within the foreseeable future, and possibly not at all. The underlying assumption is that even if rehabilitated, the disabled worker will probably not be able to return to his or her original employment, either because the work is no longer physically possible or because the job was filled during the interim. To protect themselves from paying benefits permanently to individuals who may be able to return to some alternate form of employment, insurance companies generally award benefits on the basis of "inability to perform 'own occupation' for two years and 'any reasonably suitable occupation' thereafter."[36]

The Health Insurance Association of America estimates that at the end of 1980, 21 million individuals were covered by long-term disability insurance, about 70 percent under group coverage.[37] However, this figure of 21 million does not include persons covered by plans financed through any form of self-insurance, for example, fully self-insured, administrative-services-only arrangements and minimum premium plans. Such alternative financing mechanisms are fairly common among larger companies. For example, the 1983 Hay-Huggins survey found that about 35 percent of surveyed companies at least partially self-insure their long-term disability plans.[38]

As is the case with short-term disability insurance, long-term disability benefits either exclude work-related disability or are offset by workers' compensation. The benefits are also generally reduced by primary and dependent SSDI, retirement benefits, state cash sickness plans and any other employee-sponsored benefits. Benefits from employer-purchased long-term disability insurance are taxable although, depending on income level, $5,200 may be exempted annually. Benefits purchased by the employee are not considered taxable income.[39]

Long-term disability insurance is widely used as an employee benefit for managerial employees (83 percent) and clerical staff (72 percent), and less frequently for nonoffice workers (33 percent).[40] Over half of the 791 companies with long-term disability insurance (93 percent of survey participants) responding to the Hay-Huggins survey covered all employees under their plan. Eligibility to join the long-term disability plan begins immediately upon employment in one out of three cases, during the first month of employment in 13 percent of cases and, after a longer waiting period, in 55 percent of cases. Of plans with waiting periods, 90 percent provided eligibility within twelve months of employment.[41]

The HIAA examined long-term disability income policies written in 1981, and found that more than one-half of the plans, covering one-third of employees, initiated benefits after six months of disability (Table 21.6). The vast majority of plans (82 percent), covering 88 percent of employees, provided benefits for both accidents and illness until age 65. In 14 percent of the plans, however, benefit periods were for only two years. Two-thirds of the plans, covering half of employees, replaced 60 percent of gross income. One in four employees was limited to a disability benefit equivalent to 25 percent of gross earnings. More than 90 percent of the long-term disability

TABLE 21.6
LONG-TERM DISABILITY INSURANCE

Commencement of benefit period	Percentage of cases	Percentage of employees
First day	0.9	4.6
30 days or 1 month	4.7	1.2
60 days or 2 months	2.8	4.1
90 days or 3 months	36.8	27.9
180 days or 6 months	52.4	35.3
All others	2.4	26.9
Percentage of income replaced		
25	0.5	25.9
50	7.6	7.8
60	66.0	53.4
66–67	21.2	12.5
70	4.7	0.4

Source: Health Insurance Association of America. *New Group Disability Insurance*, Washington, D.C., 1981.

income contracts specified a maximum monthly income of $1,500 or greater. However, contracts covering 70 percent of employees established an average monthly benefit between $500 and $900.[42] The HIAA survey of long-term group disability coverage also found that:

- 80 percent of employees were not insured for partial disabilities.
- Two-thirds of contracts provided coverage for rehabilitation under various conditions; generally rehabilitation services had to be prescribed by a physician and provided by a physician or rehabilitation specialist.
- The employer paid 50 to 100 percent of the premium in 90 percent of plans, covering 84 percent of employees.

The Rise in Disability Payments

Although disability transfer payments (from all programs) appear to have stabilized over recent years, with the aging of the workforce and erosion of the work ethic there is little to prevent a resurgence of what occurred in the period from 1950 to 1977. During that time, payments to disabled beneficiaries grew from $3.1 billion (1.1 percent GNP) to $42.2 billion (2.2 percent GNP). From 1965 to 1975 alone, cash payments to disabled persons grew from $9.7 billion to $33.9 billion, an increase of almost a full percent of the GNP. In 1977, the federal portion of total disability payments was 63 percent, while state and local disability programs and privately purchased disability insurance each accounted for 19 percent of total disability transfer payments.[43]

Three major factors have contributed to the rapid rise in disability expenditures:

1. Establishment of new programs, such as Social Security Disability Insurance in 1956, Supplemental Security Income (SSI) for indigent disabled in 1974, Black Lung in 1969 and private long-term disability insurance which was almost nonexistent before 1960.*
2. Higher benefits per capita, primarily through legislation. In general, per capita disability benefits grew faster than earnings, with a particularly wide disparity during the first half of the 1970s.
3. Increased use of disability programs by workers of all ages resulting from more attractive benefits (relative to earnings), wider awareness of programs and greater tendency of people to consider themselves disabled. Between 1965 and 1975 the number of SSDI beneficiaries increased by 150 percent, from 1 million to 2.5 million while the covered workforce grew by only 55 percent.[45]

Two of these important contributors to rapid growth in disability benefit expenditures, new programs and higher benefit levels, resulted from public policy decisions. Increases related to these changes were anticipated. Not projected, however, was the huge expansion in the number of individuals claiming themselves to be disabled. Lack of objective methods to measure disability has forced reliance on an individual's reported ability to perform important activities. The same physical problem may in one person be perceived as disability and in another as an impediment but compatible with work.

OPTIONS AND RECOMMENDATIONS

Employers have increasingly made available disability benefits which provide reasonable financial protection against loss of income due to a disabling illness or injury. However, as disability rates have increased, the costs of maintaining or improving disability benefits have grown much more rapidly than payroll or profits. More attention is therefore turning to how disability costs can be contained and how to assure appropriate utilization of disability benefits.

Appropriate Benefit Design

Employers who provide disability insurance have considerable flexibility in determining contents of the plan. Unlike workers' compensation, no maxi-

---

*Had these new programs not been created, it is estimated that disability transfer payments would have been less than two-thirds their level in 1975.[44]

mum ceiling on benefit levels need be specified, providing equal protection for workers at all salary levels. Regardless of the specified level and duration of benefit under long-term disability, care should be taken to effect integration of employer-sponsored programs with Social Security payments so that the desired level of total benefit is maintained. A 1981 survey showed that fewer than one-half of all surveyed long-term disability plans had integration with all other major sources of benefits. In those plans providing for coordination, direct dollar-for-dollar offset provisions are the rule (75 percent of plans).[46] If the benefit level in the employer plan is appropriately set, the dollar-for-dollar offset appears reasonable.

Long-term disability benefits as currently written terminate either at age 65 or age 70. As pension plans are rewritten and retirement at later ages becomes the norm, long-term disability plans should be reviewed to ensure that benefits continue until expected age of retirement and/or age when pension benefits are available to achieve comparable payment levels. On the other hand, it is equally important to make sure that individuals receiving generous long-term disability benefits are not eligible at the same time for private pension benefits or that the disability benefits are reduced by the amount of the pension.

As with any benefit, care should be taken to minimize abuses. In long-term disability the most common abuses tend to be claims related to an insured's job being in jeopardy, claims close to retirement age and claims for an insured who has pending litigation that hinges on whether he or she is "disabled." With greater frequency, employers are seeing long-term disability claims due to increased pressure on employees to raise productivity, to relocate or be retrained. A downturn in the economy also brings a rise in disability claims, both short- and long-term.

Ideally, short-term disability benefits should be integrated with long-term disability coverage, and those plans with benefits exceeding five months should be integrated with Social Security benefits. As of 1981 about 27 percent of surveyed companies with long- and short-term disability plans had gaps in disability coverage.[47]

Income Under Disability Plans

Disability income should be set at levels that create a work incentive but don't make it difficult or impossible for workers to support themselves and their dependents when they are unable to work due to illness or injury. Exactly what this level should be, however, is a very subjective question, and opinions vary. According to the Health Insurance Association of America, based on the tax-exempt status of most disability income* and the absence

---

*SSDI, worker's compensation and individually purchased disability insurance are all tax-free; group-purchased disability insurance benefits are subject to tax but may be eligible for $5,200 annual exemption, depending on income.

of certain work-related expenses, disability benefits which replace 65 to 75 percent of predisability gross earnings generally provide a comparable level of spendable income for the disabled individual. HIAA suggests that benefits that replace 55 to 65 percent of gross earnings create a stronger incentive to induce workers back to employment as soon as they are physically able.[48] While there is little argument that lower replacement rates provide a stronger work incentive, whether or not benefit levels set at 55 to 65 percent of predisability gross earnings provide an adequate amount to live on depends on several factors, such as level of predisability earnings, whether new medical expenses are incurred which raise out-of-pocket costs, whether special assistance is required to perform everyday activities, etc.

Contrary to what one might expect, give the multiplicity of disability compensation systems available and recent increases in benefits, low income replacement rates are more common than high replacement rates. In 1977, 70 percent of severely disabled beneficiaries received 60 percent or less of their predisability gross earnings. Half of these received below 35 percent of prior gross earnings. Fourteen percent of severely disabled beneficiaries received more than their prior gross earnings in disability cash benefits.[49] Employers should establish income levels under disability that are at least 60 percent of prior earnings if they wish to protect workers against serious financial problems associated with disability.

High income replacement rates relative to prior earnings are experienced more frequently in cases where the beneficiary is receiving disability income from more than one program. Severely disabled male beneficiaries in 1977 receiving disability income from more than one source had replacement rates above their prior earnings in 18 percent of cases compared to 5 percent of cases where the beneficiary received income from a single program.[50] However, receipt of multiple benefits does not guarantee high cumulative benefits for the disabled workers. According to a recent study by the Congressional Budget Office, approximately one-half of disabled workers receiving benefits from SSDI and other programs also receive income from some welfare program which provides benefits on the basis of a means test. *Total* benefits received by these individuals are low in comparison to average disability benefits.[51] Fewer workers are likely to be covered by SSDI, as the Social Security Administration more carefully scrutinizes eligibility under that program.

As well as discouraging some employees from returning to work as quickly as possible, higher benefit levels provided by individual compensation programs have been found to increase the number of workers filing for disability under that program. A study by the Society of Actuaries of the experience with employer-sponsored long-term disability from 1971 to 1975 found a correlation between the level of benefits and the number of claims (Figure 21.1). The same study found that when group long-term disability benefits were reduced by total SSDI payments (i.e., to the primary bene-

FIGURE 21.1
RELATIONSHIP BETWEEN LEVEL OF LONG-TERM DISABILITY BENEFIT AND NUMBER OF CLAIMS

Percentage of gross earnings replaced	Claims filed (percentage of the average)
50	90
50–60	100
61–70	106
>70	119

ficiary *and* to the dependents), claims were 91 percent of the average; if benefits were offset by primary SSDI payments only, claims were 101 percent of the average; and without any integration of SSDI and disability insurance benefits, claims were 112 percent of the average.[52]

Most employers do a good job in coordinating and controlling multiple benefits. For those that do not, a sizable payoff can be anticipated from developing an effective monitoring system.

Prevention of Disability

Preventing disability requires:

- Attention to reducing the disability toll from degenerative diseases such as heart disease, arthritis, diabetes, stroke and many cancers;
- Efforts devoted to reducing the toll of injuries, both on and off the job, and the potential causes of traumatic injury;
- A comprehensive approach to minimizing the number and severity of occupational illnesses;
- Creating an environment where stress is not a major contributor to disease;
- Dealing effectively with troubled employees, including those with alcoholism and other substance abuse problems; and
- Working to minimize the effect of specific limitations and impairments on employees' perceptions of their ability to work and on their ability to perform in an appropriate job.

Minimizing the toll of degenerative diseases on physical and mental health is treated extensively in Chapters 18 and 19. A continuum of opportunities exist from health promotion, through disease prevention, to minimization of impairment and disability. Regular aerobic exercise for example, reduces the risk of developing a heart attack while simultaneously increasing the level of energy, morale and self-reported productivity. Development of a regular exercise program after a heart attack can reduce the time period to return to work and improve confidence in the ability to perform normal work roles.[53] Finally, exercise after a heart attack improves cardiac function, and there is some reason to believe it could reduce the risk of sustaining a second heart attack.[54]

Diabetes is a condition of increasing prevalence in the workforce. Its appearance may be forestalled, in some cases indefinitely, by maintenance of normal weight. A combination of weight management and a diet high in complex carbohydrate and fiber facilitates improved control and reduces dependence on medications.[55] Aerobic exercise improves carbohydrate metabolism in diabetics, also facilitating control of blood sugar.[56] Tighter control of blood sugar is believed to reduce the rate of complications,[57] and may therefore forestall many of the impairments that can diminish productivity on the job, preclude employment and increase health care expenditures.

Osteoarthritis, a common disease associated with the aging process, appears to cause less impairment in those who are physically active. Physical activity is also helpful in preventing osteoporosis (thinning of the bones), which predisposes older people to fractures. Exercise further appears to reduce functional impairment in those who have already advanced to severe arthritis.[58]

Rehabilitation

"The best work incentive for disabled persons is the provision of rehabilitation services at the onset of disability." So began testimony from the Washington Business Group on Health to the President's Commission on Pension Policy.[59] That rehabilitation in selected cases is cost-effective appears clear. For example, Kemper Rehabilitation Management showed that in two similar cases involving paraplegics whose injuries were due to traffic accidents in their early twenties, the company will pay perhaps $150,000 for treatment (including rehabilitation services) to the individual who becomes re-employed as opposed to $1.2 million in lifetime medical benefits to his or her counterpart who does not return to work.[60] In a study of 2,106 files of disabled persons who had been successfully rehabilitated, International Rehabilitation Associates, Inc. claims to have realized a savings in disability benefits of $10.26 for every $1.00 spent on rehabilitation. The company computed that the savings amounted to over $10,500 per case and a combined total savings of more than $3.3 million.[61]

What determines whether rehabilitation is successful? Individual factors are quite important, as are the incentives for return to work. But many times what makes all the difference is *when* the rehabilitation process starts. According to one expert, "rehabilitation starts in the ambulance on the way to the hospital."[62] If no contact with the disabled employee is made early and rehabilitation efforts are not initiated until eligibility for long-term disability occurs, results are not likely to be successful. A study by International Rehabilitation Associates, Inc. of injured workers referred to the company for rehabilitation found that of those referred within three months, 47 percent returned to work, while referrals within four to six months were associated with a 33 percent return to work, and for referrals over twelve months only 18 percent returned.[63]

Employees can become comfortable with living on the income provided under disability programs and with the change in lifestyle required by their disability. At that point rehabilitation efforts are much more difficult. Personal rapport between the disabled person and the individual coordinating rehabilitation efforts is also a key ingredient to a successful outcome. Continuing contact between the supervisor and the disabled worker can also be important in helping the worker feel that there is a link to his place of employment and that others care and want him or her back as part of the work team. Willingness to invest initially, even when return to work is not assured, is also essential.

The best results are those where the person can return to the same job. If not, a different job with the same employer would be the second goal. Failing either of these, returning the person to a different job with a different employer should be tried, although placement in a company without prior experience with the employee is frequently difficult. In general, the prognosis is best for individuals who suffer catastrophic musculoskeletal injuries as the result of trauma. Working with those on disability due to low back pain is more difficult. Those with head trauma, especially if residual brain dysfunction remains, are among the most difficult problems.

Incentives need to be incorporated into both short- and long-term disability programs that encourage return to work, even work at lesser pay than pre-injury or other compensable health problems. A policy that offsets only one-half of work pay against disability payments up to a maximum level not less than predisability pay can act synergistically with structured rehabilitation efforts.

To those who may be ambivalent about how productive disabled workers are upon return to work when they still have some impairment, the results of a careful study by the Department of Labor may be illuminating. Reviewing employment records of 11,028 disabled workers and a matched group of 18,258 able-bodied workers, the study concluded:

1. As a group, impaired workers were as efficient as unimpaired workers.
2. Impaired workers lost only slightly more time through absenteeism than unimpaired workers.
3. The impaired worker was found to be as safe a worker as his or her unimpaired co-workers.
4. The record for disabling injuries—i.e., injuries that result in death,

permanent impairment, or absence from work for at least one full day—was better for the impaired than other workers.

5. In no instance had an impaired worker suffered another permanent work injury sufficiently severe to place him or her in the group of permanently and totally disabled.

6. No disabling injury to the impaired worker could be traced to his or her impairment.

7. The frequency with which workers used plant medical facilities for reasons of illness or discomfort not related to employment averaged about the same for impaired and unimpaired.[64]

Management of Disability Programs

In many large companies long-term disability and workers' compensation are managed by different departments, most often human resources and legal, respectively. Since both deal with the same health problems, employers should consider consolidating responsibility for the two into one department. Human resources is usually the best choice. If they are to remain separate, communication should be required to assure that both problems and successful approaches are shared.

NOTES

1. Colvez, A. and Blanchet, M. "Disability Trends in the United States Population 1966–76: Analysis of Reported Causes," *American Journal of Public Health,* Vol. 71, No. 5, 1981, pp. 464–471.

2. Ibid.

3. U.S. Department of Health and Human Services, Social Security Administration. *Work Disability in the United States: A Chartbook,* Washington, D.C.: U.S.G.P.O., 1980.

4. U.S. Department of Health and Human Services, Social Security Administration. *Disability Survey 72—Disabled and Nondisabled Adults,* Research Report No. 56, Washington, D.C.: U.S.G.P.O., 1981.

5. Sunshine, J. "Disability Payments Stabilizing After Era of Accelerating Growth," *Monthly Labor Review,* Vol. 104, No. 5, 1981, pp. 17–22.

6. Health Insurance Association of America. *Source Book of Health Insurance Data 1981–1982,* Washington, D.C., 1982.

7. U.S. Chamber of Commerce. *Employee Benefits 1980.* Washington, D.C., 1981.

8. 1982 Annual Report, Federal Old-Age and Survivors Insurance and Disability Insurance Trust Funds. Washington, D.C.: U.S.G.P.O., 1982.

9. National Council on Compensation Insurance. Annual Statistical Bulletin, 1981.

10. Op. cit. (Social Security Administration 1980).

11. Op. cit. (Health Insurance Association of America).

12. Op. cit. (Health Insurance Association of America).

13. "Older Men Disabled More Often Than Women," *Business Insurance,* November 2, 1981, p. 64.

14. Congressional Budget Office. *Disability Compensation: Current Issues and Options for Change,* 1982.
15. Sunshine, J. *Disability: A Comprehensive Overview of Programs, Issues, and Options for Change.* President's Commission on Pension Policy, 1980.
16. Lando, M.E., Farley, A.V. and Brown, M.A. "Recent Trends in the Social Security Disability Insurance Program," *Social Security Bulletin,* Vol. 45, No. 8, 1982, pp. 3–14.
17. U.S. Chamber of Commerce. *Analysis of Workers' Compensation Laws,* Washington, D.C., 1982.
18. Health Insurance Association of America. *Compensation Systems Available to Disabled Persons in the United States,* Washington, D.C., 1979.
19. Op. cit. (Congressional Budget Office, 1982).
20. Op. cit. (Sunshine, 1980).
21. Price, D. "Workers' Compensation Coverage, Benefits and Costs, 1979," *Social Security Bulletin,* Vol. 44, No. 9, 1981, pp. 9–13.
22. Op. cit. (Health Insurance Association of America, 1979).
23. Op. cit. (U.S. Chamber of Commerce, 1982).
24. Op. cit. (Health Insurance Association of America, 1982).
25. Hay Associates. *1983 Noncash Compensation Comparison.*
26. Price, D. "Income Replacement During Sickness, 1948–78," *Social Security Bulletin,* Vol. 44, No. 5, 1981, pp. 18–32.
27. U.S. Department of Labor. Bureau of Labor Statistics. *Employee Benefits in Industry, 1980.* Washington, D.C.: U.S.G.P.O., 1981.
28. Vaughan, E.J. and Elliott, C.M. *Fundamentals of Risk Insurance,* 2nd ed., New York: John Wiley & Sons, 1978.
29. Op. cit. (Health Insurance Association of America, 1979).
30. Op. cit. (Vaughan and Elliott).
31. Health Insurance Association of America. *New Group Disability Insurance.* Washington, D.C., 1981.
32. Ibid.
33. Op. cit. (Hay Associates).
34. "Short-term Disability Benefits at Pitney-Bowes Tied to Length of Service, Attendance, Sales Record," *Employee Benefit Plan Review,* Vol. 36, No. 2, 1982, p. 53.
35. Op. cit. (Health Insurance Association of America, 1981).
36. Op. cit. (Health Insurance Association of America, 1979).
37. Op. cit. (Health Insurance Association of America, 1982).
38. Op. cit. (Hay Associates).
39. Op. cit. (Health Insurance Association of America, 1979).
40. The Conference Board. *Profile of Employee Benefits: 1981 Edition.*
41. Op. cit. (Hay Associates).
42. Op. cit. (Health Insurance Association of America, 1981).
43. Op. cit. (Sunshine, 1980).
44. Op. cit. (Sunshine, 1980).
45. Op. cit. (Sunshine, 1980).
46. Hay Associates. *1981 Noncash Compensation Comparison.*
47. Op. cit. (Hay Associates, 1981).
48. Op. cit. (Health Insurance Association of America, 1979).
49. Op. cit. (Congressional Budget Office, 1982).
50. Op. cit. (Congressional Budget Office, 1982).
51. Op. cit. (Congressional Budget Office, 1982).
52. Op. cit. (Health Insurance Association of America, 1979).

53. Ewart, C.K., Taylor, C.B., Bandura, A. and DeBusk, R. "Immediate Psychological Impact of Exercise Testing Soon After Myocardial Infarction," *American Journal of Cardiology*, Vol. 45, No. 2, 1980, p. 421.

54. Haskell, W.L. "Physical Activity After Myocardial Infarction," *American Journal of Cardiology*, Vol. 33, No. 6, 1974, p. 776.

55. Anderson, J.W. "The Role of Dietary Carbohydrate and Fiber in the Control of Diabetes," *Advances in Internal Medicine*, Vol. 26, 1980, pp. 67–96.

56. Vranic, M. and Berger, M. "Exercise and Diabetes Mellitus," *Diabetes*, Vol. 28, No. 2, 1979, pp. 147–167.

57. Unger, R.H. "Meticulous Control of Diabetes: Benefits, Risks, and Precautions," *Diabetes*, Vol. 31, No. 6, 1982, pp. 479–483.

58. Bardwick, P.A. and Swezey, R.L. "Physical Therapies in Exercise," *Postgraduate Medicine*, Vol. 72, No. 3, 1982, pp. 223–234.

59. Washington Business Group on Health. Testimony on Disability Insurance to the President's Commission on Pension Policy. 1980.

60. Quoted in Schwartz, G. *Disability*. Washington Business Group on Health. 1980.

61. Ibid.

62. "Pay or Rehabilitation? Getting Employees Back to Work Spells Success for Employer and Worker," *Employee Benefit Plan Review*, Vol. 35, No. 5, 1980, pp. 14, 16, 18.

63. Rundle, R.L. "Move Fast If You Want to Rehabilitate a Worker," *Business Insurance*, May 2, 1983, pp. 10, 12.

64. U.S. Department of Labor. *The Performance of Physically Impaired Workers in Manufacturing*. 1948, as cited in Schwartz, G. (op. cit.).

WORKERS' 22
COMPENSATION

OVERVIEW

Workers' compensation is the oldest social insurance program in the United States. In 1908 the first state law was passed providing both medical benefits and cash payments for death and disability related to work. While operating in all jurisdictions, it is actually fifty separate state operations and several special federal programs.

Workers' compensation is expensive to employers, costing business between 0.86 and 4.34 percent of payroll, depending upon the state and frequently the experience of that firm. The national average is 2.55 percent.[1]* Costs to employers are highest in Oregon, Alaska, Hawaii, Maine and Massachusetts, where statewide average costs for workers' compensation (as a percentage of payroll) range between 4.00 and 4.34. Lowest costs (in percent) to employers are found in Indiana (0.86), Utah (1.30), Vermont (1.38), Missouri (1.40) and North Carolina (1.45). Factors contributing to the wide variation in workers' compensation costs among states include:

- differences in benefit levels
- the manner in which workers' compensation laws are administered in each state

---

*Excludes states which require employers to purchase workers' compensation from state-run funds.

- the amount of litigation over claims and the degree of attorney involvement
- different types of work injuries and illnesses based on dominant categories of employment in each state
- varying economic conditions among states, e.g., a state that is primarily agricultural will have different class rates than one which is mainly industrial
- varying attention to industrial safety programs from state to state (and often from employer to employer)
- varying wage levels by state; since the benefit awarded is generally a percentage of the gross weekly wage this will affect the average cost per case
- varying regional medical care costs, which account for approximately 30 percent of total workers' compensation costs[2]

In 1980 employers paid $22 billion directly into workers' compensation, including $15.7 billion for premiums to private insurance carriers, $4.1 billion to state funds and $2.1 billion for the cost of self-insurance benefits and administration. The Social Security Administration estimates that 79 million workers were covered by these programs during an average month in 1980, for an annual average cost per employee of $278.[3]

Benefits paid out in 1980 were $13.4 billion or 61 percent of premiums.[4] Data from prior years indicates that about 30 percent of total benefits went for medical services while 70 percent was provided as cash payments. Employer costs for a year are much higher than benefits because increasing amounts of premium income must be set aside to cover liabilities for payments in future years.

Enactment of state workers' compensation laws was fueled by both increasing social concern for workers' health and by industrialization. By 1900 there were 25,000 to 35,000 occupationally related deaths per year and 2 million serious injuries. Arguments for workers' compensation laws focused on the inevitability of industrial accidents as a byproduct of increased technology and new methods of production and, given the inevitable, it seemed appropriate to protect workers, regardless of their own negligence, by building reasonable compensation for such events into the costs of the products and services purchased by consumers. Also argued was the need to standardize the liability of the firm and minimize litigation and corporate financial exposure for work-related injuries and illnesses. Workers' compensation, therefore, has had from its inception an objective of protecting both the employer and employee from losses associated with employee injuries.

Essential principles of workers' compensation programs are:

1. Employers are obligated to pay benefits according to the schedule enacted into law regardless of whose negligence caused the injury.

2. There is no attempt to provide an injured worker with complete damages. Workers forego the right to sue in exchange for receiving statutory benefits. Usually benefits are less than could be obtained by litigation if employer negligence were found.

3. In most cases indemnity payments are made on a periodic rather than a lump sum basis.

4. Since the cost is a cost of production, employers bear the full financial burden of the program costs.

5. Insurance is required except for larger firms, which may be permitted to self-insure.[5]

PROGRAM FEATURES

Virtually every type of industrial employment is covered by workers' compensation which is compulsory in all but three states: South Carolina, Texas and New Jersey. As expected, great variation exists among state laws with respect to coverage, benefits and funding options. As of 1982 six states (Nevada, North Dakota, Ohio, Washington, West Virginia, Wyoming) required that insurance be purchased from a state-run insurance fund, while in twelve states employers had a choice of buying the insurance through commercial carriers or the state-run program. In every state except for Wyoming, North Dakota and Texas, large companies can self-insure, and in twenty-two states plus the District of Columbia, small employers are permitted to come together, usually through a trade association, to self-insure as a group.[6]

Generally workers' compensation is provided to employees whose injuries arise "out of and in the cause of employment." Typically, all well-established occupational diseases are covered. However, there are many gray areas, such as whether an employee was in the course of employment while commuting to work, exercising on company premises during lunch hours or attending a company-sponsored nonwork activity. Even greater disagreements in interpretation arise from trying to answer whether the health problem arose from the job. Heart attacks, mental health problems and cancers have been some of the major sources of litigation to establish employer liability under the workers' compensation system.

Benefits usually include seven categories:

1. *Medical Expenses* In general there are no statutory restrictions on the types of services that are covered.

2. *Total Temporary Disability* Payments to those unable to work due to work injury but expected to recover and return to work. Payment is usually based on percentage of injured worker's wage, most often

$66\frac{2}{3}$ percent subject to minima and maxima (January 1982 range of maxima $112–$942 per week). Sometimes payment level is also modified by number of dependents. Waiting periods before payments begin are common, usually three to five working days.

3. **Partial Temporary Disability** Payment to those unable to do work in their occupation but who can do other (less well paid) work. Amount of payment is usually a percentage of the difference between pre-injury and post-injury wages, subject to minima and maxima.

4. **Total Permanent Disability** Payment for workers unable to obtain any gainful employment as a result of work injury. In addition, loss of any combination of two arms, hands, legs, feet or eyes are statutorily defined as permanent disability. Under about 80 percent of programs, payments are for life, subject to minima and maxima. In other states, limitations in duration and amount of payments are made.

5. **Partial Permanent Disability** Payment for loss of a member or faculty (e.g., hearing in one ear) or for general disability. An injured worker usually receives a benefit equal to statutorily established multiple of weekly disability benefit. Generally this is given to injured worker in addition to total temporary disability payments.

6. **Death Benefit** Payment for burial and survivors' benefit to dependents. The death benefit is usually paid to surviving spouse for life or until remarriage, although some states limit payments (312 to 1,100 weeks) and about 15 percent set aggregate dollar maxima. Sometimes size of benefit varies with number of dependents.

7. **Rehabilitation Benefits** Payments for rehabilitation expenses, such as vocational training, mechanical appliances, transportation. Funds may come from an employer fund or a separate state fund.

ESTABLISHING WORKERS' COMPENSATION RATES

Workers' compensation premiums under private insurance are rated on the basis of two factors: the loss experience associated with different job classifications and individual company experience. As of 1980 there were more than 600 industry and occupation classifications, each reflecting a different loss experience. Private insurers in each state aggregate their loss figures through rating bureaus in order to establish rates that are based on the most comprehensive data available. These figures are published in rating manuals and are used to establish the premiums paid by employers. Although in some small companies employees are rated under a single classification, in most firms separate categories of employees are rated differently.[7]

Because employers vary with respect to production mechanisms, safety programs and the like, experience rating is employed for risks above a minimum size. Under experience rating, premiums set at manual rates may be altered based on a company's historical record of loss over the preceding three-year period. If the company's record is better than the class average, its premium will be lowered; if its record is worse than average, the premium will be raised.[8]

Rating bureaus collect and process loss data from private insurers and monitor the workers' compensation system to ensure that companies are appropriately classified and rated. The National Council on Compensation Insurance (NCCI), which serves forty states and provides data for independent bureaus in several other states, is the largest of these organizations. NCCI was established in 1915 after states realized that workers' compensation rating could not be accomplished by each individual state.[9] Administrative bureaus of NCCI review all policies written by member insurance companies to make sure that correct classifications and premiums are applied and that double coverage does not exist. They also provide on-site inspections to determine what classifications should be used to assign premiums. Their manual rates are standard with all insurance companies, subject to the approval of state insurance commissions.

NCCI produces a number of publications providing data on workers' compensation. For example, their *Annual Statistical Bulletin* (first edition, 1981) provides an historical review of premium and benefit level adjustments; premium, benefit and expense information; wage data; statewide average rates; statewide benefits; and statistics on the severity and frequency of work-related injuries. (Information is *not* furnished for those states which require insurance through a state fund.) Employers may find this information a useful frame of reference for comparing their own workers' compensation experience with statewide averages.

MAJOR TRENDS

1. *Accelerating Cost Increases*   As shown graphically in Figure 22.1 and detailed in Table 22.1, benefits about doubled in the 1940s, again in the 1950s and once again in the 1960s. However, the rate of rise greatly accelerated in the 1970s, with 1979 benefits 382 percent of those paid out in 1970. From the standpoint of employer contributions, the percentage of payroll remained relatively constant until the 1960s, staying within the range of .89 to .99 from 1946 to 1963. By 1970 it had risen to 1.11 and then started accelerating so that by 1979, in less than a

FIGURE 22.1
WORKERS' COMPENSATION BENEFITS 1940–1979

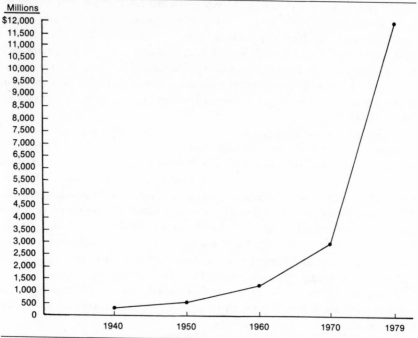

Source: *Social Security Bulletin*, Vol. 43, No. 10, 1980, p. 5 and Vol. 44, No. 9, 1981, pp. 9–13.

decade, it had reached 1.93. Employers were therefore paying about 2 percent of payroll entering the 1980s for a program that fifteen years before cost them only 1 percent of payroll.

Part of the steep cost increment has been due to the introduction of the federal Black Lung Program, whose benefits grew from $110 million in 1970 to $554 million in 1972, leveled off at approximately $1 billion/year through 1978 and grew 67 percent 1978 to 1979 to reach $1.7 billion. However, even excluding the effect of the Black Lung Program increase, 1979 benefit payments were 17 percent above the 1978 level and rose at an average annual rate of 29 percent from 1969 to 1979.[10]

2.   *Health Care Cost Inflation*   The approximately one-third of workers' compensation payments for medical expenses are as sensitive as health insurance to the major increases in price, intensity and utilization that

TABLE 22.1
TRENDS IN WORKERS' COMPENSATION COSTS AND BENEFITS

Calendar years	Estimated workers covered	Millions			Percent of covered payroll	
		Total benefits	Medical and hospitalization	Compensation payments	Cost of workers' compensation	Benefits
1940	25	$ 256	$ 95	$ 161	1.19	.72
1949	35	566	185	381	.98	.55
1950	37	615	200	415	.89	.54
1959	44	1,210	410	800	.89	.58
1960	45	1,295	435	860	.93	.59
1969	59	2,634	920	1,714	1.08	.62
1970	59	3,031	1,050	1,981	1.11	.66
1973	66	5,103	1,480	3,623	1.17	.70
1974	68	5,781	1,760	4,021	1.24	.75
1975	67	6,598	2,030	4,568	1.32	.83
1976	69.5	7,597	2,380	5,217	1.48	.88
1977	72	8,623	2,680	5,943	1.73	.93
1978	75	9,734	2,960	6,775	1.85	.94
1979	78.6	11,872	3,470	8,402	1.93	.98

Source: Social Security Bulletin, Vol. 43, No. 10, 1980, p. 5 and Vol. 44, No. 9, 1981, pp. 9–13.

TABLE 22.2
WAGE REPLACEMENT RATES (PERCENTAGE) FOR TEMPORARY TOTAL DISABILITY OF 3 WEEKS

Year	Single worker	Worker, wife and two children, in jurisdiction	
		With dependent's allowances	Without dependent's allowances
1969	68	73	60
1973	70	75	66
1977	77	83	71

Source: Social Security Bulletin, Vol. 42, No. 5, 1979. p. 15.

have occurred. During the 1970s this component cost almost tripled, excluding the effects of the Black Lung Program, as did the medical cost per covered worker.

3. **State Statutory Changes** The 1972 National Commission on State Workmen's Compensation Law recommended that benefits replace at least two-thirds of a worker's average weekly wage. Many states increased the benefits to at least this level. Constraints on the maximum weekly benefit have been progressively relaxed. For example, all states except California increased the weekly amount payable for temporary total disability in 1978 and/or 1979 and forty-two states increased the amounts in both these years.[11] Most states now provide for automatic increases in the weekly maximum benefit amount, usually based on increases in the state average weekly wage for workers covered by the unemployment insurance program. In most cases, the maximum is set at 100 percent of the state average wage.[12] Waiting periods to become eligible for temporary total disability have also declined, as have the number of days of disability required before retroactive payments can be made to cover a waiting period.

   The effects of all these changes on benefits is seen by a significant increase in the average wage-replacement rates listed in Table 22.2.

4. **Cost Differences by State** Variation in employer costs has increased over the past three decades. From 1950 to 1978 the mean weekly cost per worker increased tenfold while the standard deviation (a measure of variation among states) increased twentyfold. In 1978 the average weekly net costs of insurance per employee for forty-five types of employers ranged from $.90 in North Carolina and $1.23 in West

Virginia to $5.29 in Arizona, $6.29 in Oregon and $8.20 in the District of Columbia.[13]

5. *Declining Benefit/Cost Ratio*   This ratio was 50.9 percent in 1979, a significant decline from a high of 62.4 percent in 1975. Declines are the result of more reserves being set aside for anticipated payouts due to rapidly increasing benefits and medical costs.[14]

OPPORTUNITIES FOR COST CONTROL

Rapidly rising costs have spurred employers to seek better ways to control costs. In some cases, the correct approaches are clear, although implementation is difficult. In other cases, such as the relative costs of public or private financing for workers' compensation, the best approach is an open question.

1. *Public Versus Private Insurance*   Over 70 percent of premiums go to private carriers, whereas about 20 percent go to the state-administered funds, and the remaining 10 percent to self-insured programs. Compared to state funds, private carriers tend to have lower benefit/premium ratios, indicating lower payout per premium dollars collected. In 1978, for example, private carriers collected $2.32 for every dollar paid out compared with $1.96 for state funds.[15] An analysis in *Business Insurance* by an insurance analyst in the risk management department of Kellogg Co. concluded that state-run funds are more efficient. He reports, based on private industry published figures and information from seventeen of the eighteen state funds, that for 1979 "both systems [were] paying approximately the same amount in benefits to workers for each premium dollar paid by employers." However, based on his analysis, industry administrative expense (including commissions) of 16.8 percent greatly exceed state-run fund administrative expense of 5.7 percent.

   While striking, this analysis does not explain why much lower costs have not led to state-run funds getting almost all the business in states where they operate. Insurance companies also claim that they provide loss-control services, unavailable from state-operated funds. Also, the analysis in *Business Insurance* did not take into account issues of how quickly claims were processed, error rates, ease of filing claims, numbers of claims leading to litigation, claimant satisfaction and employer satisfaction. Nonetheless, the article concludes that, "the leading market share held by eleven of the twelve competitive [state-run] funds indicates their savings are being recognized by employers.[16] At a minimum, this study suggests the importance of employers in states with competitive systems shopping around carefully to determine if they are getting the best deal. Included in the inquiry should be not only premium rates, but also experience with dividends, if offered, so that net costs can be compared. Assessment of the quality of the services provided is also important prior to making a decision.

2. **Participation in Group Compensation Funds** The 1982 U.S. Chamber of Commerce report, "Analysis of Worker's Compensation Laws" shows that twenty-two states allow group self-insurance workers' compensation funds. In Michigan, the first state to allow such arrangements in 1974, twenty-one funds with an excess of $120 million in annual premiums were operating in 1981. One article on this subject claims that "cost savings of 35 to 45 percent are not uncommon."[17] In general those considering establishing such a fund should have combined annual premiums of at least $1 million. Frequently, programs are set up by trade associations for their members, and in some states funds are restricted to employers in the same industry. For smaller businesses, group compensation systems provide an opportunity for greater participation in control of claims administration, expenditures of assets and development of fund policies and optional activities such as safety programs.

3. **Learning From the Federal Experience** The Federal Employees' Compensation Act (FECA) of 1916 created a system for providing benefits to injured federal employees. In 1974 two new provisions were enacted, "continuation of pay" and "free choice of physician." Continuation of pay eliminated the waiting period and specified that a federal employee sustaining a disabling job-related injury would be entitled to full salary (minus usual deductions) for up to forty-five calendar days after injury. Free choice of physician permitted an employee to choose the physician he or she wished to treat a job-related injury, superseding the previous requirements to be treated by a government physician or a private physician selected by the Department of Labor.

The result of these changes has been a great increase in claims and total compensation costs (Figure 22.2). In 1976, the first calendar year postenactment, claims rose 567 percent over 1974 levels and continued to increase in subsequent years, despite a relatively constant federal civilian work force.[18] Disbursements rose from $270 million in 1974 to $700 million in 1979. In the postal service, reported accidents from 1970 to 1976 went from 84,290 to 55,467, but the number of injury claims more than doubled from 16,838 to 35,282. A General Accounting Office (GAO) 1979 report to Congress concluded that, based on its investigations, 46 percent of a random sample of 410 time-loss injuries were minor or frivolous. GAO estimated that the existence

FIGURE 22.2
LOST-TIME TRAUMATIC INJURY CLAIMS, FEDERAL CIVILIAN WORKFORCE

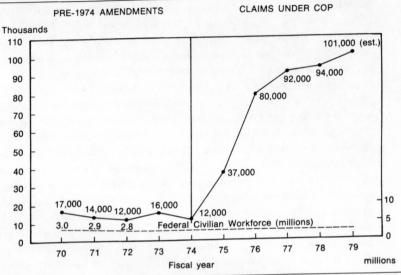

Source: Report to the Congress by the Comptroller General, Multiple Problems with the 1974 Amendments to the Federal Employees' Compensation Act, HRD-79-80, June 11, 1979, as cited in LaDou, J. and Whyte, A.A. "Workers' Compensation: The Federal Experience," *Journal of Occupational Medicine*, Vol. 23, No. 12, 1981, p. 824.

of a three-day waiting period would have reduced the number of short-term continuation of pay claims by 50 percent.

The free choice of physician provision invites shopping around on the part of some workers for a physician who will provide a very liberal estimate of disability leave required. GAO cited a case when a postal clerk, whose big toe was injured by a falling tray, had a bruise diagnosed by a public health physician, who suggested four days' rest. The employee saw several private physicians until one agreed she needed forty-five days off the job. In another case, a postal employee on short-term disability for a work-related knee injury won a tennis tournament.

While there are doubtless many employees who have appropriately benefited, at least from elimination of the waiting period, the 1974 changes created strong economic incentives to exaggerate the impact of minor work-related injuries. Because abuse has become so widespread due to the changes, many employees are apparently embarrassed to file claims for valid injuries.

4. **Setting Benefit Levels** In general, compensation levels should be set so that the aggregate remuneration to the injured worker does not impose a financial hardship nor provide a financial incentive for making questionable claims or inappropriately prolonging the period of disability. Gross wages are typically the base upon which benefits are figured. Workers' compensation benefits are tax-free. The congressional-appointed Commission on State Workmen's Compensation Laws recommended that state laws should provide a weekly cash benefit to disabled workers equal to at least two-thirds of the worker's weekly gross wage.[19] However, the percentage of a gross wage which translates into take-home pay depends upon federal income tax withholding deductions, Social Security contributions or other voluntary payroll deductions. A worker with a gross weekly wage of $350 could have take-home pay of $275, or $250, or $225 or less. An alternative approach is to provide a fixed percentage of net earnings. Maximum benefit levels create further inequities, since better-paid workers are likely to obtain a small portion of their pre-injury spendable income. If a higher paid worker is not covered by a supplemental policy, due to the maximum allowed weekly benefit, he or she will receive a smaller proportion of previous salary and will have a stronger financial incentive to return to employment.

Workers' compensation programs might well be advised to provide incentives for "permanently disabled" workers to return to some form of employment. If less than a dollar-for-dollar offset against workers' compensation payments is taken for the first portion of income, a strong motivation would be provided for a return to the workforce.

One ominous portent is the possibility of including employee benefits in the base of workers' compensation benefits. The U.S. Court of Appeals for the District of Columbia ruled that jurisdiction benefits must be considered as wages in calculating a workers' compensation award.[20] While the District is unique in that it operates under the 1927 Longshoremen's and Harbor Workers' Compensation Act, states have been concerned that this issue is being increasingly raised in appeals and litigation. It has become enough of a concern that Michigan passed a statute specifically excluding benefits from consideration in determining workers' compensation benefits.[21]

5. **Attempts at Reform** In Michigan, passage of legislation which took effect January 1, 1982 required insurers to reduce workers' compensation rates 20 percent in 1982 and 20 percent further in 1983. At the same time a package of twelve reform bills was enacted that tightened and redefined eligibility. Some prominent features of the legislation are a ninety-day limit to file a claim after injury, requirement for

coordination of workers' compensation with Social Security, workers' compensation from other states and with other related benefits, limitations on attorney and physician fees and a requirement that workers undergoing rehabilitation accept jobs they are capable of performing as a condition of benefits payment. It also increased from 5 to 12 percent the interest rate that must be figured in calculation of lump-sum payments, and required insurers to pay 12 percent interest if claim payments were not timely.[22]

Florida passed a wage-loss law in 1979 under which an injured worker does not receive permanent partial disability except for extreme impairment. Benefits are paid only for proven lost wages after an accident. After enactment, premiums dropped 15 percent (1979), then 11.4 percent early in 1981, followed by a further 15.6 percent in July 1981. These reductions were attributed to improved loss experience and reductions on the amount of money spent on litigation. Coupled with tighter policies were increases in the weekly benefits to $223 (as of January 1982) from $130 in August 1979. Although the premium declines are striking, the fact that the reforms occurred midway in a six-year transitional program to base premiums on total wages rather than the first $100 of each employee's weekly earnings makes it likely that individual employers have experienced differential effects on their workers' compensation costs.[23]

Each state has a different system and its own opportunities for improvement. It seems rational to attempt reforms that move in the direction of providing a more reasonable net replacement benefit with higher maxima while simultaneously limiting adverse incentives by adopting stringent criteria for award, maintaining a panel of knowledgable physicians and other professionals to reduce shopping around, and probably retaining a short waiting period. Establishing maximum weekly benefit levels at about 80 percent of *spendable* earnings (gross wages less federal and state income taxes and Social Security), already in place as of 1982 in Michigan and Iowa, appears a promising approach to improving equity of benefits for disabled workers.[24]

6. *Moving Toward More Open Competition* Most states require rate filings and approval by the state insurance commission to sell workers' compensation insurance. As of 1982, however, five states (Michigan, Minnesota, Oregon, Kentucky and Rhode Island) had passed open rating laws that permit marketplace competition, and advocates claim it can lead to significant savings. An advisory committee of the National Association of Insurance Commissioners investigating open rating felt that it would probably most benefit medium-sized employers whom they feel are perceived by insurance companies as the most profitable segment of the business. Large employers' costs would be the least

effected, they suggested, because many competitive techniques are already used to minimize costs for this class of purchasers. However, they also felt that movement toward this system could increase the costs to small business (always more expensive to service), reduce the trend toward self-insurance and lead to more unbundling of safety and loss prevention programs.[25]

On balance, it appears likely that truly competitive approaches would reduce total costs. Illinois insurance director Philip R. O'Connor estimates that Illinois employers could have saved $50 million in premium costs in 1981 if there had been competitive rating.[26] Competitive rating would also provide a market test of the efficiency of state funds versus private insurance carriers. If, as predicted, the smaller employers face higher prices, they may be compelled to participate in group self-insurance programs to an increasing degree.

7. **Reducing Litigation** Provisions of workers' compensation law which require subjective judgments invite litigation. In many current programs there is considerable uncertainty as to whether some disabled employees will receive benefits and the amount that will be received if an award is made. It is not surprising that in a sample of closed cases throughout the United States in 1975, one-fifth had been contested overall and one-third to one-half of the serious cases involving permanent, partial and total disability or death had been contested.[27] In many states, courts at all levels are inundated with worker's compensation cases. In Illinois, for example, they comprise about 20 percent of all cases going to the State Supreme Court.[28]

While the no-fault principle underlies workers' compensation, suits have been successfully brought against employers on the basis of product liability. In 1976–77, $86 million was paid for work-related injury product liability cases.[29] And an appellate court in Illinois ruled that an employer who owns a worksite can be sued as a negligent landowner under common law.[30]

A major yet unsolved problem which fosters litigation is the difficulty in determining whether a worker's disease deserves benefits under the workers' compensation statutes. Rapid resolution of many of these problems is unlikely because a) the evidence linking the disease category to worksite exposures may be equivocal, and b) even if the link is unassailable it may be impossible in many instances to know if the disease in a *particular* person was caused by occupational exposure. For example, exposure to iron oxide increases the risk of cancer of the lungs and larynx by two- to fivefold.[31] However, the cancer caused by this is indistinguishable from the cancer caused by smoking and the cancers appearing in nonsmoking workers without occupational exposures.

TABLE 22.3
CONTESTED WORKERS' COMPENSATION CLAIMS

	Percentage	
	Injury	Occupational disease
Claims contested	9.8	62.7
Reasons for contesting:		
Compensability	20.6	72.5
Percent of disability	55.8	12.0
Other	23.6	15.5
Awards initially contested	10.0	60.0

Source: Brown, T.C. "Denial and Compromise of Workers' Compensation Claims for Occupational Disease," paper presented at American Public Health Association, Oct. 16, 1978, reproduced in Price, D.N. "Workers' Compensation in the 1970s," Social Security Bulletin, Vol. 42, No. 5, 1979, p. 23.

The relative frequency of contested claims for injury and occupational diseases are displayed in Table 22.3 and underscore the major role litigation plays in determining compensability for the latter category. The frequency of litigation is also increased by the fact that workers' compensation usually provides greater net income than disability plans.

Percent of disability is the major litigated issue for work-related injuries. Particularly difficult are determinations of permanent partial disability in the absence of specific statutory schedules. In Florida, before a law change, attorneys were involved in 70 percent of the permanent partial disability cases. However, adopting a law specifying that compensation is to be based on lost wages takes cognizance of the widely differing effect that the identical injury can have on different workers. As a result of the law, a study of changes from the year preceding its passage to the first year after enactment found a decrease in the percentage of permanent partial disability from 30 to 6 percent of total claims. Attorney involvement decreased from 34 to 20 percent in permanent partial loss cases and from 13 to 6 percent in all cases. Savings from the Florida wage-loss provision were estimated for 1980 at over $300 million.[32]

8. **Cardiovascular Disease and Stress** An increasing number of claims are being filed alleging that acquired cardiovascular disease was work-related and therefore a compensable disease. This is a part of a trend to claim degenerative diseases, including common osteoarthritis, as work-related. Since cardiovascular disease is still the most common

cause of death among middle-aged and older Americans, attributing a significant amount to occupational factors could quickly swamp the workers' compensation system. Ironically, long-term repeated strenuous physical exercise, of the type sometimes claimed as the cause of heart disease, has been well-established as protective against heart disease.[33] It is possible that strenuous physical work in some individuals with underlying heart disease may result in heart failure sooner than if the effort had not occurred. However, a member of the American Heart Association's Committee on Stress, Strain and Heart Disease stated the committee's feeling that ". . . it is not possible to determine precisely when cardiac insufficiency would have occurred during the natural history of the underlying disease or from normal 'wear and tear of life' without that stress."[34] Despite this, a number of states specifically provide workers' compensation benefits for designated categories of workers (e.g., policemen, firemen) who develop heart disease.

A growing amount of recent evidence has related stress, including occupational stress, to increased incidence of a number of psychiatric and somatic problems. Work-related factors include the physical environment, management style, job characteristics and the network of supportive social ties with co-workers and supervisors. (See Chapter 19.) In one of the few careful studies in this murky area, over a six-year period employees working in fast-paced, psychologically demanding jobs were found to have an increased risk of developing symptoms of coronary heart disease and of dying of heart and blood vessel disorders. Individuals working in jobs with minimal decision-making requirements or low personal freedom also had a greater risk of developing cardiovascular disease.[35] However, when all studies on this subject are considered together, there is no clear-cut evidence that would permit ascribing a particular instance of cardiac disease or other conditions to specific work environments or job characteristics.

It is doubtful, however, that lack of clear evidence relating work stress to specific ailments will entirely eliminate claims for workers' compensation based on stress-related illnesses. This category of claim is likely to involve legal representation and extensive litigation, with frequent appeals. States are reluctant to attempt to deal with this problem legislatively. Many state laws contain provision for coverage of work stress-related problems. A summary of the more than 100 laws enacted in 1981 dealing with workers' compensation does not mention a legal attempt to define stress-related injuries or to set new standards for compensability based on job stress-related health problems.[36]

Courts and administrative agencies have issued opinions that are striking in their diversity and sometimes break new ground in most disconcerting ways. In 1981, a state appeals court in Indiana awarded the state maximum death benefit of $71,500 to the widow of a businessman who collapsed and died of a heart attack while watching his retail store burn. The court ruled that watching the conflagration constituted work-related emotional stress that aggravated his previously known heart condition. The chairman of the Indiana Industrial Board, which administers the state's workers' compensation laws, indicated concern that the result would be more emotional-stress claims, "which are difficult to resolve because it's [stress is] very nebulous." He further mused, "If somebody hits a fellow worker on the back and says 'Good morning' and the person dies of a heart attack, is that compensable? How far do you go?"[37]

Perhaps the most difficult cases for workers' compensation boards and, all too frequently, courts to resolve are those where it is claimed that work stress has precipitated mental health problems. In an extreme but poignant case, a police officer claimed he could no longer work due to fear of injury precipitated when he saw a fellow officer shot in the head by a motorist stopped for speeding. That a police officer's job frequently puts him or her in perilous situations is unquestioned, but should the effects of increasing vulnerability which may come as an inevitable part of a job be considered a compensable work-related injury? In many cases it is alleged that stress has exacerbated mental health problems which the claimant had previously been able to control. However, the natural history of mental illness is so variable that it is impossible to determine whether the exacerbation leading to an inability to function on the job would have occurred in the absence of work stresses. A partial solution to this dilemma was reached by the Supreme Court of Ohio when it held that disabilities caused solely by emotional stress without contemporaneous physical injury or physical trauma are not compensable under workers' compensation.[38]

PREVENTING AND REDUCING WORK-RELATED DISABILITY

Work-related disability is a special category of disability because it is financed through the workers' compensation system. Almost 25 percent of severely disabled adults participating in a 1978 Social Security Administration survey on disability reported a condition that was caused by an injury or accident which occurred while working, and 13 percent reported a condition resulting from a poor working environment, e.g., noise, heat or smoke.[39] In 1980 approximately 5.6 million cases of occupational injury and illness were reported by private sector businesses with more than ten employees.* Two

---

*Occupational injury is defined by the Department of Labor to be any injury which results from a work accident or from exposure involving a single incident in the work environment. An occupational illness is considered to be any abnormal condition or disorder, other than one resulting from an occupational injury, caused by environmental factors associated with employment, including both acute and chronic illnesses.

and one-half million of those cases involved days away from work, days of restricted activity or both, while the other 3.1 million were nonfatal cases which occurred without lost workdays. These figures translate into an incidence of 8.7 cases of occupational illness and injury per 100 full-time employees, and account for a total of 42 million lost workdays.[40] Total economic losses, including lost wages, medical expenses, production declines and other indirect costs are estimated to be about $21 billion annually.[41]

Reducing the number, severity and consequences of injuries to workers both on and off the job requires attention to a number of different health problems. Occupational safety programs, both mandated and voluntary, have helped to reduce greatly the frequency of occupational injuries from 10.5 per 100 full-time workers in 1972 to 8.5 in 1980. Occupationally related fatalities have also dropped markedly from an incidence of 0.98 per 10,000 workers in 1974 to 0.71 per 10,000 workers in 1980.[42,43] The perceived importance of accident prevention programs is reflected in the finding that 65 percent of California employers with at least 100 employees at one or more sites reported active programs in 1981.[44] Despite these efforts, 4,400 workers died in 1980 as a result of an occupational injury or illness, and nearly 5.5 million work-related injuries were reported, more than half of which resulted in lost workdays.[45]

Reliable estimates of the percentage of worksite accidents related to the use of alcohol or other drugs are unavailable as of 1984. However, at least two major studies have shown the accident rate of alcoholics to be three-and-a-half times that of a control group.[46] High sobriety rates and improved performance among employees referred to an employee assistance program suggest that such programs should help to reduce accident rates. Epidemiological analysis of injury cases by site, time of day, use of equipment, type of activity, body position and other potentially relevant conditions can assist in pinpointing causes of work-related injuries and devising remedies to minimize risk of recurrence.

In many companies musculoskeletal problems are the most common cause of both temporary and permanent work-related disability. Prime among this class of problems is low back pain. In 1979 the National Safety Council reported 460,000 disabling occupational back injuries, up 60,000 from the previous year and constituting about 20 percent of all occupational injuries and 33 percent of compensation dollars.[47] In 1978 $14 billion was spent on the treatment of industrial back injuries, and low back injuries were estimated to account for 25 to 40 million lost workdays.[48,49] According to the National Council on Compensation Insurance, the average lost time and medical cost of a disc injury, $23,250 in 1979, is more than four times the average cost of lower-back sprains ($5,515) and five times the average cost of lower-back strains ($4,203).[50]

Unfortunately, while preventive techniques of various kinds to avoid back injuries have been advocated, evidence to support the effectiveness of any is at best limited. Perhaps the most intuitively sensible approach is to screen prospective workers by medical history and not to hire those with previous history of back injury for jobs involving material handling, since recurrence rates for back problems are high. Low back x-rays as part of pre-employment physicals, once performed routinely for those applying for specific types of work, are no longer recommended. These x-rays do not appear to indicate workers who will have back problems; they also expose workers to unnecessary radiation.[51] Physical examination of the back likewise has not been validated as an effective method of determining those at risk of back injury.[52]

Although training workers to lift with bent knees and a straight back has long been advocated as an effective measure, there is no clear evidence that this training alone is effective. Researchers at Liberty Mutual studied at random 192 compensable back injuries reported to their loss prevention representatives throughout the United States. They were unable to find fewer back injuries than expected from those policyholders who were trained in safe lifting procedures. Nor did companies using as selection procedures one or more of 1) medical histories, 2) medical examinations or 3) low back x-rays have a lower back injury rate than those using none of these procedures. However, they did find a strong and statistically significant correlation between those jobs involving manual handling tasks requiring overexertion (for at least 25 percent of the average work force) and the frequency of low back injuries. The acts or movements associated with low back injury in this study are summarized in Table 22.4. The researchers concluded that "the only currently effective control for low back injuries is the ergonomic approach of designing the job to fit the worker," and they estimated that about two-thirds of back injuries could be prevented utilizing that approach.[53] Training that specifically addresses how to avoid job-specific maneuvers that increase risk of back injury may also prove effective. A new organization, Back Systems, Inc. of Dallas, Texas has developed targeted back injury prevention training and ergonomic design programs specific to each type of job and preliminary indications suggest that this approach can reduce the frequency of back injury problems.

While occupational diseases only constitute about one to two percent of claims for workers' compensation, compensation per case is generally more expensive than for occupational injuries except in the small number of injury cases involving permanent disability or death.[54] Requirements under OSHA and TOSCA for extensive data collection efforts to relate worksite exposures to disease will undoubtedly increase greatly the number of conditions considered to be job-related.

Currently, the best ways to prevent occupational diseases and work-related injuries are to:

1. Ensure compliance with NIOSH and OSHA standards through an accurate and thorough monitoring system.

TABLE 22.4
ACTS OR MOVEMENTS ASSOCIATED WITH LOW BACK INJURY

Act or movement	Percentage of injuries
Lifting	49
Twisting/turning	18
Bending	12
Pushing	9
Pulling	9
Lowering	7
Sudden change in posture	5
Carrying	5
Slipping/tripping	4
Standing	4
Falling	3
Squatting	2
Catching	2
Sitting	1
Throwing	0.5
Walking	0.5
Shoveling	0.0
Other	9

Source: Snook, S.H., Campanelli, M.S. and Hart, J.W., "A Study of Three Preventive Approaches to Low Back Injury," *Journal of Occupational Medicine*, Vol. 20, No. 7, 1978, pp. 478–481.

2. Where appropriate, enlist the aid of a safety or loss prevention expert available through insurance carriers or private organizations.

3. Maintain an active safety program.

4. Have a qualified employee or outside consultant repsonsible for keeping current on the occupational health and safety literature, especially regarding emerging occupational health risks and improved control measures that might be applicable.

5. Invest in ergonomics. Careful design of job to fit the worker is a proven loss prevention technique. Also consider investing in job-specific back injury prevention programs.

6. Have available qualified medical expertise to help determine early signs of occupational illness and to take steps, both in terms of treatment and personnel redeployment, to minimize the severity of the condition.

7. Promote programs that increase general resistance to disease and minimize nonwork behaviors that increase risk of occupational disease. While our understanding of what makes a person hardy or more resistant to the effects of insults to the body is limited, good health habits in general clearly decrease the risk of illness and death. In the often-cited study by Dr. Lester Breslow and colleagues in Alameda County,

habits such as obtaining seven to eight hours sleep per day; eating breakfast regularly; never smoking; not snacking between meals; engaging in regular physical activity; using alcohol moderately or not at all; and maintaining proper weight reduced the risk of dying of a large number of diseases.[55] Employer encouragement of employees acquiring and/or maintaining good health habits through health education and health promotion programs may yield significant dividends in reduction of occupational disease burden (Chapter 18). However, it is also clear that such programs are not substitutes for direct attention to worksite hazards.

8. Be aware of the possible synergistic interrelationships between health practices and occupational exposures. Smoking in asbestos workers has been shown to further increase the risk of dying of lung cancer over fivefold. Asbestos workers who smoke less than one pack a day have a risk of 50.82 times that of a nonsmoking, nonasbestos worker. For asbestos workers smoking more than one pack a day, the corresponding figure is 87.36 times.[56] Smoking acts synergistically in this situation to increase greatly the risk of a disease which in this group of workers is almost always accepted as the result of a worksite exposure. Armed with this information, Johns Manville took the unusual step of refusing to hire smokers in any of their asbestos mines or plants which use asbestos in the manufacturing process.[57]

9. Provide a cushion for employees confronted with unavoidable stress overloads. Promoting a strong network of social support, teaching employees to improve the way they react to stress, including developing better coping skills and learning relaxation techniques are all important (see Chapter 19).

10. Consider operating an employee assistance program. An EAP can be an effective means of providing outlets and referral channels for personal problems that individuals are not able to deal with effectively alone. Counseling and appropriate treatment can prevent continuing declines in an individual's productivity and can help turn many workers whose performance is poor enough to warrant termination into productive and valued employees (see Chapter 20).

The final set of strategies revolves around preventing an established health problem from becoming a disability. Equal impairment in two workers may result in one active, productive worker at his or her original job and in one ex-worker who has been awarded benefits for total permanent disability. Impairments are determined by objective assessment, such as a careful physical examination or simulation of activities necessary to perform a job. But many other factors determine degree of disability, including motivation, encouragement from others, financial incentives, cultural norms regarding what degree of impairment justifies not working and environmental factors, such as willingness of employers to redesign jobs, to facilitate accessibility to the job site, and to purchase special equipment to reduce the functional impact of impairments. Blindness is usually considered a clear-cut criterion for receiving disability payments. Yet IBM won an award in 1982 for employing a significant number of blind people at all job levels.

Efforts that may enhance the ability to retain an impaired employee in the workforce should be directed at preventing the employee from feeling isolated in his or her work environment. Supervisors in particular should maintain close communication with the injured or ill worker, indicate how he or she is missed and that the supervisor and co-workers look forward to his or her return and are anxious to help in whatever way possible. Early attention to rehabilitation is mandatory.

NOTES

1. National Council on Compensation Insurance. *1981 Annual Statistical Review.*
2. Ibid.
3. "Comp Costs Jump 10 Percent in 1980," *Business Insurance*, November 8, 1982, p. 55.
4. Ibid.
5. Vaughan, E.J. and Elliott, C.M. *Fundamentals of Risk and Insurance*, 2nd ed. New York: John Wiley and Sons, 1978.
6. U.S. Chamber of Commerce. *Analysis of Workers' Compensation Laws 1982.*
7. "Rating Job Hazards," *Journal of American Insurance*, Vol. 56, No. 3, 1981, pp. 23–26.
8. Ibid.
9. Op. cit. (U.S. Chamber of Commerce).
10. Price, D.N. "Workers' Compensation: Coverage, Benefits and Costs, 1979," *Social Security Bulletin*, Vol. 44, No. 9, 1981, pp. 9–13.
11. Ibid.
12. Price, D.N. "Workers' Compensation Program in the 1970s," *Social Security Bulletin*, Vol. 42, No. 5, 1979, pp. 3–24.
13. Elson, M.W. and Burton, J.F. "Workers' Compensation Insurance: Recent Trend in Employers' Costs," *Monthly Labor Review*, Vol. 104, No. 3, 1981, pp. 45–50.
14. Op. cit. (Price, 1981).
15. Price, D.N. "Workers' Compensation: 1978 Program Update," *Social Security Bulletin*, Vol. 43, No. 10, 1980, pp. 3–10.
16. Martin, B.T. "Comparing Comp Systems: Study Shows State Funds Are More Efficient," *Business Insurance*, January 18, 1982, pp. 19–20.
17. Grieves, P.C. "Pooling Your Risks: Group Comp Funds Offer Big Savings To Small Firms," *Business Insurance*, December 7, 1981. p. 25.
18. La Dou, J. and Whyte, A.A. "Workers' Compensation: The Federal Experience," *Journal of Occupational Medicine*, Vol. 23, No. 12, 1981, pp. 823–838.
19. Op. cit. (Vaughan and Elliott).
20. "Ruling Adds Benefits To Comp Calculation," *Business Insurance*, February 8, 1982. pp. 30–31.
21. "Michigan, Florida Employers Receive Work Comp Rate Relief," *Business Insurance*, January 11, 1982, pp. 1, 28.

22. Ibid.
23. Ibid.
24. Tinsley, L.C. "Workers' Compensation: Key Legislation in 1981," *Monthly Labor Review*, Vol. 105, No. 2, 1982, pp. 24–30.
25. Norris, E. "Disadvantages of Open Rating Outlined," *Business Insurance*, June 14, 1982, pp. 3, 36.
26. Norris, E. "Illinois Business Face Many Battles over Work Comp," *Business Insurance*, May 24, 1982, p. 16.
27. Op. cit. (Price, 1979).
28. Op. cit. (Norris, 5/24/82).
29. Op. cit. (Price, 1979).
30. Op. cit. (Norris, 5/24/82).
31. Barth, P.S. and Hunt, H.A. *Workers' Compensation and Work-Related Illnesses and Diseases*. Cambridge, Massachusetts: MIT Press, 1980.
32. Sherwood, S. "Florida Wage-loss Program Cuts Work Comp Rates 28 Percent," *Business Insurance*, June 22, 1981. pp. 1, 22.
33. Paffenbarger, R.S. and Hale, A.B. "Work Activity and Coronary Heart Mortality," *New England Journal of Medicine*, Vol. 292, No. 11, 1975, pp. 545–550.
34. "Question Liability for Heart Attacks: Doctor," *Business Insurance*, October 26, 1981, p. 19.
35. Karasek, R., Baker, D., Marxer, F., Ahlbam, A. and Theorell, T. "Job Decision Latitude, Job Demands and Cardiovascular Disease: A Prospective Study of Swedish Men," *American Journal of Public Health*, Vol. 71, No. 7, 1981, pp. 694–704.
36. Op. cit. (Tinsley).
37. Macs, J. "Stress Ruling May Raise Work Comp Costs," *Business Insurance*, June 14, 1981, pp. 3, 25.
38. "Workers' Comp Denies Stress Without Injury," *Business Insurance*, April 20, 1981, p. 34.
39. U.S. Department of Health and Human Services. Social Security Administration. *Work Disability in the United States: A Chartbook*. Washington, D.C.: U.S.G.P.O., 1980.
40. U.S. Department of Labor. Bureau of Labor Statistics. *Occupational Injuries and Illnesses in the United States by Industry, 1980*. Washington, D.C.: U.S.G.P.O., 1982.
41. "Tracking Workers' Compensation Costs," *Journal of American Insurance*, Vol. 57, No. 1, 1981, pp. 6–9.
42. Op. cit. (U.S Department of Labor).
43. U.S. Department of Labor. Bureau of Labor Statistics. *Occupational Injuries and Illnesses in the United States by Industry, 1978*. Washington, D.C.: U.S.G.P.O., 1980.
44. Fielding, J.E. and Breslow, L.J. "Health Promotion Programs Sponsored by California Employers," *American Journal of Public Health*, Vol. 73, No. 5, 1983, pp. 538–542.
45. Op. cit. (U.S. Department of Labor, 1980).
46. Observer and Maxwell, M.A. "A Study of Absenteeism, Accidents and Sickness Payments in Problem Drinkers in One Industry," *Quarterly Journal of Studies on Alcohol*, Vol. 20, No. 2, 1959, pp. 302–312 and Pell, S. and D'Alonzo, C.A. "Sickness Absenteeism of Alcoholics," *Journal of Occupational Medicine*, Vol. 12, No. 6, 1970, pp. 198–210. As cited in Shain, M. and Groeneveld, J. *Employee Assistance Programs*. Lexington, Massachusetts: D.C. Heath, 1980.

47. Antonakes, J.A. "Claims Cost of Back Pain," *Bests Review*, Vol. 82, No. 5, 1981, pp. 36–40.

48. Goldberg, H.M., Kohn, H.J., Dehn, T. and Seeds, R. "Diagnosis and Management of Low Back Pain," *Occupational Health and Safety*, Vol. 49, No. 6, 1980, pp. 11 and 24–30.

49. "High Costs of Back Pain Fuels Interest in Prevention," *Employee Health and Fitness*, Vol. 2, No. 10, 1980, p. 115.

50. Norris, E. "Study Tracks Cost of Back Injuries," *Business Insurance*, March 12, 1984, p. 72.

51. Rowe, M. "Are Routine Spine Films on Workers in Industry Cost- or Risk-Benefit Effective?" *Journal of Occupational Medicine*, Vol. 24, No. 1, 1982, pp. 41–43.

52. Snook, S.H., Campanelli, M.S. and Hart, J.W. "A Study of Three Preventive Approaches to Low Back Injury," *Journal of Occupational Medicine*, Vol. 20, No. 7, 1978, pp. 478–481.

53. Ibid.

54. Op. cit. (Barth and Hunt).

55. Breslow, L. and Enstrom, J.E. "Persistence of Health Habits and Their Relationship to Mortality," *Preventive Medicine*, Vol. 9, No. 4, 1980, pp. 469–483.

56. Hammond, E.C., Selikoff, I.J. and Seidman, H. "Asbestos Exposure, Cigarette Smoking and Death Rates," *Annals of the New York Academy of Sciences*, Vol. 330, 1979, pp. 473–90.

57. U.S. Department of Health and Human Services. Public Health Service. *Cardiovascular Primer for the Workplace*. Washington, D.C.: U.S.G.P.O., 1981.

WHAT SMALLER 23
EMPLOYERS
CAN DO

"*I* have the same health care cost problems as the big boys, but I can't do much about it. I'm better off spending my time on other problems where I feel I can make an impact on the bottom line." So spoke Tom Wright, the president of a seventy-five-employee welding company, when asked what he was doing about the costs of ill health in his company. His reaction is typical, a combination of anger, frustration and resignation.

Smaller employers, regardless of their business, frequently feel that it is hard for them to have the same impact on most personnel-related problems as the larger companies with at least one experienced professional responsible for the general area of human resources and more often a sophisticated staff of personnel professionals. Everyone defines "smaller" differently, but certainly companies with fewer than 300 employees are likely to have limited personnel to deal with corporate health management. Yet over half of the workforce in the United States works for employers with fewer than 100 employees, and nearly 70 percent work for employers with fewer than 250 employees.[1] Therefore, if progress is going to be made in improving health and reducing the rate of rise of health-related costs, both by direct and indirect measures, smaller employers must be involved.

SPECIAL PROBLEMS OF SMALLER EMPLOYERS

Smaller companies are different. Most of what is written and talked about concerning corporate health management focuses on the experience of larger employers. In some ways the world of smaller employers is very different from that of the Fortune 500 or even the Fortune 1500. While the diversity of smaller employers almost defies generalization, a few characteristics are shared widely enough to be mentioned. Most smaller employers do not have an expert in health-related issues on staff. Frequently there is no personnel manager, no director of compensation and benefits, no health and safety officer, and no nurse or other health professional. As a rule everyone wears many hats. Frequently the company president is the one who deals with health-related problems, whether it is deciding which insurer to choose or how to handle a disability case. In other cases the chief financial officer or store or plant manager is assigned this responsibility. Sometimes the responsibility is split, simply because no one person has time to deal with safety and health requirements, health insurance, disability (if any exists), workers' compensation and other related issues. Whoever is assigned responsibility for each of the insurance areas is usually assisted by a broker who is often well trusted and has a long-standing relationship with the company. Virtually all contact between the company and the insurers may be done through the brokers. These characteristics have very significant limitations and reduce the flexibility of the organization with respect to health management.

However, smaller employers often are not aware of their inherent advantages and therefore fail to capitalize on some of the good aspects of being smaller. The principal advantage is flexibility. Smaller companies can make changes much more quickly than larger organizations. In addition, smaller employers are less likely to be bound by collective bargaining agreements which often limit options for larger employers. Decisions can be made much more quickly in small organizations. Usually it is the president who listens to the alternatives or recommendations made by the person with responsibility for a health-related problem and makes a decision on the spot, or over a day or two. Contrast this with the efforts of larger companies, which have established channels for recommendations, usually involving a large number of different staff and line departments, extensive written briefings and several presentations at different levels. In most cases, a decision to change a health program or introduce a new benefit takes a minimum of a number of months. Implementation time can also be much more rapid in a smaller organization. Smaller companies are usually privately held and the person who runs the company often is the major or sole stockholder and can make the decision to proceed without having to consult anyone else.

STRATEGIES FOR SMALLER BUSINESSES

Health Insurance

Despite their inherent advantages there is still a great tendency for smaller employers to consider health care costs as uncontrollable and to limit their attempts at control activities to frequent switching of insurance carriers in the face of 30 to 50 percent increases, settling for a marginally smaller increment. However, many options for a more aggressive posture are available, including:

1. **Self-insuring** Any company with more than 200 to 300 employees should carefully investigate the potential advantages of self-insuring for most health care services and buying reinsurance to cover excess costs to insurees and to the company. Some insurers offer minimum premium plans, good credits for reserves and deferred payments. These may constitute a package which is as attractive as self-insurance, but the two options should be compared before a decision is made. Employers with 1,000 or more employees are at lower risk for totally self-insuring and may decide that the certainty that their payments will not exceed a certain figure is not worth the price of the reinsurance policy.

2. **Use of Third-Party Administrators** A number of claims administrators cater to the needs of smaller employers. Some have developed efficient on-line claims processing operations. Many perform more aggressive claims review than may be routinely available from many of the larger carriers. Some have a data base management system which permits a more flexible and more complete analysis of data than is usually offered. Such systems can be very helpful in pointing to the specific problems in costs and utilization which can be ameliorated through adoption of well-targeted cost-containment programs. A number of third-party administrators are able to offer reinsurance.

3. **Experience Rating** Often, smaller employers are considered as part of a larger group of insurees, and their rates are not strictly based upon the experience of their own group. If the claims experience of a smaller employer is particularly good and appears well below the levels of premiums being paid, it is worthwhile to shop for another carrier who may be willing to provide credit for favorable experience over a number of years.

4. **Multiple Employer Trusts (METs)** Multiple employer trusts are trust funds which collect insurance premiums from a number of smaller

employers and pay for health care services from these funds, either directly or through a claims processing company. METs generally are not considered subject to state insurance laws, and have therefore been unregulated, without statutory reserve or underwriting requirements. To some small employers METs have a bad name, and not without reason. Some METs have failed spectacularly, primarily due to inept and sometimes even corrupt management. However, the majority of METs function well, use reasonable underwriting and accounting principles and have adequate reserves and conservative management. Often METs are available for companies within a specific industry and deserve exploration. Since they combine the insuree population of a number of smaller employers into a large group, they are able to obtain lower rates for administration than could any of their members individually. Before joining, however, no matter how attractive the rates, it is essential to check the ratio of claims paid to premiums over the years, review whatever audits have been conducted and look carefully at the background of the management, the organizers of the MET and what rewards to the managers are built in for effective cost-containment activities.

5. *Use of Several Brokers* Most smaller employers tend to develop a relationship with one broker and continue it over a long period of time. At the time of contract renewal for health insurance, the broker usually requests bids from a number of carriers and presents them to the employer. Surprisingly, however, in several instances when an employer has asked two or three brokers to seek the best bid, they come back with different bids, with the lowest sometimes differing considerably from broker to broker. In two typical examples, the same carrier, given the same information at about the same time by two different brokers, returned substantially different bids. While the reasons for such differences are unclear, there are potential advantages in obtaining bids through several brokers.

6. *Cost-containment Features* Whether entirely self-insured, in a shared-risk arrangement or self-insured with some reinsurance, the smaller employer should pay attention to the cost-containment features offered by different carriers and third-party administrators. What kind of utilization review have they been doing? Do they have experience using model treatment profiles or other aggressive prepayment reviews? Are they organized to run a mandatory second opinion surgery program? Reviewing the savings they have achieved for other companies of the same size and type provides a better measure of success than claimed savings.

7. *Preferred Arrangements*   Smaller employers may wish to take advantage of the preferred arrangements that have been negotiated by larger employers or by major insurers or other carriers in their area. If the area is replete with such arrangements, it may be imperative to piggyback onto some of these as a defense against having to pay prices which have been inflated to cover some of the shortfall from the preferred arrangements. In some cases the carrier from whom the employer buys insurance or administrative services will already have negotiated preferred arrangements on behalf of its subscribers. Before deciding that a significant savings will accrue to the smaller employer from these arrangements, review Chapter 14.

8. *HMOs*   Smaller employers tend to have leaner health insurance benefit packages than larger employers. Therefore, there will be fewer instances where the employer saves substantial dollars by promoting HMO enrollment, which will usually cost as much or more than the indemnity plan. However, the HMO alternative may look increasingly attractive over time as the health benefit coverage broadens and if insurance costs under the indemnity plan rise faster than the HMO premiums. It is important to remember that the insuree saves considerable out-of-pocket expenses by enrolling in almost any HMO. For low-salary employees, not belonging to an HMO may be a barrier to receiving necessary care due to lack of funds to pay the co-insurance and deductible.

Disability and Workers' Compensation

The same principles for managing the costs of disability and workers' compensation apply to both smaller and larger employers (Chapters 21 and 22). Setting realistic payment levels under disability, giving priority attention to early rehabilitation, maintaining close contact with the disabled employee and being willing to let the employee return to a less demanding job, either temporarily or permanently as required, can all help to limit the level of employee disability and associated employer costs. Preventive approaches, such as fitting the job to the physical capabilities of the worker, requiring the wearing of safety belts when driving company vehicles, encouragement of good health habits and maintaining an environment which minimizes work stress, can also make substantial contributions to minimizing disability.

   In those states which permit it, small employers may accrue significant savings by taking advantage of group worker's compensation funds. Group compensation funds offer small employers an opportunity to be more active in expenditure of assets, control of claims administration and establishment of fund policies. In addition, group compensation systems can facilitate the development and implementation of worksite safety programs and other optional activities.

Mental Health

A company with a limited number of employees is in the best position to create a positive environment where each worker feels that he or she is an important member of a team that is working together toward clearly defined goals. Each employee can be made to feel that the company cares about him or her. Smaller organizations have the best opportunity to develop quality circles or other activities which increase worker participation and pride in the products. Attention can be given to assure that authority and responsibility are consonant. In short, the personal touch is easiest to maintain in a smaller business. A feeling of belonging to and being comfortable with an organization can translate into lower turnover, lower levels of stress and reduced rates of illness and disability.

Few small companies have employee assistance programs (EAPs). However, the benefit to the employee of a small company is no less than to an employee of a larger business organization. For the company that does not have the resources to administer its own program, contract services are available from a variety of sources, including mental health clinics, psychologists, large organizations that run these EAPs at a number of sites in the area, local mental health councils and nonprofit alcohol treatment organizations. Local hospitals may also offer assistance services. In general, any size employer should look for organizations which do not have other services to sell, especially inpatient mental health services or inpatient alcohol or other drug abuse services. Since in smaller companies employees tend to know each other, it is usually advantageous to locate the program off-site so that employees can maintain a sense of privacy and confidentiality.

Health Promotion

The majority of smaller employers already have launched some activities to promote health and prevent disease. Among the most popular are hypertension screening, cardiopulmonary resuscitation (CPR), accident prevention and smoking cessation.[2] Unfortunately, these activities often are intermittent, of limited impact and of less than desirable quality. While developing an in-house program may be impractical due to resource constraints, there are a number of good options available:

1. Take advantage of programs offered by local voluntary organizations. For example, the Heart Association, Lung Association and Cancer Society all have smoking cessation programs available, at low or no cost. In most locales, the Heart Association offers free hypertension

screening and follow-up. The YMCAs offer a variety of programs on nutrition, exercise, stress control, back injury prevention, and other subjects.

2. Obtain help from one or more local hospitals. Many have health promotion programs, and some of these are of excellent quality. Some hospitals are willing to send their health educators, nurses and others to the business site.

3. Consider the programs offered by some insurance carriers. A number of insurers offer health risk appraisals and others offer on-site health promotion services to their clients at reasonable costs in a number of areas. At least one company, Pilot Life, offers a comprehensive guide appropriate for intermediate and smaller employers on how to develop a program, procure resources to run it and evaluate it.

A few communities have a wellness council of businesses and health organizations. These councils are usually the source of information on what has worked and not worked at other companies. In addition, they may distribute materials and have seminars where companies can learn how to effectively plan and establish a program. One of the most impressive organizations of this type is the Wellness Council of the Midlands, in Omaha.

Recently the National Center for Health Education, a well-respected organization with considerable experience in developing and mounting health promotion programs, has made helping smaller businesses their highest priority. Based in New York, they are developing networks of smaller employers as well as materials that may be very useful. (The address of the National Center for Health Education is 30 East 29th Street, New York, NY 10016.)

Two other potential sources of high-quality health promotion and disease prevention services are HMOs and universities. Some HMOs are offering these services to employers, regardless of the number of their employees enrolled in the HMO. Universities in many parts of the country have developed cooperative relationships with businesses that have a strong interest not only in mounting good programs but also evaluating their effects and effectiveness.

In addition, smaller companies can give their employees and their families a health newsletter which includes not only well-written articles for the layman on how to stay healthy, but also can incorporate employer-specific information on the availability of health promotion activities, reports of progress that employees have made and special company-sponsored or employee-sponsored activities. New York-based Healthvoice is an example of one of these publications.

Smaller organizations are also conducive to contests between different small units that foster competition in the direction of health. For example, contests can be held for the greatest improvement in exercise, the most weight lost and kept off over a several-month period and the greatest reduction in the number of smokers. Each person who is part of the winning unit should be rewarded. Alternatively, individuals can be given incentives for making sustained changes in health habits. Smaller companies often have greater flexibility in what can be provided as incentives.

Finally, smaller employers can take advantage of the opportunities to offer a health-oriented benefit package that reimburses for effective preventive procedures, such as mammography, Pap smears, examinations for blood in the stool, well-baby care, amniocentesis and others. By covering these items, employers tell employees that their company has a serious interest in helping them and their dependents safeguard their health.

Employer As Role Model

Particularly in a smaller organization, the health habits and attitudes of the executives are known to all employees. Executives are role models, regardless of their behavior. If they smoke, this fact says something about the commitment of the company to improving health. If they exercise regularly and eat a prudent diet, this will be noted and will have an effect on the habits of other employees. If they are always seen to buckle their safety belt, it is likely that others will be influenced to do the same. If they learn to manage their stress and set a less frenetic tone, others will see this as desirable and seek to do the same. Therefore, top managers have the potential, through their own actions, to cause improvement in the health of the other employees.

Innovation

Finally, the flexibility enjoyed by most smaller companies can give them an advantage in corporate health management. Innovation should be strongly encouraged. It is the innovators who will be the first to actively and successfully practice corporate health management.

NOTES

1. U.S. Department of Commerce, Bureau of the Census. *County Business Patterns—United States, 1981.* CBP–81–1. Washington, D.C.: U.S.G.P.O.
2. Fielding, J.E. and Breslow, L. "Health Promotion Programs Sponsored by California Employers," *American Journal of Public Health,* Vol. 73, No. 5, 1983, pp. 538–542.

WHERE TO START: 24
DEVELOPING A HEALTHFUL CORPORATE ENVIRONMENT

Trying to improve health and reduce health-related costs is a formidable challenge, especially since the problem has so many dimensions. To get the maximum effect, however, a coherent program must be developed to bring the pieces together. A considerable portion of the efforts by companies to date have been focused on taking direct steps that can reduce the rate of rise of health care costs. These efforts are directed primarily toward reducing the cost of care once a person has entered the health care system and/or shifting a greater portion of the cost to the employee. However, the costs of ill health encompass much more than the costs paid for health care services. And there are at least three other important opportunities to reduce overall ill-health costs: 1) reduce how often employees and dependents have serious health problems; 2) reduce how often those with health problems seek professional care; and 3) shorten the period of disability associated with a health problem.

REDUCING THE FREQUENCY OF SERIOUS HEALTH PROBLEMS

Health Promotion Activities, Policies and Opportunities

Overall, the single most promising approach to reducing the frequency and severity of health problems is to actively help employees maintain their health and reduce those preventable risks that increase their chances of getting ill in the future. Organized health promotion and disease prevention activities such as smoking cessation, blood pressure detection and control, nutrition, weight management, aerobic exercise, back injury prevention and stress management are essential to this effort.

Of equal importance is the encouragement that an employer provides for adoption and maintenance of healthful behaviors. Part of creating a healthful environment is making good options available: for example, an attractive and well-lit stairwell to encourage use of stairs for exercise during the working day; low-fat, low-calorie, low-salt foods in the cafeteria and in vending machines on premises; and showers and lockers, or maybe a parcourse if the setting permits, to encourage exercise at noontime or before or after work. Another part is creating policies that lead to health as an organizational norm. An enforced smoking policy, which, in recognition that the majority of employees are nonsmokers establishes nonsmoking as the norm, may do more to encourage smoking cessation than the most elaborately promoted, high-quality smoking cessation program in a company without a policy and in which the top executives all smoke. Support for on- and off-the-job safety programs is also part of creating the right environment. For example, are supervisors serious about identifying and remedying potential safety hazards? Does the company routinely stay well ahead of OSHA requirements, or is it always concerned about what might be found if a spot inspection were made? Is the environment conducive to work? Is it well-lighted, reasonably ventilated, without excess noise, at a comfortable temperature and in good repair?

Addressing Personal Problems That Affect Job Performance

What is discouraged may be as important as what is encouraged. Is it a firm and consistently enforced policy that no alcohol or other drugs are permitted on-site? Do managers have training to deal with personal problems which interfere with job productivity? There should in all cases be a high-quality helping resource to whom employees with these problems can be directed, either on- or off-site. The same resources should be available to dependents of employees, since their health problems affect the employees' performance and their health insurance is partially or fully paid by the employer.

Creating Incentives for Health Improvement and Maintenance

Providing direct incentives can assist employees in the development and maintenance of a healthful and productive environment. Employees who quit smoking and a year later still don't smoke can be given a bonus, especially since the company will save considerable money in health insurance and in reduced absenteeism from that lifestyle change. Since fit employees tend to be absent less and more productive, money can be paid for sustained exercise habits, pounds taken off that don't come back, or for seat belt use observed in the company parking lot. Another direct incentive is to pay employees for allowed sick days not taken. Although it is apparent that not all sick days taken are illness-related, paying for sick days not used says to employees: "The company benefits from your taking care of yourself. If your physical and mental health is so good that you don't need to take those days, we want to share the benefit which the company derives from your presence on those days."

Incentives can also be used to reward personal attention to health maintenance. For example, under the dental plan, co-payment for restorative services can be reduced in increments each consecutive year that an enrollee has routine preventive dental care.

Incentives for health are sometimes more indirect. For example, there are a number of companies where the word is out: "Managers who exercise, control their weight, manage their stress well and do not smoke have a greater chance of advancement here." In at least one company, the chairman will not permit smoking at any meeting and is obvious in registering his disapproval of managers who do not utilize the company health facility.

Limiting Work Stressors

Stress management is a skill from which every worker can benefit. But managing unavoidable stress is not an adequate approach to avoid the ill effects of many frequent stressors. Psyche and soma have a very close relationship. If stress is omnipresent and strongly felt, the body and/or mind will have reactions which are costly to individuals, their families and their employers. Whether the result is ulcers, a migraine, an irritable colon, anxiety attacks, transmitting stress to family and friends or eventual burnout, productivity will suffer, insurance bills will rise and disability and workers' compensation cases will escalate.

One of the most important ways employers can reduce stress is to provide workers with the feeling that they are secure in their jobs. This is not to suggest that nobody should ever be fired, but rather to have a company policy which limits firing to situations where an employee has consistently done a poor job which is recognized by boss and co-workers alike. If employees feel that they can be fired at the whim of their immediate supervisor, they will always have a high level of stress and a negative sense of their employer. A second way to reduce stress is to make sure that employees feel that they are important to the organization. Regular two-way communication between all employees and top management about the company's plans and problems, company-sponsored activities that build company loyalty and a policy of open communication which permits employees to have access to top management if they feel unfairly treated all contribute to a healthful corporate environment.

Job responsibility, authority and resources must go hand in hand if excess stress is to be avoided. There are few greater job stressors than having a responsibility without the resources to be able to do what is expected of you. Work styles also have the ability to either increase or decrease stress. In some organizations, everybody feels pushed all the time, communication is limited to tersely delivered statements expressing expectations of what should be done, only limited time is available to discuss problems with the supervisor and almost impossible deadlines are the rule. This management style can achieve desired results in the short run but in the longer term promotes high turnover, burnout, lack of creativity, and high absentee and illness levels.

One of the best ways to cushion stress is to have a strong social support network of family, friends and groups with whom personal problems and ways to solve them can be discussed frankly. The proper work environment facilitates the development of a network of co-workers, an important source of support. Some company activities which help foster this environment are recreational programs, employee special interest groups that are allowed to use the facilities at work and the establishment of work teams that encourage close working relationships and the development of close friendships among team members.

Health Benefits That Benefit Health

Employer-sponsored health benefit plans can contribute to health. Services which have been well established as effective preventive measures should be covered in the benefit plan. Examples are immunizations, well-child examinations, some forms of cancer screening (Pap smears, mammography), amniocentesis and genetic counseling, and hypertension screening and control. To provide strong incentives for their use, co-insurance and/or deductibles might be waived for these services. In addition, coverage should be provided for services, which while not strictly preventive, are known to reduce the overall utilization and costs of health insurance, disability or workers' compensation. For example, providing ambulatory mental health visits on a sliding scale of reimbursement whereby the first few visits are totally reimbursed and then reimbursement gradually decreases to 50 percent has been shown to reduce overall health care costs and is likely to improve productivity on the job. Another example is paying 100 percent for preventive dental services, which is known to minimize more expensive restorative, prosthetic and periodontal work in the future.

REDUCING THE USE OF THE HEALTH CARE SYSTEM

At some time almost everybody suffers from some ailment. But the majority of health problems never get to the doctor's office or to any other health professional. They are dealt with by the owner of the body or mind they are afflicting, by two principal means, self-care and appropriate use of the health care system.

Everybody provides some self-care, whether putting on a bandage, drinking only liquids for a day after have a severely upset stomach or taking some pills that he or she was given by the doctor last time for the same problem. Self-care includes getting help from a pharmacist, an underutilized health professional who can provide good advice on dealing with a number of minor ailments. The availability of many medicines over the counter that were previously obtainable only through a doctor's prescription has broadened the number of options for self-care. Self-care can also include looking up your symptoms or condition in one of the several good self-care books that suggest both how to treat the problem and when to see the doctor.

Most health problems go away in time, with or without professional help. Diarrhea, muscle strains, a tension headache that doesn't vanish after two aspirin or Tylenol, a child with colic or a scraped knee—all are best handled by time and patience.

Many problems can't be cured by the best of modern medicine. Many cannot be adequately controlled to eliminate pain or restore normal function. Over time, most people accept what medicine can or can't do and appropriately reduce their use of the health care system with respect to their chronic health problem. Examples are low back pain, migraines, a sinus condition, bursitis, or reduced hearing despite a hearing aid.

Employers have many opportunities to reduce their employees' reliance on the health care system. Equally important are the dependents and retirees, who together frequently account for over one-half of employer health care costs. One approach is to provide classes on self-care, supplemented by a self-care book and suggestions on how to make better use of the pharmacist for treatment of minor ailments. While studies on the effect of self-care have shown inconsistent effects on health care utilization, self-care courses have shown promise, they are inexpensive, and employees register high levels of satisfaction with them.

Whether a person decides to see a health professional for a problem is not simply a function of the nature of the problem. If two persons have the same identical problem, one may hardly talk about it and merely wait it out, while the other may promptly call for an appointment saying that the pain is terrrible. Individual differences in the way people use health services will always exist, but the differences can occur in the same person, but on two different days. For example, the first time Mary's problem arose, she was feeling otherwise good, enjoying her job, in good physical shape and looking forward to spending the weekend with her family. The second time the problem arose, she was much less satisfied with her job, was out of shape, and generally depressed about her interpersonal relationships. It is no surprise that the pain seemed much worse the second time and that she wanted to see her doctor. Considerable research has shown that general physical condition and mental attitude have not only a great impact on the perception of a health problem but also on the outcome.

In the majority of health visits to a primary care physician's office, no somatic problem can be found. Visits often serve other purposes: to talk about a personal problem, to obtain reassurance, to find out whether fatigue is due to a physical problem. People who are feeling energetic and happy are much less likely to have these kinds of physician visits. Employer-sponsored health promotion programs and other efforts to improve worker morale and self-esteem can contribute to positive attitudes and feelings.

In some cases, increasing utilization in the short term can reduce it overall. Screening for cervical, breast and colon cancer, checking blood pressure elevation, receiving regular prenatal examinations and, if indicated, amniocentesis, all reduce the long-term utilization and costs of health care. The presence of good employee assistance programs and good ambulatory mental health coverage that eliminates barriers to seeking prompt professional help reduce overall utilization.

DECREASING THE PERIOD OF DISABILITY

Attention to all of the above opportunities will reduce the frequency of disability, but many employees will still have serious illnesses and accidents that require professional help. The third focus of employer efforts should be to reduce the impact of these problems on employees, their families and the company. In the case of serious health problems, it is essential to maintain contact with the employee during his or her illness. The manager is the best person to maintain this contact and to reinforce the notion that the worker is missed by his or her co-workers and that everyone is anxious to help make the return to work as easy as possible. Consultation with appropriate health professionals experienced with disabilities in workers can help to minimize the time the employee is away from work. In one company, a cardiologist helped reduce the time from heart attack to return to work from three months to five weeks, with a noticeable positive effect on self-confidence in returning employees and on the disability costs to the company. Flexibility in work assignment and scheduling are important ingredients in limiting disability absences to no longer than what is absolutely required. Also important, however, is to have the disability plan provide economic incentives for early return to work.

KEEPING TRACK OF HEALTH

How does a company know how it is doing in its attempts to maximize the health of its employees? In most cases, companies are still trying to determine what are the real costs of illness and finding that even this is very complicated. Nonetheless, as a number of employers have found, there are relatively inexpensive ways of obtaining and periodically updating information about employee health. Employee surveys already often cover some aspects of health, most commonly morale and feeling about their job. It is not difficult to add well-tested questions on attitudes about employees' health, the degree to which employer programs are helping them safeguard their health, and what types of modifications in existing health programs or new health programs they would like the company to sponsor.

Most likely, the company medical department already has information on some health problems of employees. Use of in-house health services by employees can be monitored over time and the frequency and types of problems seen analyzed to discern trends. Many employers also maintain records of occupational exposures, some via sophisticated data systems that correlate occupational exposures to health problems seen on examinations and reported on claims for health care, disability and workers' compensation. Screening of employees in conjunction with health promotion programs at the worksite or as part of company-sponsored or federally mandated periodic examinations can include questions about health attitudes, health behaviors and health risks.

With appropriate attention to assure confidentiality and privacy, these various sources of health-relevant data can be summarized and sometimes combined to provide a clear picture of the health of the workforce. This information can help employers identify current and future problems, based on known risks, that are soluble through new action programs or changes in existing programs. A health data bank is also required to assess the effect of programs that have been put in place and to start to appreciate the dividends to employers and employees from investment in creating a healthful environment. Over time, the dividends should accrue not only to employees, but to dependents and to retirees as well.

INDEX

Absenteeism, causes of, 266, 348. *See also* Sick leave plans

Administrators, third-party, 42–43, 158, 395

Aetna Life and Casualty, 42, 100, 213

AFL–CIO, 226

Age, and health care costs, 132–133, 134

Aid to Families with Dependent Children, 243

Alcoholics Anonymous, 86

Alcoholism, 83, 88, 97–101

Alcoholism programs, 332, 333, 340. *See also* Employee assistance programs.

All-payer plans, 241

AMA–PAC, 245

American Association of Fitness Directors in Business and Industry, 294

American Dental Association, 68, 74

American Heart Association, 384, 398

American Hospital Association, 40, 226

American Management Association, 324

American Medical Association, 40, 226, 255

American Medical Peer Review Association (AMPRA), 183

American Psychiatric Association, 92

American Psychological Association, 92

Anesthesia, 2

Arizona Coalition for Cost-Effective Quality Health Care, 227

Arizona Hospital Association, 227

Armer, John, 93

Arnett, R. H., 72

AT&T employee assistance program, 334

Back injuries, 387, 388

Back Systems, Inc., 387

Bailit, H. L., 70

Bank of America, 55

Bell, Donald R., 69

Bell System, 68

Benefit packages, 58

Benefit utilization, 114–120, 129–130, 131

Berol Corporation, 58

Black Lung Program, 360, 374

Blindness, 390

Blood pressure, 268, 272–273, 277–279, 301

Blue Cross/Blue Shield, 20–21, 40–41, 50; activism of, 230; alcohol and drug abuse coverage by, 98; coalitions and, 225, 226, 227; competition and, 258–

259; dental plans, 68, 69, 76, 77, 79; health promotion programs and, 294, 295; HMO investments of, 204; home maternity program, 107; medical service adviser system, 63; mental health services, 85, 89, 90, 92; preferred arrangements and, 210, 212; regulation and, 244
Bone marrow transplants, 49
Bonne Bell, 293
Breslow, Lester, 388
Burnout, 314–317
Business and Health: A Report on Health Policy and Cost-Management Strategies, 233
Business Insurance, 377
Business Roundtable, 226

Caffeine, 328
Cain, Carol, 59
California Dental Service (CDS), 67, 77
California Hospital Association, 211
California Psychological Health Plan, 93
California Wellness Plan, 93
Campbell Soup Company, 298
Cancer, 264, 265, 267, 268, 271, 277, 279, 283
Cardiac bypass, 167–168
Cardiovascular disease, 3–4, 6, 264–265, 266, 267, 269, 270, 271, 277, 283, 321, 383–385
Carpenter, Robert, 225
Carroll, M. S., 72
Catastrophic coverage, 48
Center for Disease Control, 4
Certificate-of-need (CON) legislation, 239
CHAMPUS, 92
Charter Medical Corporation, 202, 204
Chemical New York Corporation, 58
Children, causes of hospitalization among, 8, 9
Cholesterol, 268, 269, 270, 272, 283–284
CIGNA, 42
Claims analysis, 110–154; benefit utilization, 114–120, 129–130; cost-control strategies, 150–154; cost increases, 130–140; hospital costs, 120–127; incurred claims, 140–141, 142; mental health benefits, 145–150; multiple carriers, 141–145; physician costs, 127–128; reasons for, 110–114
Claims rejection, reasons for, 113
CNA Financial Corporation, 202, 204

Coalition Action Program, 230
Coalitions, 224–235; activities of, 227–232; future of, 233–234; goals of, 224–225; membership of, 225–227; national and regional, 232–233
Coal Mine Safety Act of 1969, 12
Co-insurance, 42, 47–48
Commercial carriers, 41–42, 212
COMPETE, 214
Competition, 253–262; barriers to, 258–259; consumer choice and, 256–258; consumer knowledge and, 259; regulation and, 261–262; tax laws and, 253–255
Comprehensive coverage, 46–47. *See also* Medical benefits.
Congressional Budget Office, 254, 256
CON legislation, 239
Connecticut General Insurance, 42, 67, 175, 201, 204
Consumer Product Safety Commission, 246
Control Data Corporation Staywell program, 290–291
Converse Rubber Company, 326
Co-payments, 54
Coping skills, 327–328
Corporate health management, history of, 1–5
Costs: benefit utilization and, 115, 116, 131; claims analysis and, 111–112; hospital, 117, 120–127; sharing of, 51–54; strategies for control of, 150–154
Coverage limits, 48–49
Craig, T. J., 90

Davis, Carolyne, 241
Death, causes of, 6, 7, 8, 9, 264, 265. *See also* Terminal illness.
Deductibles: dental, 71; medical, 41, 42, 47, 53, 54
Deere, John, 202, 205
Delta Dental Plans, 68, 79
Dental benefits, 67–81; controlling costs and quality, 75–79; cost-effective, 79–80; costs of, 71–72, 73; deductibles, 71; dental practice and, 72–75; employee contributions, 71; growth in, 67–68; incentives for preventive care, 69, 70; maximums, 71; payments for services, 116, 128; reimbursement principles of, 68–70; surgical, 71
Dental disease, 266

Dental Service Corporations, 68
Depression, 83
Diabetes, 345, 364
Diagnosis-related groups (DRGs), 123, 126–127
Diagnostic tools, 2
Dialysis, costs of, 24–25
Diet, 283–284
Disability: defined, 347; prevention of, 363–364; rehabilitation and, 364–366; sex differences in, 351; work-related, 371–372, 385–390 (*see also* Workers' compensation).
Disability compensation, 345–368; benefit design, 360–361; controlling cost of, 351–360; income under, 361–363; increasing claims for, 345–349; long-term, 357–360; rise in payments, 359–360; short-term, 353–357; for small employers, 397–398
Disabling conditions, 10–11, 345–348
Diseases, preventable, 264–277
Dodson, J. D., 311
Dow Chemical Company, 69
Drug abuse program, 340. *See also* Employee assistance programs.
Drug abuse treatment, 88, 97–101
Drugs. *See* Prescription drug benefits
DuPont, 57
Dychtwald, Ken, 310

EAP. *See* Employee assistance programs.
Early periodic screening and diagnostic testing (EPSDT), 243
Elderly, causes of death among, 8
Employee assistance programs (EAPs), 332–344; alcoholism, 332, 333; components of, 335–336; evaluation of, 341–343; features of, 338–341; management of, 336–338; results of, 334–335; successful, 338–341
Employee benefits, cost of, 13–14. *See also* Medical benefits, Dental benefits.
Employee Benefits Survey, 47–48
Employee burnout, 314–317
Employee coalitions. *See* Coalitions.
Employee education programs, 63–64
Employee expense accounts, 58–59
Employee incentives. *See* Incentives.
Employee turnover, cost of, 15
Employers: attitudes toward health care benefits, 30–38; attitudes toward

HMOs, 197–200; preferred arrangements and, 211, 216–221; small, 393–400
Employers' Health Care Costs Coalition, 225, 230
Environmental Protection Agency, 246
Epidemiology, 267–268
Equitable Insurance, 42, 327
ERISA, 41, 42
Exercise, 280–283, 294, 328
Eye care, 104–105

Far-West Administrators, 214
Federal Administrative Procedures Act, 246
Federal Employee Health Benefit Program, 57, 90
Federal Employees' Compensation Act (FECA), 378
Federal Mutual Benefit Association, 354
Federal Office of Health Maintenance Organizations, 202
Fisher, C. R., 10
Fitness programs, 280–283, 294
Fitness Systems, 287
FMC Corporation, 175
Ford Motor Company, 278
Foundations for Medical Care (FMCs), 182

General Motors' employee assistance program, 334
Gibson, R. M., 72
Glasser, J. H., 89
Goldbeck, Willis, 105
Government regulation. *See* Health care regulation.
Group Health Association of Washington, 88

Harris, Louis, 197, 198, 200
Hay–Huggins survey, 46, 47, 48, 49, 68, 354, 356
Health: direct costs of, 12–14; employer responsibility for, 11–17; indirect costs of, 15–17, 31–32; new definition of, 5–6
Health Alliance Plan, 226
Health care, misconceptions about, 22–24
Health Care Financing Administration, 10
Health care regulation, 236–252; capital expenditure controls, 239; competition and, 261–262;

construction standards, 237–238; effects of, 248–250; facilities operation standards, 237–238; HMOs, 240; Medicare and Medicaid, 242–243; organization of, 247–248; politics of, 244–246; private health insurance, 244; process of, 246–247; state rate setting, 240–242; utilization review, 238–239

Health care system, 19–29; competition in, 253–262; consumers and, 256–258, 259; future of, 28–29; hospital billing, 19–20; hospital reimbursement, 20–21; physicians as consumers, 21–22; reducing use of, 404–406

Healthfax, 292

Health insurance, small business strategies, 395–397

Health Insurance Association of America (HIAA), 49, 226, 351, 354, 356, 358, 359, 361–362

Health insurance carriers, 39–45; Blue Cross/Blue Shield, 40–41; commercial, 41–42; future of, 43–44; third-party administrators, 42–43, 158, 395

Health insurance claims. *See* Claims analysis.

Health Maintenance Life Insurance Company, 184

Health Maintenance Organizations (HMOs), 28, 59–60, 193–208; alcoholism and drug abuse services, 98; coalitions and, 230; competition and, 256, 258; consumer attitudes and, 200; costs of care, 194–197; defined, 193–194; employer attitudes and, 197–200; enrollment in, 194; evaluation of, 202–203; financial viability of, 201; future of, 206–207; mental health services, 89; private investment in, 204–206; regulation of, 240; for small employers, 397, 399

Healthnet, 233

Health Plans Inc., 205

Health problems: preventable, 264–277; reduction of, 401–404

Health promotion programs, 287–304, 401–402; evaluation of, 297–304; examples, 288–293; fitness programs, 294; incentives, 293, 402–403; private insurers and, 294–295; for small

companies, 398–400; success in, 295–297

Health Research Institute (HRI), 87, 106

Health risks, 266–287; appraisal, 268–276, 294; controllable risk factors, 267; costs of, 277; epidemiology, 267–268; high blood pressure, 277–279; lack of exercise, 280–283; nutrition, 283–284; obesity, 284–287; smoking, 279–280

Health Services Council, 228

Heart disease, 3–4, 6, 264–265, 266, 267, 269, 270, 271, 277, 283, 321, 383–385

Heart transplants, 25, 49

Hewitt survey, 64

Hewlett-Packard, 201

High blood pressure, 268, 272–273, 277–279, 301

HMO Act of 1973, 236, 240

HMOs. *See* Health Maintenance Organizations.

Home health care, 106–108

Hospice care, 105–106

Hospital admissions, 121

Hospital benefits, 49

Hospital Corporation of America, 293

Hospital costs, analysis of, 117, 120–127

Hospitals: billing by, 19–20; Blue Cross/Blue Shield and, 40, 41; preferred arrangements and, 211; reimbursement of, 20–21; utilization rate for, 27–28

Hotlines, 336–337

Hypertension. *See* High blood pressure.

IBM, 86–87, 313, 390

IBM Plan for Life, 292–293

INA Corporation, 42, 202, 204

Incentives, 51–61; cost sharing, 51–54; employee expense accounts, 58–59; for health improvement, 293, 402–403; optional levels of coverage, 57–58; prepaid plans, 59–61; rewards for low use of services, 54–56

Indemnity plans, 42

Independent practice association (IPA), 60, 193, 206, 214; costs of, 194, 195; employer perceptions of, 199; utilization of, 195. *See also* Health Maintenance Organizations.

Infants: causes of death among, 7, 9; intensive care for, 8

Insurance carriers. *See* Health insurance carriers.
International Longshoremen's and Warehousemen's Union-Pacific Maritime Association, 67
International Rehabilitation Association, Inc., 364
InterStudy, 194, 202

Job performance, 309, 312, 326–327, 402
Job security, 313–314
Johns Manville, 389
Johnson & Johnson Live for Life program, 288–290
Joint Commission of Accreditation of Hospitals (JCAH), 100, 237–238

Kaiser Family Foundation, 197, 199
Kaiser Foundation Health Plan, 202
Kelly, Curt, 225
Kemper Insurance Companies, 100
Kemper Rehabilitation Management, 364
Kidney transplants, 25, 49
Kimberly Clark health promotion program, 291

Laboratory services, 49–50
Legislation. *See* Health care regulation
Levin, B. L., 89
Levinson, Harry, 329
Liberty Mutual Insurance Company, 387
Limits, 48–49
Live for Life program, 288–290
Liver transplants, 49
Longshoremen's and Harbor Workers' Compensation Act, 380
Long-term disability insurance, 357–360
Lung cancer, 264, 268
Lutheran Hospital Society of Southern California, 213

McCarthy, Eugene, 178
McGuire, T. J., 91
Machinist's Union, 70
McKinlay, J. B., 3
McKinlay, S. M., 3
Major medical coverage, 46. *See also* Medical benefits.
Manic depressives, 83
Massachusetts Mutual Life Insurance Company, 278

Mattel Health Enhancement Program, 291–292
Mechanic, David, 86
Medicaid: cost control and, 21; dental benefits and, 79; hospital costs and, 120; prescription drugs and, 103; regulation and, 237, 240, 242–243, 247
Medi-Cal, 212
Medical benefits, 46–66; broadening of, 26–27; co-insurance, 47–48; cost of, 13, 14; coverage limits, 48–49; deductibles, 47; employers' attitudes toward, 30–38; existing features, 46–49; hospital, 49; incentives, 51–61; in-hospital doctor visits, 49–50; laboratory tests, 49, 50; office visits, 50; redesign of, 50–65; surgical, 49; union concerns about, 64–65; x-rays, 49, 50
Medical care, limits on, 24–25
Medical service adviser system, 63
Medicare: costs and, 21, 25, 27; diagnosis-related groups and, 123; hospice care and, 106; hospital costs and, 120, 123; regulations and, 237, 238, 240, 242–243, 247, 262; second opinion programs and, 176, 177; supplements to, 15–16
Medications. *See* Prescription drug benefits.
Medserco, 205
Mendocino County Stay Well Program, 54
Mental health, 309–331; employee burnout, 314–317; job performance and, 309, 312, 326–327; stress and, 310–317; stress management, 317–330. *See also* Employee assistance programs.
Mental health benefits, 82–96; analysis of costs, 145–150; cost effectiveness of, 88, 91–94; coverage, 87–88; historical perspective on, 82–84; outpatient coverage, 83–84; prepaid, 93; providers of services, 84–88; psychotherapy, 83–84, 92; recommended approach to, 93–94; for small companies, 398; users of, 82–84; utilization of, 89–91
Mental health professionals, 85–86
MET. *See* Multiple employer trusts.

Metropolitan Life Insurance, 42, 67, 205, 282, 294, 351
Michigan Health Economics Coalition, 226
Michigan Hospital Association, 227
Midwest Business Group on Health, 233
Model Treatment Program, 43
Montgomery Ward, 74
Multiple employer trusts (METs), 395–396

National Association of Insurance Commissioners, 100, 381
National Center for Health Education, 399
National Center for Health Statistics, 7, 8, 9
National Commission on State Workmen's Compensation Law, 376, 380
National Council on Compensation Insurance (NCCI), 373, 386
National Health Planning and Resources Development Act of 1974, 239
National HMO Census, 194
National Medical Care Expenditure Study, 65
Neonatal intensive care, 8
Neuroses, 83
Newhouse, J. P., 52, 72
New York Business Group on Health, 224–225
New York Teamsters, 72
New York Telephone, 298, 327
Nutrition, 283–284

Obesity, 272, 278, 284–287
Occupational Safety and Health Act (OSHA), 11, 12
O'Connor, Philip R., 382
Olin Corporation, 47
Omnibus Budget Reconciliation Act of 1981, 201, 353
Operations. *See* Surgery.
Options: benefit packages, 58; level of coverage, 57
Organ transplants, 25, 49
Orthodontic services, 69, 71
Outpatient procedures, 56

Patterson, D. Y., 90
Peer Review Organizations (PROs), 120, 176, 243
Penjerdel Council of the Greater Philadelphia Chamber of Commerce, 229
Pennsylvania Blue Shield, 76, 77

Phelps, C. E., 72
Physicians: analyzing costs of, 127–128; Blue Cross/Blue Shield and, 40; in-hospital visits, 49–50; office visits, 50; oversupply of, 27; preferred arrangements and, 211
Pioneer Hi-Bred International, 293
Pitney-Bowes, Inc., 356
Planning Research Corporation, 107
Political action committees (PACs), 245
Potomac Electric Power Company's (PEPCO) Employee Advisory Service, 334
Preferred Provider Organizations (PPOs), 209–223; defined, 209–210; employer considerations with, 216–221; employer savings with, 216–217; reasons for participation in, 211–214; recommendations on, 221–222; for small businesses, 397; types of, 214–216
Prepaid plans, 59–61
Prescription drug benefits, 101–104
President's Commission on Mental Health, 82, 84
Preventive services, 54
Productivity, and stress management, 326–327
Professional Standards Review Organization (PSRO), 183, 238–239, 249
Prudential Insurance Company, 42, 70, 202, 204
PSRO. *See* Professional Standards Review Organization.
Psychiatric social worker, 85, 86
Psychiatrists, 85–86
Psychologists, 85–86
Psychotherapy, 83–84, 92. *See also* Mental health benefits.
Public Health Service, 4, 6, 73

Quaker Health Incentive Plan, 58, 59

Ramsey, Doris J., 212
Rand Health Insurance Experiment, 23, 52, 91–92
Rate setting, 240–242
Regulation. *See* Health care regulation.
Rehabilitation Act of 1973, 12
Republic Steel Corporation, 225
Reynolds, R. J., 202, 205

Risks. *See* Health risks.
Rohr Industries, Inc., 212

Safeco, 205
Safeco's Health Action Plan for Everyone (SHAPE), 293
Safety programs, 387–389
San Diego Foundation for Medical Care, 230
Schizophrenia, 83
Sears Department Stores, 74
Second opinion surgery programs, 166–180; consumer education and, 176; design of, 177–178; future of, 179; growth of, 174–177; history of, 170–174; voluntary, 176–177
Segal, Martin E., 213
Self-insurance, 155–165; administrative costs of, 157–158; choosing plan administrator, 160–161; cost savings of, 159; effects of, 156–158; future of, 164; history of, 156; legal considerations of, 163–164; limiting risk in, 159–160; plan size and, 159, 160; reserve requirements for, 157; self-administration of, 161–163; for small businesses, 395; third-party administrators for, 42–43, 158
Selye, Hans, 310–311
Short-term disability, 353–357
Sick leave plans, 354–356
Small employers, 393–400; special problems of, 394; strategies for, 395–400
Smoking, 264, 267, 268, 272, 277, 279–280, 293
Social Security Disability Insurance (SSDI), 14, 347, 352–353, 360, 362–363
Social support, 317–319
Spencer, Charles D., and Associates, 197
SSDI. *See* Social Security Disability Insurance.
State Mutual Life Assurance Company of America, 279
Staywell program, 290–291
Stockham Valves and Fittings, Inc., 74–75
Stop loss, 48
Stress, 272, 278, 309–331; characteristics of, 310–312; disorders related to, 313; employee burnout and, 314–317; employers and, 312–314; job performance and, 309, 312, 326–327;

limiting, 403–404; management of, 323–330; social support and, 317–319; Type A behavior and, 314, 322; work conditions and, 316, 317, 319–323, 324, 328–330, 403–404
Stroke, 263, 264, 265, 267, 271, 277, 283
Supplemental Security Income (SSI), 360
Supplementary Medical Insurance (SMI), 242. *See also* Medicare.
Surgery: costs of, 134; increase in, 167–169; second opinion for, 166–180; unnecessary, 166–169
Surgical benefits, 49
Survey of Disabled and Nondisabled Adults (1972), 347

Tax Equity and Fiscal Responsibility Act (TEFRA), 10
Tax laws, restructuring of, 253–255
TELANSWER, 63, 259
Terminal illness, 105–106
THE (The Health Enhancement) Program, 291–292
Third-party administrators, 42–43, 158, 395
Tibbitts, Samuel J., 213
Tosco Corporation Health Enhancement Program, 291–292
Toxic Substances Control Act (TOSCA), 11, 12
TPA. *See* Third-party administrators.
Transplants, 25, 49
Travelers Insurance Company, 42, 294
Type A behavior, 272, 314, 322

UCLA Center for Health Enhancement Education and Research, 291
Unions: dental plans and, 67, 78, 70, 72; medical benefits and, 64–65
United Auto Workers (UAW), 68, 91, 226–227
U.S. Administrators (U.S.A.), 43
U.S. Corporate Health Management, 63, 259
U.S. Department of Health and Human Services, 73
U.S. Department of Health, Education and Welfare, 6
U.S. Department of Labor, 88
Utah Health Cost Management Foundation, Inc., 228

Utilization review (UR), 181–192; future of, 191; history of, 182–183; private, 187–191; types of, 183–191

Vinick, Alan, 201
Vision care benefits, 104–105

Waldo, D. R., 72
Washington Business Group on Health (WBGH), 85, 232–233, 245, 250, 287, 342, 364
Wausau Insurance Company, 204
Weight Watchers, 286
Wellness Council, 399
Winegar, Marshall, 263
Women: disability and, 351; health care costs of, 132, 133

Work conditions, 316, 317, 319–323, 324, 328–330, 403–404
Workers' compensation, 353, 369–392; contested claims for, 383; cost control for, 377–385; features of, 371–372; increasing costs of, 373–376; preventing work-related disability, 385–390; principles of, 370–371; rates for, 372–373; for small employers, 397–398; trends in, 373–377
Wright, Tom, 393

Xerox Corporation, 47, 294
X-ray costs, 49, 50

Yerkes, R. M., 311

Zenith Corporation, 63